BALTIC PILING

PROCEEDINGS OF THE CONFERENCE ON BALTIC PILING DAYS 2012,
TALLIN, ESTONIA, 3–5 SEPTEMBER 2012

Baltic Piling

Editors

M. Mets & R. Raudsepp
Estonian Geotechnical Society, Tallinn, Estonia

CRC Press
Taylor & Francis Group
Boca Raton London New York Leiden

CRC Press is an imprint of the
Taylor & Francis Group, an **informa** business
A BALKEMA BOOK

Published by:
CRC Press/Balkema
P.O. Box 447, 2300 AK Leiden, The Netherlands
e-mail: Pub.NL@taylorandfrancis.com
www.crcpress.com – www.taylorandfrancis.com

ISBN 13: 978-0-415-64334-4 (hbk)

Visit the Taylor & Francis Web site at
http://www.taylorandfrancis.com

and the CRC Press Web site at
http://www.crcpress.com

Typeset by MPS Limited, Chennai, India

Baltic Piling – Mets & Raudsepp (Eds)
© 2013 Taylor & Francis Group, London, ISBN 978-0-415-64334-4

Table of contents

Baltic Piling – Mets & Raudsepp (Eds)
© 2013 Taylor & Francis Group, London, ISBN 978-0-415-64334-4

Committee members

Organising Committee

Chairman	P. Ilves
Co-chairman's	V. Celminš
	L. Furmonavičius
	T. Tarkkio
Scientific Secretar	M. Mets
Secretaries	R. Raudsepp
	J. Pello
	T. Liblik

Advisory Committe

Jean-Louis Briaud
Neil R. Taylor
Roger Frank
Heinz Brandl
Harry Poulus
Renato P. Cunha
Bengt H. Fellenius
Paul W. Mayne
Donald A. Bruce
Zbigniew Sikora
V.M. Ulitski

Scientific Comittee

Charles W.W. Ng
Melvin England
Jorgen S. Steenfelt
Alessandro Mandolini
Flor De Cock
Diemar Adam
E.N. Bellendir
M.J. Nikitenko
M.L. Zotsenko
Ahmed Elgamal
Mark Randolph
Kazimierz Gwizdala
V. Fedorovsky

Registration and Secretariat (Enquiries)

Baltic Geotechnics'
Tallinn Technical University
Ehitajate tee 5
19086 Tallinn, Estonia
Phone: +372 620 2408
Fax: +372 620 2405
E-mail: jpello@edu.ttu.ee

Baltic Piling – Mets & Raudsepp (Eds)
© 2013 Taylor & Francis Group, London, ISBN 978-0-415-64334-4

Sponsors

Baltic Piling Days Main Sponsors

Bauer

BAUER MASCHINEN

BMGS

BMGS

GIB

GIB AS GEOTEHNIKA INSENERIBÜROO

Baltic Piling Days Sponsors

Merko

merko

Fort Ehitus

Fort Ehitus

Skanska

SKANSKA

Skanska EMV AS

Savi

Nordecon

Riverside

Pile Dynamics Europe

Infralock

Transiidikeskus

Geomachine

Geoengeneering

Ruukki

Eesti Energia AS

Geokon

Fundex

Baltic Piling – Mets & Raudsepp (Eds)
© 2013 Taylor & Francis Group, London, ISBN 978-0-415-64334-4

A brief overview of the historical developments of the foundation engineering in Brazil

R.P. Cunha
University of Brasília, Brasília, Federal District, Brazil

P.J.R. Albuquerque
State University of Campinas, Campinas, São Paulo, Brazil

ABSTRACT: This paper presents a summary of the historical developments that took place in Brazil since its early beginning, from the colonial times to modern age. It relies on published (most in Portuguese) information pinpointed from historical and reference books widely available today that were, on the other hand, written with basis on oral and written recounts & documents that survived through the time. It does not have the intention to present an insightful discussion on the subject, but rather illustrate how, in common terms, the foundation engineering technology has been upgraded from the early arrival of the Portuguese colonizers to nowadays society. It finalizes with a broad, and updated, contextualization of the foundation types that are normally found in Brazil at its different topographic, geotechnical and demographic regions.

1 INTRODUCTION

With the evolution of the civil engineering, and the high level of demand from the construction of structures with increasingly slender and heavier buildings, there was also the development of the foundation engineering in Brazil. In addition to the structural elements of the project, another recent factor of extreme importance for the development of the techniques for foundation execution and design was the limited (or lack of) urban space and the choice of location for the implementation of the new constructions in a fast developing country. In many cities another aspect took place, i.e., previously (historically, for many reasons) disregarded regions of the "Brazilian continent" began to be populated and this fact imposed technical difficulties that necessarily needed to be overcome.

Therefore, it is briefly presented herein the history of the foundation engineering in Brazil, narrowing part of the facts at its most developed and populated cities, Rio de Janeiro and São Paulo (and surroundings). It does not intend to make a state of the art on the subject, but rather highlight the most interesting and relevant points in this evolutionary process of design and construction techniques. Within this scope, as it will be shown, some historical cases are emphasized together with the main techniques that have been employed in the country since the early beginning of its development stage. It shall be pointed out that most of the presented data comes from a compilation (and translation) of information submitted in the chapter 1 of the renowned book "Foundations, Theory and Practice", published in Brazil by its Soil Mechanics Society (ABMS, 1996), added to pertinent passages from the recently published book on the "History of Geotechnical Engineering in Brazil", by Sayão (2010). The paper finalizes with a technical overview of solutions gathered by a survey on few, but prominent, foundation companies that develop technology and execute foundations on the distinct regions of the country.

2 HISTORICAL DEVELOPMENT

2.1 *Colonial period (XV–XVIIIth centuries)*

In the colonial period of the XVI century, around 1530, the buildings could be classified into three groups according to Katinsky (1994). The first group corresponded to works/factories associated with the sugar industry and exporting port facilities, whereas in the second group one could find military buildings like forts and barracks, plus churches, monasteries, public buildings and related commercial chains, and finally the third group gathered civil works and houses of any dimensions plus business stores, warehouses, flour mills and so on. According to this author, there are just few reports on the foundation characteristics of these buildings, being possible to confirm a common knowledge that, since such remote period, foundations were made of excavated trenches filled in with "compacted" stones.

Around the XVII century, in the range of 1684, Silva Nigra and Clemente (1950) state that for the construction of the Monastery of St. Benedict, in the city of Rio de Janeiro, it was a written a "Statement of

Construction" which contained specifications for the construction of the monastery. In one of the items presented by such document, one can find the techniques for the construction of the base foundation. In this paper the author demonstrates that concepts of "tension" and "resistance of the material' where already known at the time, being associated to the need of reaching a foundation level prone to exist a subsoil material of sufficient hardness to support the superstructure building. This document also mentioned that the presence of water in the subsoil could ruin the trenches dragged for the construction of this particular foundation.

2.2 *Monarchic period (XIXth century)*

The monarchic period lasted around 1 century, and by its mid course, in the range of 1850, Brazil was ruled by an aristocratic and conservative society, driven by an agricultural economy that depended exclusively on slave labor. This was the period in which the coffee culture granted economic stability to the empire, allowing some few entrepreneurs, like Mr. Ireineu Evangelista de Sousa[1], or the Baron de Mauá, to open a bank and to give financial incentives to large and diversified private enterprises, as, for example, the construction of the first Brazilian railway road in Rio de Janeiro. The construction of this railroad was an impressive achievement for the time since this construction was made up of bridges and tunnels, the largest of such with over 2000 meters in span. There are unfortunately few reports on the foundations of such bridges, nonetheless such engineering work can be highlighted as a true benchmark for the geotechnical engineering in Brazil. From then on, many other railroads and associated engineering works were built along the vast domains of the empire.

At the same time the construction of civil houses began to use panels made of bricks allied to beams and iron columns, due to the collapse of existing constructions built solely on "taipa"[2]. It emerged, since then, the execution of industrial buildings, markets, warehouses, railway stations and associated works, as the "Light Station" in São Paulo. This construction, among others erected between 1884 and 1896 in the cities of São Paulo, Rio de Janeiro, Recife and Manaus, associated the use of imported bricks, steel structural elements, and a metal roof. In view of such developments, particularly in the structural field, it became also necessary to leverage and improve current construction (and design) techniques for the foundations. The trenches filled with "compacted" stones were soon replaced by shallow foundations, strip foundations, or rigid blocks of brick masonry or carved rock, manually laid down at foundation base and compacted inside the ditches.

Figure 1. Historic pictures from the Pedregulho reservoir (1880 Marc Ferrez, from Gilberto Ferrez collection and Public Archives from the city of Rio de Janeiro).

According to Vargas (1998), in the late XIX century, already after the installation of the first engineering schools in the country, witnesses could observe that the Brazilian civil engineers held some understanding of the "technology of soil".

It is worthy to mention the interesting case of the water reservoirs from the Pedregulho station in Rio de Janeiro (Figure 1), where two tanks were constructed to storage water for this (expanding) city, one upstream of the other with a differential topographic level of 5 meters.

After several days of tropical rain, the reservoir located on lower level, built near the slope base, had several cracks in the walls and had to be immediately emptied. According to this author, it is nowadays

[1] *Irineu Evangelista de Sousa* (1813–1889), was a merchant, shipowner, industrial and Brazilian banker. Given the thorough contribution to the industrialization of Brazil in the Empire period (1822–1889), he was granted the nobility status of Baron (1854) and later Viscount of Mauá (1874). He was a pioneer in several areas of Brazil's economy, among which the establishment of the first iron melting plant and shipyard in the country, the construction of the first railway, the railroad Mauá, the installation of a gas lighting system in the city of Rio de Janeiro, the creation of the first private bank in Brazil, and the installation of a submarine telegraph cable between South America and Europe (Wikipedia citation).

[2] *Taipa* is a constructive technique imported from the Portuguese colonizers, still used nowadays in the poor inner rural areas of Brazil, which is based on the manual compaction of a mixture of local (predominantly clayey) soil, water and gravel, to form and erect vertical walls (adapted from Wikipedia citation).

known the formal transcripts of a meeting dated July 12, 1880, i.e., over two months after the construction of the reservoir. The records reveal that engineer Honório Bicalho[3] diagnosed the cause of the accident as due to differential settlements that were produced by "the lack of homogeneity of the terrain, related to the decomposition of the primitive rock, gneiss, which originated a residual soil of compressible and collapsible characteristics".

During the nineteenth century the interest in mining works, particularly iron mining, started to gain strength in Brazil. From then on, it started a large demand for geological surveys in such related works, and a close relationship between geology and (geotechnical) civil engineering was established in the country. Then, in 1907, initiates the "Engineering Geology" field of expertise, which early works lie in the geological survey for the implementation of railways and roads. The engineering geology also significantly contributed at such time (as nowadays) to the development and implementation of the projects of the foundations of embankments and dams, or the art works (bridges, tunnels) of road/railway constructions.

With the expansion of the empire, enlargement of the cities and population growth, and associated natural evolution of the engineering problems, new challenges have emerged by the mid to the final period of this century. Many problems were related to pier foundations designed for the construction of ports, in regions comprised by geologically characterized quaternary deposits, that means with a high frequency of sedimentary saturated layers of soft soil materials.

It is also reported that even though wooden piles were known and openly used at such time (as for "classical" example the Municipal Theater of Rio de Janeiro, erected in early XX century), the lack of knowledge on the behavior of the subsoil, allied to the inexistence of settlement control techniques, led local contractors to avoid constructing new buildings in areas of soft deposits. On the other hand, it was precisely the construction of a foundation comprised of timber piles, driven by an innovative (for the time) steam fuelled hammer, which promoted the successful construction of deep foundations on such coastal sedimentary strata. The particular case of the design and construction of the Customs Docks in Rio de Janeiro, which started in 1866, is an example of that.

2.3 *Republican period (XXth century)*

In the early twentieth century the designers and contractors started to count with one of the best construction materials of the modern civil engineering industry: The reinforced concrete. It is of knowledge that concrete multi layer structures of this era could be built over shallow footings of reinforced concrete or on top of massive rigid blocks of non reinforced concrete. Besides, deep foundations could be represented by reinforced concrete types in addition to the already in use wood ones. This era was also initially marked by the entry in the market of the first specialized company on seaport construction company and pile foundations in Rio de Janeiro, and by the expansion of the port of Santos, where the first reinforced concrete pier supported by concrete piles and retained by a sheetpiling wall was executed. Between 1926 and 1927 the dock of Rio de Janeiro was also expanded and the foundations consisted of pressurized piers executed with driven shaft elements of reinforced concrete.

However, until the end of the '20s, empiricism was predominant in geotechnical engineering. "Experimentation, however little, was carried out directly in the stonemasonry of the work and in a restrained fashion", according to the former ABMS president, the engineer, Antonio Nápoles Neto, in a lecture given in 1970 (Sayão, 2010). Possibly the first Brazilian geotechnical undertaking, was described by Domingos da Silva Cunha in the Brazilian Journal of Engineering (October, 1920) with the title "Experiments on terrain for the study of foundations". In this paper, Cunha warned of the variations in resistance and deformation of the soil as a function of chemical content variations in the terrain's humidity and depth.

Nevertheless, it was precisely in the next decade that the first academic investigations were carried out by technological research institutes of the young Republic, and by the Polytechnic School of São Paulo (USP), the latter one more focused on the particular problems of the foundation engineering. In 1934, the Laboratory for Material Testing of the city of São Paulo was transformed into the Institute of Technological Research of São Paulo (IPT), attached to the academic activities of the Polytechnic School. Early work in IPT started in 1938 under the coordination of engineer Odair Grillo[4] and aimed both the correct paving of earth roads and the study of foundations of bridges and buildings. In regard to foundation engineering, which is the scope of this paper, the first step to solving engineering problems was the development in 1939 (Grillo, 1939) of a drilling/sounding equipment for the exploration and analysis of the subsoil. Thus, IPT standardized a soil exploratory device method, later published by Vargas (1945), which consisted of the introduction of a standard thin walled steel sampler by a percussion method with water circulation, simultaneously to the counting of the required number of strokes of a falling impact hammer of 60 kg in weight, from a gravity drop height of 75 cm. The number of blows necessary to penetrate this sampler to a standard length of 1 feet (30 cm) was termed as $N_{(IPT)}$. Besides

[3] *Honório Bicalho* was an engineer graduated by the Military School, later Polytechnic School of Rio de Janeiro (nowadays known as the Federal University of Rio de Janeiro, UFRJ), who has actively participated in the construction of public works of ports and railways in the Brazilian monarchic empire (Wikipedia citation).

[4] *Odair Grillo* is recognized as the first Brazilian, graduated in engineering in 1934, to use in design the soil mechanics with practical and research grounded geotechnics in the manner as it is understood nowadays. In 1947 founded the first course on soil mechanics at the USP Polytechnic School (Wikipedia citation).

this number, another standardization was created by a private company of research and project in soils and foundations. This latter one also started to be largely adopted in the country, with a distinctive nomenclature of $N_{(MG)}$. Subsequently, Terzaghi and Peck (1948) suggested the worldwide method commonly practiced today, known as Standard Penetration Test (SPT), with its corresponding $N_{(SPT)}$ number. The high variability of the testing results from both "Brazilian" techniques at time, given differences in procedure and especially due to distinct "standard" N quantifications, posed serious problems of comparison and foundation design (via empirical rules, as still used nowadays). Therefore, over time, persuaded by periodic discussions on technical congresses, the IPT's number N variable was replaced by the international SPT quantification method of Terzaghi and Peck (1948), which approach was later on regulated and standardized for the whole country by the Normative NBR 6484 in its first version of 1980.

Among the numerous construction sites and foundation works carried out in the country during the 20th century, some are worth mentioning and deserve to be described, especially given their innovative adopted solutions and the engineering difficulties they faced with the available technology of the time.

A first case that deserves mention in the city of São Paulo, around 1925, is the construction of the first skyscraper in Latin America, the Martinelli building (Figure 2). The foundations were laid down using wooden piles and spread footings up to 16 m depth. It required lowering the water level, which resulted in structural damage to a neighboring building, leading to a temporary suspension of the work. The 25 storey, reinforced concrete Martinelli skyscraper was considered the highest in the world at the time.

Figure 2. Martinelli building in São Paulo (after Sayão, 2010).

Another case was the construction of the foundations of the State Bank Building. It shall be initially mentioned that, given the experience coming from previous geotechnical investigation campaigns in São Paulo, designers and contractors already had some knowledge beyond the local geotechnical parameters, i.e., they could also anticipate somehow the foundation behavior of the common buildings in the region. Nevertheless, in the present case, the execution process of the piles of this building encompassed the dynamic penetration of closed end tubes through a stiff clay stratum. This execution feature caused the lifting of the surrounding ground level by values around 70 cm, leading to breakage of surrounding piles that have been (previously) molded in place. It was observed by load tests carried out in the site that such piles had compressive bearing capacities as low as 30% of the expected values. The problem was then solved by the installation of additional piles, summing up to a total of 400 piles in this particular work, all of them partially excavated around the shaft after the casting period to prevent damage during the driven stage of the new piles.

Another example that also deserves mention is the case of the Building of the Paulista Insurance Company in São Paulo. This is a 26 storey building construction that began in 1940, and in which the investigation of the subsoil was partially conducted inside the construction area given the existence of an old building that was about to be demolished (during further construction stages). Thus, the piles were designed based solely in one of the sides of the site, which allowed those foundations to behave as end bearing due to an existing shallow stiff clay stratum. However, it happens that on the other (non surveyed) side existed a layer of soft silt. This misconception in design was further aggravated by the (satisfactory) load tests solely carried out on piles that were executed on the investigated side. After three years, in 1943, the building was opened and soon afterwards, the IPT Institute detected in the monitoring records an accelerating tendency of column settlements from the building side located in the non investigated part of the site. Diary differences, or settlement increases of up to 1 mm from morning to afternoon readings, were noticed, together with a steadily tilting of the building. This edification is presented in Figure 3.

According to what was presented by Villares (1956), the solution of the problem was achieved by an overall freezing process of the subsoil in 162 freezing steps, in

Figure 3. Historic picture from the end-of-construction building of the Paulista Insurance Company (after Sayão, 2010).

which the final temperature was kept at −20° Celsius. It was followed by the drilling of injections wells, where galvanized metallic pipes were installed up to 14 m in length. With the hardened ground and transitory suspension of settlements, injection of concrete was carried out through these pipes, transforming them into cast-in-situ reinforcement piles. The building was finally realigned in the vertical direction by the temporary insertion and pumping of hydraulic jacks onto existing columns.

Still around the '40s two other studies conducted by the IPT stand out. One of them is related to the foundations of the bridge over the Rio Grande river during the construction of the railway Corumbá-Santa Cruz de la Sierra (in Bolivia), which crossed a terrain filled with quicksands and swelling clays. The other one refers to the design of the foundations of the Steel plant in Volta Redonda, where for the first time it was noticed foundation problems associated with soils derived from rock alteration, or saprolites (Nápoles Neto and Lorraine, 1956).

In the '50s, and with continuing studies of the previous decade, both IPT and the private company Geotécnica SA monitored the settlements in tall buildings founded on shallow foundations at the coastal zone of Santos, an area typically known (later on) by its thick layers of saturated soft clays. The results

Figure 4. Construction of the foundations of the ministry buildings (1959) (Source: Public Archives of the Federal District).

of the investigation in Santos were henceforth published by Teixeira (1959a, 1959b and 1959c), by Machado (1961) and by Vargas (1961), the latter ones related to the settlement of foundations over soft sand deposits. Burland et al. (1977) during the state-of-art of an international conference on Soil Mechanics presented arguments and conclusions about the Brazilian experience in forecasting and monitoring the settlements in buildings.

Within this same decade, the crossing project of the Guaíba river in Rio Grande do Sul deserves a special note, as this project highlights the use of bridges founded on deep foundations of the Franki type with relatively long lengths. In Rio de Janeiro the project and construction of the building Marquês de Herval, conducted by Prof. Costa Nunes[5], also calls for attention. This building was constructed by subsoil excavation with the simultaneous water level drawdown to up to 9 meters of depth, in a region surrounded by older constructions founded on shallow foundations.

It was also in the '50s that the construction of today's capital of the Federative Republic of Brazil, the city of Brasilia, has started. Figures 4 and 5 present photos of the construction phase of the ministry buildings, in the heart of the new city. Rio de Janeiro's companies were responsible for conducting the site investigation of the subsoil, which is composed of sandstones and siltstones covered by a layer of residual soil, originated from the same typical metamorphic type rocks of the region. Faced with such distinctive (from the coastal zone) geological conditions, the contractors opted for deep foundations with Franki type piles and uncased manually excavated caissons.

[5]*Antônio José da Costa Nunes* was one of founders of the Brazilian Association of Soil Mechanics (ABMS), being its second President in 1952. A renowned Professional at both academic and industry levels, he founded in 1957 one of the largest geotechnical private companies in Brazil, the Tecnosolo SA. After him is named one of the most prestigious medals conceded biannually by ABMS to the best DSc Thesis of the country.

Figure 5. Construction of the ministry buildings in Brasilia (1959) (Source: Public Archives of the Federal District).

Later on in Rio de Janeiro, between 1960 and 1964, the Duque de Caxias[6] oil refinery was built by employing several types of deep foundations, among them pressurized drilled caissons, which led to a total length that exceeded 400 km of piles. Other relevant information of this work is the consumption of concrete and steel that reached 120 cubic meters and 10 tons respectively. Since then, one could affirm that the foundation engineering was finally consolidated in the country.

In 1963, the José Bonifácio[7] steel plant, belonging to the Paulista Steel Group (Cosipa) was inaugurated close to Santos. The region is characterized by the presence of wetlands with soft clay deposits at surface extending to up to 25 m in depth. Faced with such geological conditions the solution by shallow foundations proved to be unattainable, allowing the exclusive use of deep foundations which totaled 110 thousand piles varying from centrifuged precast concrete types, metallic H shaped piles and mixed (composite) piles. The main problems encountered in implementing the foundations of the plant were the execution of the jointing sections of the long piles, the enhanced corrosion by the marine environment, the excess pore pressures generated during the dynamic insertion of the piles and negative friction between the shaft and the surrounding soft clay by consolidation phenomena. To understand and manage the behavior of the foundations, the IPT conducted over 300 pile load tests.

Still in the '60s, the construction of the highway Piaçaguera-Guarujá, now known by Cônego Domênico Rangoni, has started close to Santos. About 12km of the road lies in a region of very soft organic soils up to 40 meters in thickness. Six bridges were necessary to be built in this highway, with particular emphasis on the one crossing the Bertioga channel, that required 8 pressurized drilled caissons in the central span and dozens of metallic piles, with varying lengths from 10 to 60 m on the edges. The road was opened in 1970 as a result of the implementation of pioneering techniques of geotechnical engineering.

Figure 6. Example of pressurized drilled caisson platform and site work over a river in the interior part of Brazil (Source: Personal databank from the authors).

The '70s was marked by several works of great technical challenge and financial impact for the country, taking advantage of the good winds in which Brazil was sailing on at the time (nowadays recalled as the "economic miracle" era). Among several works it is highlighted the Transamazônica highway (1972) through the rainforest, The Lagoa-Barra highway (1971), the Steel railway (1973), the Immigrants highway (1976), the undergrounds at both Rio de Janeiro and São Paulo cities, and some of the astonishing in size, for the time, hydroelectric power plants (Itaipú, Tucuruí) that still serve the country nowadays.

Regarding the work of foundations, the literature highlights the Rio-Niterói Bridge (1974), where a tragic accident took place during a load test over the Guanabara Bay, causing the death of two workers. This bridge is 13.3 km long and 26.6 m wide, with six lanes of traffic, and is still considered nowadays a national landmark of engineering, given all the executive difficulties and the geotechnical particularities. As for the bridge foundations themselves, in the dry (land) sections metallic and Franki piles were used, whereas in the wet (sea) section pressurized 1.8 m diameter drilled caissons were constructed with lengths that varied according to the thickness of the superficial organic soft clay of the bay. This was considered a pioneer solution for the time. Altogether 1138 caissons were executed, of these 462 were pressurized by air, 199 were of mixed type and 477 used the "Bade-Wirth" technology. The pressurized caissons with enlarged

[6] Luis Alves de Lima e Silva, the Duke of Caxias, (1803–1880), nicknamed "The Peacemaker" and "The Iron Marshal", was one of the most important statesmen of the military and history of the Empire of Brazil.

[7] Jose Bonifacio de Andrada e Silva (1763–1838) was a naturalist, poet and statesman. It is known by the epithet of "Patriarch of Independence" for being a decisive one for the Independence of Brazil.

Figure 7. Angra dos Reis nuclear power plant, which construction started in 1970 (Source: Portal Brasil, internet governmental site of free access).

base (Figure 6) were adopted whenever it was possible to support them in the sedimentary compacted sandy stratum located underneath the soft clay, up to a total length of 30 meters. The mixed types were executed with the dynamic insertion of a steel pile into a young residual soil and the subsequent submerged concrete molding of the caisson's shaft. Finally, the Bade-Wirth types were founded on rock and were mechanically drilled with the support of a metallic shaft, following the subsequent structural reinforcement with a steel cage that was lowered down just before the final molding stage.

The '80s began with works of great impact that in the vast majority have been started in the preceding decade. As key examples one can recall the first nuclear power plant in Brazil (Angra dos Reis 1 – Figure 7) and the conclusion of the hydroelectric power plants. This decade is also marked by the construction of the Carajás railroad, the expansion of the Rio de Janeiro international Airport of Galeão (also known by Tom Jobim[8] airport) and the construction of the international Airport of Guarulhos in São Paulo.

In 1982 it was opened the Port of Sepetiba in Itaguaí, renamed in 2006 to Port of Itaguaí. The region of the implementation of the port has a subsoil typically composed by an organic silty clay of high compressibility and low bearing resistance, with thicknesses of up to 12 meters. Large diameter and length piles were designed and implemented for the storage yard of the port, and the final solution also encompassed the dredging of the soft clay layer and its replacement by a hydraulically compacted sandy fill.

In the '90s the magnificence of the Brazilian economy slowed down, given international external

Figure 8. Foundation reinforcement at Paulista Mall (after Sayão, 2010).

problems, as the "petrol crises", and national internal financial calamities, as hyperinflation and recession (nowadays recalled as the "lost" era). Nevertheless some milestone works can be recalled upon, as the excavation and foundation reinforcement of the Sears store in São Paulo in 1989, to what would be known as the Paulista Shopping Center. Located in an area of large commercial appeal and high acquisitive economic power, this work was marked by its great difficulty (see some of the few remaining pictures on Figure 8).

The original conception was the deployment of three additional basements to the original Sears Mall, built in 1949. It is reported that the original building consisted of three floors and a basement with a level difference of 4.3 m to the street. The first building block had shallow foundations founded at 2 meters below the basement level, and the second block was supported by drilled caissons of 14 meters long, which were constructed in 1972 during the initial expansion project of this mall.

The foundation solution adopted for the first building block was the substitution of the shallow foundations to deep ones, allowing the excavation of extra 9 meters of soil, or the 3 additional required basements. For the second block, the strategy was the simultaneous reinforcement of the existing caissons along with the excavation. At the end, the caisson's bases were enlarged and concreted with additional steel reinforcement cages. Besides of the foundations, retaining structures made of (secant and non secant) pile curtain walls with caissons of enlarged (and non enlarged) bases were constructed to safely sustain the excavation process. It is worthy to mention that

[8] *Antonio Carlos Brasileiro de Almeida Jobim* (1927–1994), better known as Tom Jobim, was a composer, conductor, pianist, singer, arranger and Brazilian guitarist. It is considered the greatest ever exponent of Brazilian music, and one of the creators of the Bossa Nova movement. Composer of the song "The Girl from Ipanema" (adapted from Wikipedia citation).

Figure 9. Updated picture from the on going construction of the new sports arena in Brasília (*Brasilia2014* official internet site).

the commercial activities of the Sears Mall were not interrupted during the execution of such services.

2.4 *New millennium (XXIst century)*

Similar to the readjusting tides, so are the economies of the world. After an era of depression and internal readjustments, which painfully demanded many personal sacrifices, political compromises and a new economic order, Brazil was able to turn over the table and reestablish itself in the path of sustained development and growth. Being part of what is called nowadays the BRIC[9] group, the engineering possibilities, challenges and achievements for the new millennium are undoubtedly broad and large, as well as the encouraging mood and optimistic feeling of the people inside and outside the country. The coming Football World Cup event (in 2014) and the Olympic Games (in 2016) have called the attention for a general readjustment of key urban elements, as transportation, airports, housing, hotel and entertainment venues, energy, security and, off course, stadiums and sport arenas of several types (for instance, see Figure 9).

Therefore, among the various geotechnical works which debuted together with the starting millennium, one shall highlight the full renovation of the (tilted) buildings in Santos, the expansion of the Porto Alegre airport, the conclusion of the express South highway of Florianópolis, the execution of the piled embankment of the SESC, close to the future installations of the Olympic villa in Rio de Janeiro and the diverting beltway around São Paulo.

Also, it is worthy to mention the constructions of refineries and pipelines in the northeastern and northern regions of Brazil, the expansion of the petrol industry by the discovery of unknown pre-salt[10] oil basin, the renovation and construction of new sport arenas throughout the 12 base cities of the coming World Cup, and the (fast growing) residential urban expansion of interior cities all over the country, among other things.

The fast development of the country involves not only the use of more sophisticated techniques for foundations to support the increasing magnitude of the loads from superstructures that become more slender and tall everyday, but it also involves the formation of a skilled working force to safely deal with, design and construct such demanding structures. This point is valid for all fields of engineering, from foundation, to civil, mechanical, electrical, naval, energy and so on. Besides, the increasing demand for sustainable construction, space optimization and green areas have obliged urban constructions to use "less noble" subsoil areas for parking, transportation, storage and related items, leaving open environments for the more noble activities of living, entertaining and working.

Therefore, there is a potential challenge to develop and sustain an increasing demand for technologies capable of producing deep foundations with longer, slender and vibration/noise free piles, that would intervene and disrupt to a minimum the surrounding environment, and would eliminate the production of undesirable (unfriendly environmental) by-products, as contaminated soil, soil-bentonite mixtures, etc. The retaining structures have been increasingly important in all aforementioned engineering solutions, and shall (and will surely) be enhanced together with the technological developments of the foundations.

It is with such spirit that one must realize that the foundation engineers of the future (those who are still working or studying today) will face a new reality in which the foundation problems will be engaged and solved altogether with other perspectives, i.e., from the geological and geotechnical point of view to the environmental restrictions, special construction characteristics, optimization procedures, and for sure the sustainable attitudes of the new age.

---

[9] In economics, *BRIC* is a grouping acronym that refers to the countries of Brazil, Russia, India and China, which are all deemed to be at a similar stage of newly advanced economic development. It is typically rendered as "the BRICs" or "the BRIC countries" or "the BRIC economies" or alternatively as the "Big Four". Goldman Sachs has argued that, since the four BRIC countries are developing rapidly, by 2050 their combined economies could eclipse the combined economies of the current richest countries of the world. These four countries, combined, currently account for more than a quarter of the world's land area and more than 40% of the world's population (Wikipedia citation).

[10] In 2007, the Brazilian oil company Petrobrás announced the discovery of oil in the so-called *"Pre-Salt"* layer, which later turned out to be a large oil field, extending over 800 km off the Brazilian coast, the state of Espírito Santo to Santa Catarina, below thick layers of salt (rock salt) and encompassing the sedimentary basins of Espirito Santo, Campos and Santos. The first pre-salt oil was extracted in 2008 in the Tupi platform and some wells are being tested and should begin commercial production in this decade. It is estimated that there are stored about 80 billion barrels of oil and gas, which would put Brazil in the privileged position of sixth largest holder of reserves in the world – behind Saudi Arabia, Iran, Iraq, Kuwait and UAE (Wikipedia citation).

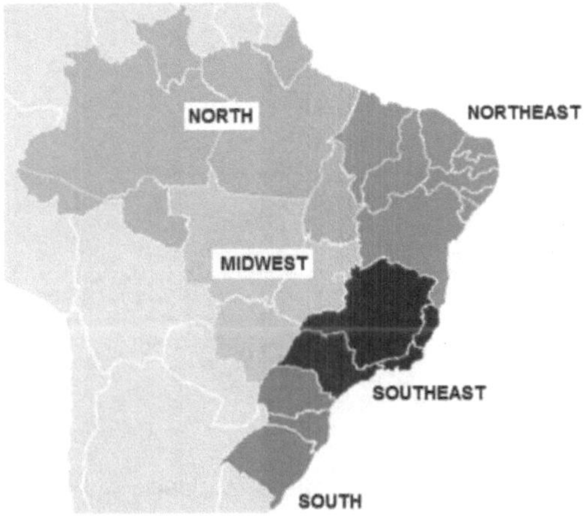

Figure 10. Regions of Brazil (adapted from the *Wikipedia* site).

3 TECHNIQUES AROUND THE COUNTRY

3.1 *General overview of techniques*

The foundation techniques employed in the Brazilian territory have close relationship with the differential development at regional level, and their inherent geological, pedogenetic, geographic and, off course, geotechnical characteristics.

Brazil is a continental country of 8.5 millions km^2, that means around 82% of the total area of the continental Europe. The country is vast and regionally distinct throughout its five federative main regions (North, Northeast, Midwest, Southeast and South – see Figure 10), comprising 26 states plus the Federal District, which is a special unit of the federation.

Although large and habitable in many parts (with the exceptional living and environmental difficulties of the areas covered by rainforest or floodplains) the country has its residents disproportionately distributed, since it has almost 70% of its population living in a narrow band of land no wider than 200 km from the coastline (*brazilazul* internet site). This is so perhaps because it has a long coastal region that stretches across the Atlantic Ocean, covering around 7400 km in length. Maybe due to historical reasons of the closer proximity to existing commerce points (as sea ports), or simply given geographical facilities, i.e., along most of the coastline there is a lack of abrupt geographical features of large magnitude and complexity, and the land and sea harmoniously coexist.

Brazil has indeed passed through a large array of geological processes that gave rise to a variety of rocks, whose geographical distribution is complex and also decisive in the history of occupation and development of the country. The distribution of different rock units in the country is characterized by the existence of two geotectonic provinces, the Basement and the Sedimentary Basins. Moreover, it is an old and stable (given seismic records) territory in the South American tectonic plate.

The geologic formations distributed on the surface of the continent went through extensive weathering processes that formed soil profiles which are typical in certain regions, such as, for example, the presence of evolved soils in the South, (parts of the) Southeast, and Midwest sections of the country. These soils are generally structurally porous with lower water levels, being predominantly residual in nature. In the coastal region (parts of the South, Southeast & Northeast), or within the rainforest (North), there is a predominance of distinctive soil types, given the abundance of deep sedimentary strata formed under alluvial, lacustrine, or marine environments, which led to alternating layers of saturated clays and sands.

Given aforementioned geotechnical characteristics of each location, simplified and generalized in terms to the reader, the inner and coastal areas of Brazil, from north to south and from east to west, adopted particular, distinctive foundation execution techniques to deal with their own characteristics. This allowed feasible, economical and safe "typical" mainline foundation types in each region, that could be locally built under the general Brazilian guidelines for allowable limiting states of failure and service.

The more recent evolution of techniques for the implementation of foundations in Brazil dates from the late '90's, with large-scale employment of continuous flight auger (CFA) piles throughout the regions, something which was then restricted to the Rio-São Paulo area of influence. Actually this technique has spread throughout the country given the emergence of manufacturers of equipment for this type of foundation, the progressive price decrease of labor and parts for this technique, and the designers' advancement of knowledge (and confidence) on the behavior of this type of foundation under different geotechnical environments.

Up to the mentioned period, say around the early '90's, Franki, precast concrete, metallic and bored type piles with or without bentonite mud were extensively employed at large civil engineering ventures. But with the advent of increased foundation loads due to the emergence of taller buildings with multilayer basement floors (as parking spacing has become crucial in nowadays overpopulated urban cities), there was a natural need to change somehow the design conception of the foundation projects. Thus, this aspect allowed the preferable use of "environmentally friendly" executive techniques that lead to lesser impacts in terms of noise, vibration, excavation and material usage (as the use of bentonite mud, or river sand).

Nowadays there is a strong tendency of using both the CFA and the bored (non displacement) pile types in large constructions of modern urban centers of the country, as, for instance, São Paulo, Rio de Janeiro, Belo Horizonte, Brasília, Salvador, and so on. Other types of foundations, as the (displacement) precast concrete driven and the Franki pile types, or the (non displacement) drilled caissons are still used, but preferably in industrial regions or zones away from large urban conglomerates (personal experience with colleagues from the Foundation industry). Exceptions for

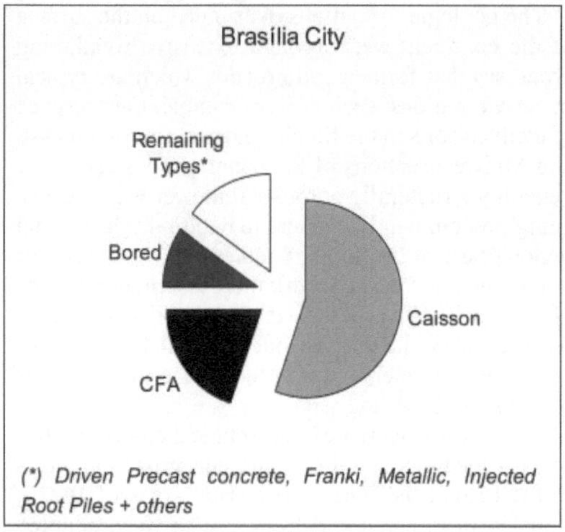

Figure 11. Illustrative example of the approximate distribution of pile types currently executed in the city of Brasília.

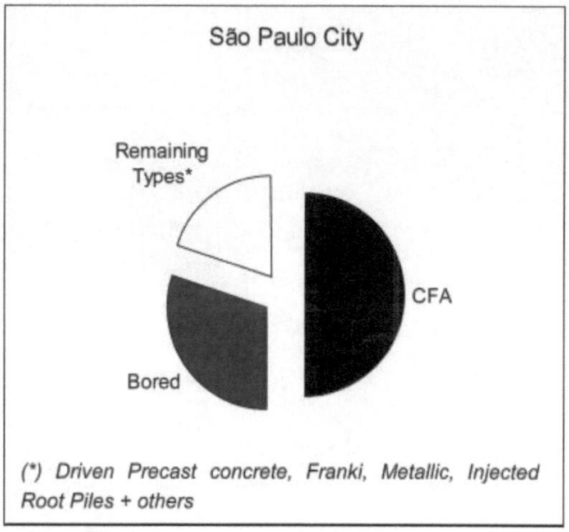

Figure 12. Illustrative example of the approximate distribution of pile types currently executed in the city of São Paulo.

this rule do exist, given the particularities of each of the numerous cities of the country. For instance, the capital Brasília is a typical example. Given its geological and geotechnical characteristics, with a well-known stable, structured and unsaturated superficial tropical deposit with the locally famous "porous" clay (around more than 85% of the Federal District area), Brasília has always been the place where open case, non pressurized drilled caissons could be competitive. Nevertheless, this tendency is slowly fading away given legal and union issues allied to the escalating usage of CFA piles in the Brazilian market, and it may well soon disappear in the future, at least for the urban area of Brasília.

This particular aspect is clearly visible when one compares the distribution of piles types (also based on personal experience and discussion with industry professionals) from both cities of Brasília and São Paulo, as respectively depicted in Figures 11 and 12.

As noticed before, each city in the country has its "typical" tendencies and preferences, nevertheless an overall trend can be established – as it will be tentatively done in the following section.

3.2 *Generalized tendency for the regions*

As previously stated, Brazil is currently going through a period of systematic growth of its economy, by virtue of a sustainable and responsible economic policy (that started way back a decade ago), where the focus is generally on large infrastructure investments (petrol and hydropower energy, transportation, exportation facilities & sport arenas) or the construction of numerous housing units to reduce the (large) existing habitational deficit of the country. Hence, in all regions of the country it is clearly detectable many engineering sites that adopt high standard techniques to execute modern foundation types, with the basic purpose to ally security and environmental issues to more condensed working timetables.

Having said that, one could perhaps generalize the usage of the onshore foundation techniques in Brazil (simply) in accordance to the distinctive geological and geotechnical aspects of the regions, as follows:

- Shallow foundations (flexible, rigid blocks or rafts), cast-in-situ and precast driven deep foundations are generally adopted in *areas of evolved or residual (tropical) soil deposits*, saturated or not. In urban or nearby areas, the monitored continuous flight auger (CFA) pile type tends to be the preferable solution from designers, as among other advantages, it can reach diameters up to 1.2 m and maximum lengths of 35 m. In addition to these solutions, it is also common the use of pre-stressed centrifuged (or full section) precast driven concrete piles, inserted into place by hydraulic or diesel hammers or simply by gravity drop hammers adapted to crawler cranes. The employment of enlarged base caissons, or bored piles socketed into the local rock, is also an alternative especially in regions without water level. In the sites where there is a reasonable distance to concrete plants or to precast pile factories, bored piles or Franki type piles are frequently used. It must also be noted the use of "root" (injected) pile types in specific projects, given their own executive peculiarities as lightweight equipment, possibility of embedment into the rock, feasibility of execution in deposits with boulders, and allowance to inclined piles to withstand eventual loads in the horizontal direction. The root piles have been particularly used in such regions to bear energy transmission towers, underpin existing structures, or to allow foundations through deposits of complex hard-soft stratifications. One shall also observe the lack, or limited presence, in this paper of commentaries on metallic piles (H, I, circular shaped, used railway guides, etc.). Actually, they are generally used when the remaining solutions cannot be employed for one reason or another. Although the metallic piles do have several advantages, the ductility being one of them, they are unfortunately still expensive in the country. More recently other types of (large scale) deep foundations started to be used in limited projects of these regions, as for example the "displacement" continuous flight auger pile type (previously named as Omega pile), barrette type piles with hidromill cutters (still used when combined with underground excavation works for subways), large diameter bored piles with "bucket" type barrels, that stabilize the shaft with the use of a polymeric fluid or metallic casing, and large diameter bored piles (cased or not) drilled to fresh rock by special drilling bit coupled machines. The technique of piled rafts[11] have also advanced in Brazil, specially as solution of tall edifications in the Midwest region, where one takes advantage of the combined support of the superficial raft (founded on hard stratum) with the slender long deep piles. Nevertheless, this solution is still considered "too" innovative in the country, and its usage restricted to few areas where expertise and work conditions allow its deployment;

- The coastal region and some parts of the north region (along the rainforest), as previously mentioned, present subsurface *deposits of fluvial-marine origin*, with levels of sandy marine terraces, mangrove deposits, lacustrine-fluvial sediments, and recent alluvial deposits. It can be found layers of soft and compact sandy materials interspersed with layers of thick soft marine clay/silt stratums. In the southeast and northeast regions two distinct but combined landscapes are generally found: hills and plains. Thus, the urban occupation with large buildings has been developed (and is under way) within a confined stretch of space confined between the hills and the coastline (typical case of Rio de Janeiro, for instance). The predominant presence of superficial dense sand layers in most of the surface area of the lowlands allows the adoption of shallow foundations for these areas. In cases where such deposits are characteristically formed with loose stratums, shallow foundations continue to be used when coupled to dynamic compaction techniques (mainly in the northeast). Sometimes the deposits are previously reinforced with the use of compaction gravel/sand piles, so that such foundations continue to be the preferred design solution. Nevertheless, in the absence of the "pre treatment" alternatives, particularly for large edifications, the design returns to the normal deep foundation alternatives, being the CFA, the precast concrete, and the Franki piles the preferred option in most of the coastal cities. In cases of compressible and saturated soft (fine) deposits, which generally account for around 50% of the plain lowland areas, the decision is made exclusively in terms of deep foundations. For such cases, the designers favor the use of precast concrete driven piles, Franki piles, rotary-injected and root piles, CFA piles, and metallic type piles, with normal or telescoped sections. The latter type is characterized by the use of a reducing cross section with depth, which allows a decrease in the pile's weight while maintaining similar bearing capacities. Depending on both design and combination of different steel sections that make up the pile, one can reach savings between 15% and 25% on the total weight (likewise, financial savings). Bored piles by standard techniques are still used, although the requirement of a temporary metallic casing can turn this solution non-competitive in face of the other alternatives. The stabilization of the hole by bentonite mud can be also used sometimes, although it is slowly being replaced by methods (solutions) that are more acceptable. Finally, though environmentally out of fashion, timber or wooden piles are still used in the country, however confined to specific areas (north especially), particular projects (small edifications and docks or ports), and peculiar conditions (bridges in the rainforest for military or civilian use). In such cases, the wooden piles are driven into position by using regular cranes and gravity drop hammers.

[11] According to Janda et al. (2009), piled-raft is a "foundation system in which both structural components (piles and top raft) interact with each other and with the surrounding soil to sustain vertical, horizontal or moment loads coming from supported superstructures."

4 CONCLUSIONS

Despite the fact of being one of the South American countries that got at a later stage its independency from the colonial power (1822), Brazil was able to accompany and even surpass the urban development of many of its neighbors. This aspect is not only reflected in the foundation engineering field but in many others, say from energy generation to underwater petrol extraction, and generally relates to both design and construction points of view.

At present time, the conditions and perspectives are propitious to the adoption of new foundation technologies in the country, aiming the modernization of the companies, the optimization of the construction processes, and off course the embracing of environmentally friendly procedures. Nevertheless, the challenges ahead are still great and shall demand a new mentality from part of contractors, designers, site workers, clients and the government itself.

The continuing investment on new machines and technology, the formation of a skilled work labor at distinct levels (from technicians to Ph.D. Professors), the appreciation of the high education value, the national and international outsourcing of exceptional professionals, the free exchange of knowledge between multi-level institutions, and the urgent reconfiguration of many of the outdated Brazilian standards are some of the complex (interrelated) items that must be especially valued by the Engineering Societies and public in general.

Notwithstanding such large mid-term necessities, this manuscript has somehow attested the capacity that this country has, under unfavorable foundation engineering conditions, to fully overcome its own inherent difficulties of size, labor, equipment, transport, information, communication, landscape and geotechnics. The paper demonstrates in its generalized and simplified manner through historical time that Brazil was undoubtedly able to overcome the pitfalls over its path, and therefore has the will, the means, the technical capacity and surely the high motivation to successfully continue pursuing its own objectives in the near future.

ACKNOWLEDGEMENTS

This paper wouldn't be feasible without the meticulous hard work carried out by MSc Engineer Osvaldo de Freitas Neto, who carefully conducted most of the bibliographic survey of the historic information presented herein. The authors therefore express their deep gratitude for Mr. Neto's accomplishment that allowed an invaluable source of data that enriched the manuscript.

Traveling funds for international events kindly provided to the first author by the University of Brasília (DPP/UnB), and by both CNPq and CAPES governmental organizations, is also acknowledged and valued.

REFERENCES

ABMS. 1996. *Foundations: Theory and Practice*. 2nd edition, PINI Editors, São Paulo. 751p. (In Portuguese).

Burland, J. B., Broms, B. B. and de Mello, V. F. B. 1977. Behavior of foundations and structures. State of Art Report. *Proceedings of the 9th International Conference on Soil Mechanics and Foundation Engineering*, Tokyo, Vol. 2, pp. 495–546.

Grillo, O. 1939. The Soil Mechanics at the Technical Road. *Bulletin of the DER*. Year V, Vol. 5, No. 4. São Paulo (In Portuguese).

Janda, T., Cunha, R.P., Kuklík, P. and Anjos, G.M. 2009. Three dimensional finite element analysis and back-analysis of CFA standard pile groups and piled rafts founded on tropical soil. *Soil and Rocks*, Vol. 32, No. 1, pp. 3–18.

Katinsky, J.R. 1994. *Colonial Construction systems. History of Technique and Technology in Brazil*. UNESP. CEETEPS, São Paulo. (In Portuguese).

Machado, J. 1961. Settlement of Structures in the City of Santos, Brazil. *Proceedings 5 th Int. Conf. Soil Mechanics and Found. Eng.*, Paris.

Nápoles Neto, A.D.F and Lorraine P. 1956. The Foundations of the plant at Volta Redonda. Publication No. 539. *Technology Research Institute (IPT)*, São Paulo. (In Portuguese).

Sayão, A. 2010. *History of Geotechnical Engineering in Brazil. 60 years of the Brazilian Association of Soil Mechanics and Geotechnical Engineering*. ABMS Editors, São Paulo, 252 p.

Silva Nigra, D. and Clemente, M. 1950. *Three Benedictine artists. Typography*. Ministry of Education and Culture. Salvador. (In Portuguese).

Teixeira, A.H. 1959a. Case History of Building underlain by Preconsolidated Unusual Condition of Clay Layer. *Proceedings of 1st Panam. Conf. Soil Mechanics*. Mexico.

Teixeira, A.H. 1959b. Contribution to the study of the Time-Settlement Relation of Actual Structures. *Proceedings of 1st Panam. Conf. Soil Mechanics*. Mexico.

Teixeira, A.H. 1959c. Typical subsoil conditions and settlement Problems In Santos, Brazi. *Proceedings of 1st Panam. Conf. Soil Mechanics*. Mexico.

Terzaghi, K. and Peck, R.B. 1948. *Soil Mechanics in Engineering Practice*, John Wiley and Sons, New York.

Vargas, M. 1945. Exploring the Underground for the Purpose of Civil Engineering. Publication No. 242. Year 41, No. 149, October. *Technology Research Institute (IPT)*, São Paulo. (In Portuguese).

Vargas, M. 1961. Foundations of Tall Building on Sand in Sao Paulo, Brazil. *Proceedings of the 5th International Conference on Soil Mechanics and Foundation Engineering*, Paris.

Villares, A. 1956. The Underpining of the 26 story Building Cia de Seguros – Sao Paulo, Brazil. *Geotechnique*, Vol. 6, Issue 1, March, pp. 1–14.

Baltic Piling – Mets & Raudsepp (Eds)
© 2013 Taylor & Francis Group, London, ISBN 978-0-415-64334-4

Deep excavations and deep foundation systems combined with energy piles

R. Katzenbach, S. Leppla & T. Waberseck
Technische Universität Darmstadt, Institute and Laboratory of Geotechnics, Germany

ABSTRACT: The current discussions about the use of fossil or nuclear resources and the development of Renewable Energies, like Geothermic etc., cause a lot of new enormous challenges. Therefore economically efficient and technically reliable energy sources are needed. The geothermal use of subsoil can be a solution, since it reduces the use of fossil resource tremendously, which are limited in their availability, and thereby also sustains the reduction of CO_2 emissions and other greenhouse gases. Geothermal Energy is renewable and absolutely sustainable. With this contribution the authors present fundamental construction aspects of deep excavations and deep foundation systems and how to use these structural elements not only for load transmission into the subsoil but as energy exchanging elements as well. With several major project examples practical applications are presented.

1 INTRODUCTION

Economic and environment-friendly design focuses on a reduction of construction material used, construction time spend and energy consumed within the buildings construction and service time.

In geotechnical engineering particularly deep excavations and deep foundation systems offer a high potential for an economical and ecological optimization of the design and the construction process.

Combined Pile-Raft Foundations (CPRF) proved to be both reliable and cost-effective deep foundation systems. The design concept of the CPRF is based on advanced calculation methods and on the monitoring of the foundation performance within the concept of the observational method. The CPRF has been successfully used for many high-rise building projects. The optimised design of deep foundation systems relies on the knowledge of shaft and base resistance values, which should be obtained from load tests.

The use of deep foundation elements as well as the use retaining structures of deep excavations is not limited to their load bearing capacity but may be extended by incorporating them in a geothermal energy supply system. By thermally activating earth-contact concrete elements in order to exchange heat with subsoil the fossil energy consumption of a building can be reduced significantly.

2 DEEP EXCAVATIONS

2.1 *General aspects*

For the type and the design of retaining structures depends on several aspects:

- geometry,
- depth,
- type of foundation,
- properties of the subsoil,
- ground water level,
- loads,
- buildings close to the pit and
- construction steps.

Retaining systems can be classified into permeable and impermeable and resilient and low grade deformation systems. Figure 1 shows the different kinds of retaining structures for not water permeable systems.

The resilient systems are girder systems, sheet pile walls and separated bored pile walls.

Girder systems consist of vertical steel beams with a concrete foot and horizontal wooden beams. Permeable sheet pile walls are made of steel without any blinding between the sheet piles. Separated bored pile

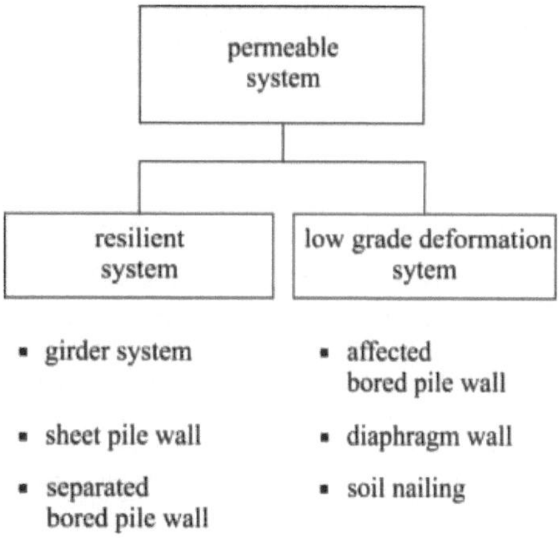

Figure 1. Water permeable retaining systems.

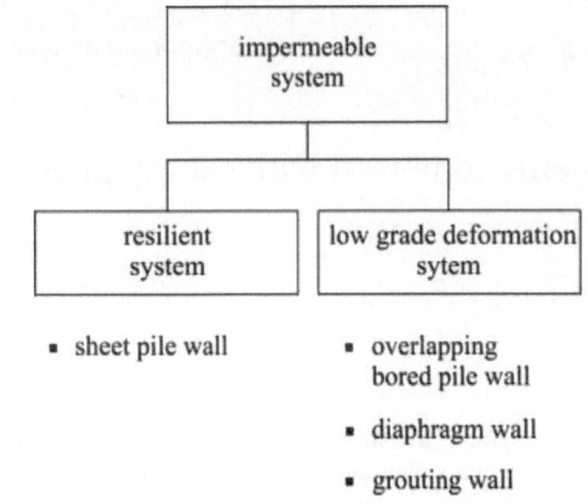

Figure 2. Water impermeable retaining systems.

walls normally have a sprayed concrete layer between the piles.

The low grade deformation systems are affected bored pile wall, diaphragm wall and soil nailing.

At an affected bored pile wall the piles only touch each other. The water permeable diaphragm wall is a reinforced concrete wall without any blinding between the elements. The soil nailing is a dowelled layer of sprayed concrete.

Figure 2 shows the different kinds of water impermeable retaining structures.

The resilient system is the sheet pile wall with blinding between the sheet piles.

The low grade deformation systems are overlapping bored pile walls, diaphragm walls with blinding between the elements and grouting walls. For the construction of grouting walls normally cement suspensions are injected into the subsoil.

2.2 *Definition of deep excavations*

The term "deep excavation" is established in the technical terminology of civil engineering (Moormann & Katzenbach 2002, Duncan & Bentler 1998, Nussbaumer & Wolffersdorff 1997, Peck 1969). For the quantitative demarcation the "Deep Excavation Index T_{EI}" (Equation 1).

$$T_{EI} = \frac{H_{res}}{E_S} \cdot 100 \quad [\mathrm{m}] \tag{1}$$

The deep excavation indes T_{EI} is a relation of the horizontal forces H_{res} and the oedometric modulus E_S. The resulting horizontal forces H_{res} conclude the excavation depth, the active earth pressure and the ground water pressure. The oedometric modulus E_S describes the soil properties.

With $T_{EI} > 0.4$ m an excavation can be classified as a deep excavation.

Figure 3 shows some examples for the deep excavation index $T_E > 0.4$ m. Excavations are defined as deep

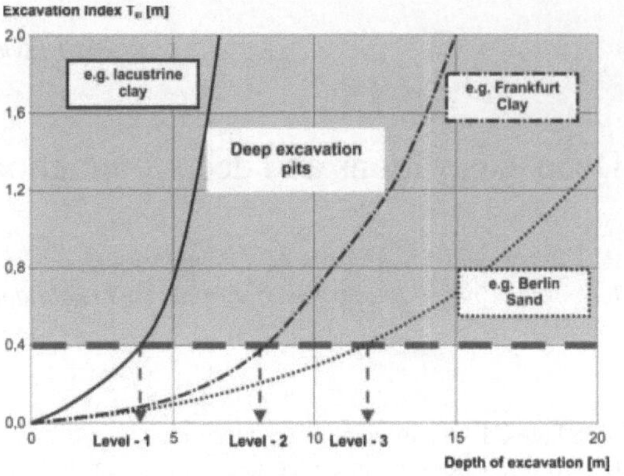

Figure 3. Deep excavation indexes.

when the excavation depth is less than 4 m in lacustrine clay with depth of less than 4 m, in Frankfurt Clay less than 8 m and in Berlin Sand less than 12 m.

3 COMBINED PILE-RAFT FOUNDATIONS

Designing Combined Pile-Raft Foundations (CPRF) requires the qualified understanding of soil-structure interaction, shown in Figure 4 (Randolph & Clancy 1993, Poulos 2001, Hanisch et al. 2002, Katzenbach et al. 2009). According to its stiffness the CPRF transfers the total vertical load of the structure R_{tot} into the subsoil by contact pressure of the raft (R_{raft}) as well as by the piles ($\Sigma R_{pile,i}$) (Equation 2).

$$R_{tot} = \sum R_{pile,i} + R_{raft} \tag{2}$$

In comparison to a conventional foundation design of a pile group for CPRFs a design philosophy with different and more complicated soil-structure interaction is applied. Piles are used up to a load level which is much higher than the permissible design values for bearing capacities of comparable single piles because the performance of the entire foundation structure is taken into account.

The distribution of the total building load between the different bearing structures of a CPRF is described by the CPRF coefficient α_{CPRF} (Equation 3) which defines the ration between the amount of load carried by the piles $\Sigma\ R_{pile,i}$ and the total load of the building R_{tot}.

$$\alpha_{CPRF} = \frac{\sum R_{pile,i}}{R_{tot}} \tag{3}$$

A CPRF coefficient of zero describes a raft foundation without piles, a coefficient of one represents a classic pile group, neglecting the existence of a raft. Positive effects of the CPRF are:

- reduction of settlements, differential settlements and tilts,

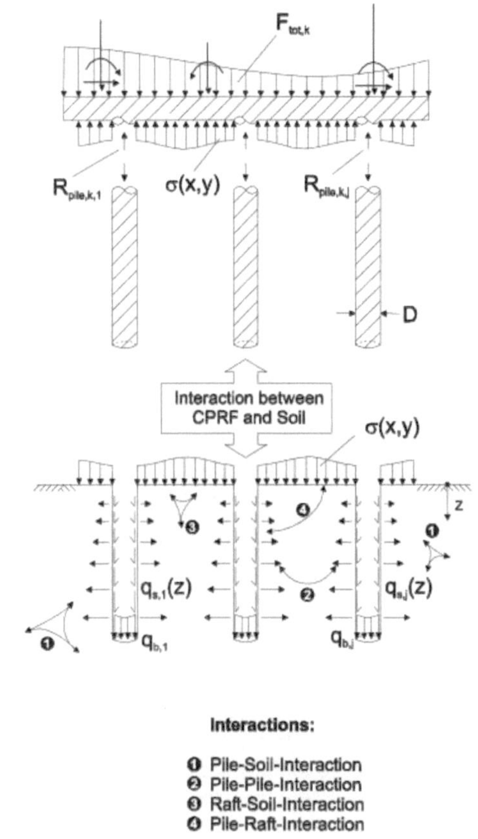

Figure 4. Soil-structure interaction of a CPRF.

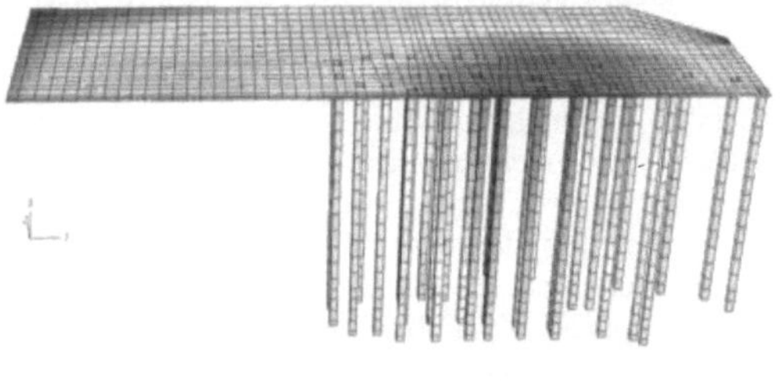

Figure 5. Exemplary finite element model of a CPRF.

- increase of the overall stability of the foundation,
- reduction of the bending stress for the foundation raft and
- cost optimisation of the whole foundation.

In order to assess the load-settlement behaviour and to calculate the internal forces of a CPRF three-dimensional finite element simulations are usually required (Figure 5). These simulations allow the consideration of complicated geometric shapes and take into account all relevant interaction effects. They provide a valuable tool to perform simulations with different pile configurations in order to optimise the foundation structure.

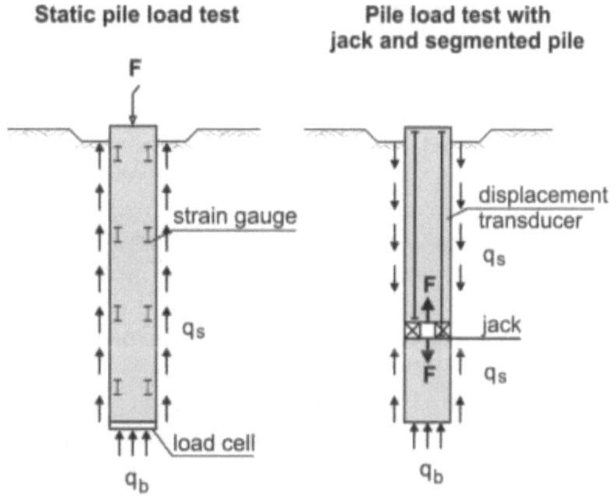

Figure 6. Scheme of pile load tests.

A constitutive model used for simulations should provide a reasonable good simulation of the stress-strain behaviour of soils, which depends on the stress path and the previous stress history. The material behaviour of the piles and the raft are simulated as linear-elastic in the finite element analysis. The soil is modelled with an elasto-plastic constitutive law.

The bearing behaviour and the load transfer within a CPRF have to be monitored due to the requirements deriving from the soil, the superstructure and the foundation according to the concept of the observational method on the basis of the monitoring program set up in the design phase.

4 LOAD TESTS

Based on project- and site-related soil investigation with core drilling and laboratory tests the soil parameters are determined. Those tests are important and essential for the initial definition of soil mechanical properties of the soil layers, but usually not sufficient for an entire and realistic capture of the complex conditions, caused by the interaction of subsoil and construction (Katzenbach 2005).

In order to reliably determine the ultimate bearing capacity of piles, load tests need to be carried out (Briaud et al. 2000). In general, static load tests of piles in rock require very high loads (Figure 6). In many cases the rocky layers are covered by loose rock material, which is not foreseen for the load transfer and therefore should not influence the test results. Due to these boundary conditions conventional load tests can hardly fulfil the requirements. By using the Osterberg method high loads can be reached without installing anchors or counter weights. Hydraulic jacks induce the load in the pile using the pile itself partly as abutment. The results of the field tests allow a calibration of the numerical simulations.

5 OBSERVATIONAL METHOD

The design of geotechnical constructions requires a good knowledge of the relevant soil parameters. But even if a comprehensive soil investigation campaign is performed, the risk of undetected but relevant subsoil conditions exists and cannot be eliminated during the design process. In order to nevertheless guarantee a safe construction the observational method can be applied. The observational method is an integrated part of the safety concept and needs to be planned in the design phase of the project.

In Eurocode 7 "Geotechnical design – Part 1: General Rules" the following rules for the application of the observational method are stated:

"When prediction of geotechnical behaviour is difficult, it can be appropriate to apply the approach known as "the observational method", in which the design is reviewed during construction. The following requirements shall be met before construction is started:

- *acceptable limits of behaviour shall be established;*
- *the range of possible behaviour shall be assessed and it shall be shown that there is an acceptable probability that the actual behaviour will be within the acceptable limits;*
- *a plan of monitoring shall be devised, which will reveal whether the actual behaviour lies within the acceptable limits. The monitoring shall make this clear at a sufficient early stage, and with sufficiently short intervals to allow contingency actions to be undertaken successfully;*
- *the response time of the instruments and the procedures for analysing the results shall be sufficiently rapid in relation to the possible evolution of the system;*
- *a plan of contingency actions shall be devised, which may be adopted if the monitoring reveals behaviour outside acceptable limits.*

During construction, the monitoring shall be carried out as planned. The results of the monitoring shall be assessed at appropriate stages and the planned contingency actions shall be put into operation if the limits of behaviour are exceeded. Monitoring equipment shall either be replaced or extended if it fails to supply reliable data of appropriate type or in sufficient quality."

In the German National Standard DIN 1054:2005 "Subsoil – Verification of the safety of soil engineering and foundations" the application of the observational method is recommended for the following types of construction projects:

- *"very complex construction projects (construction projects classified in Geotechnical Category GC 3);*
- *construction projects with pronounced soil-structure interaction, e.g. mixed (shallow and deep) foundations, raft foundations, flexibly anchored retaining walls;*

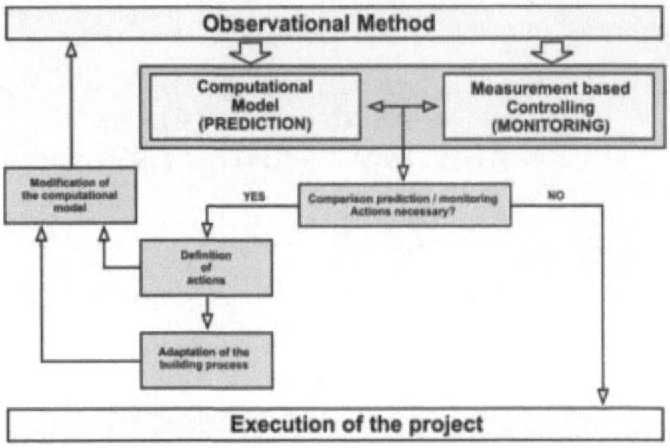

Figure 7. Principle of the observational method.

- *construction projects with substantial and variable water pressure action, e.g. trough structures or waterfront structures in tidal areas;*
- *complex interaction systems consisting of ground, excavation structure and neighbouring buildings;*
- *construction projects in which pore-water pressures may reduce stability;*
- *construction projects on slopes."*

The observational method is always a combination of the common geotechnical investigations together with the theoretical modelling and a plan of contingency actions (Figure 7). Monitoring alone is not sufficient to ensure the stability and the serviceability of the structure and, according to the standardisation, not permitted for this purpose. Overall the observational method is an institutionalised controlling instrument to verify the soil or rock mechanical modelling (Rodatz et al. 1999, Katzenbach et al. 2005).

The identification of all potential failure mechanisms is essential. The measurements need to be designed in that way, that all these mechanisms are observed. The measurements need to be reliable and should be processed in a direct and defined way. The measurement devices need to be of an adequate accuracy to allow the identification of critical tendencies. The required accuracy needs to be identified within the design phase of the observational method.

Contingency actions need to be planned in the design phase of the observational method as well. The construction method or design of the project needs to be adjustable (Peck 1969). The applicability of the observational method depends on the ductility of the system (Gudehus 2004). It is essential that contingency actions can be applied. The observational method cannot be seen as a potential alternative for a comprehensive soil investigation campaign. A comprehensive soil investigation campaign is in any way of essential importance.

To establish the above mentioned "acceptable limits" before the construction has started is a challenging task and requires powerful calculation tools which rely on suitable constitutive laws for the subsoil. As a secondary effect the observational method is a tool of quality assurance and allows the verification of the parameters and calculations applied in the design phase. The observational method helps to achieve an economic and save construction.

6 GEOTHERMAL USE OF SUBSOIL

6.1 *Theoretical principles of heat transport in subsoil*

Heat is energy, transported or transferred on the basis of a temperature gradient. The heat transport or transfer takes place as material-bound or non-material-bound transport towards the lower temperature level (Brandl 2006):

Material-bound transport:

- heat conduction
- convection
- dispersion

Non-material-bound transport:

- thermal radiation

Further heat transport could be released by radioactive effect, viscous dissipation and the work done by pressure changes (Nield & Bejan 1999).

6.1.1 *Heat conduction*

The transfer mechanism heat conduction describes the transfer of energy from higher energetic molecules to molecules of a lower energy level. Heat conduction takes place within a body, a resting liquid or gas with areas of different temperatures, or between two neighboring bodies mutually in touch and of different temperatures in subsoil. Heat conduction occurs in the solid as well as in the fluid phase.

The connection between the heat flux density $\dot{q}$ which is the amount of energy Q transported through the area A within the time unit t:

$$\dot{q} = \frac{Q}{A \cdot t} = \frac{\dot{Q}}{A} \tag{4}$$

and the temperature gradient is described by Fourier's principle of the molecular heat transport:

$$\dot{q} = -\lambda \frac{\partial T}{\partial n} \tag{5}$$

with the temperature gradient:

$$\frac{\partial T}{\partial n} = \frac{\partial T}{\partial x} i + \frac{\partial T}{\partial y} j + \frac{\partial T}{\partial z} k = \text{grad}\, T \tag{6}$$

where Q = amount of heat [Ws], $\dot{Q}$ = heat flux [W], A = area [m^2], t = time [s], $\dot{q}$ = heat flux density [W/m^2], λ = thermal conductivity [W/(m · K)], T = temperature [K], x, y, z = Cartesian co-ordinates [m], i, j, k = unit vectors [–].

According to the first law of thermodynamics the increase of internal energy of a volume element dV equals the gain in energy from the heat sources minus the heat flux density via the volume element's surface:

$$\rho c\, dV \frac{\partial T}{\partial t} = \dot{Q}_i\, dV - \begin{pmatrix} \left(\dot{q}(x+dx) - \dot{q}(x)\right) dydz + \\ \left(\dot{q}(y+dy) - \dot{q}(y)\right) dxdz + \\ \left(\dot{q}(z+dz) - \dot{q}(z)\right) dxdy \end{pmatrix} \tag{7}$$

where c = specific thermal capacity [J/(kg · K)], ρ = density [kg/m^3], V = volume [m^3], $\dot{Q}_i$ = heat sources [W/m^3].

Within an infinitesimal volume element the heat flux densities are developed in a Fourier row, given here as an example for the x-direction:

$$\dot{q}(x+dx) = \dot{q}(x) + \frac{\partial \dot{q}_x}{\partial x} dx \tag{8}$$

From this relation Fourier's differential equation for the transport by heat conduction arises to:

$$\rho c \frac{\partial T}{\partial t} = \dot{Q}_i - \left(\frac{\partial \dot{q}_x}{\partial x} + \frac{\partial \dot{q}_y}{\partial y} + \frac{\partial \dot{q}_z}{\partial z} \right) \tag{9}$$

After the elimination of the heat flux densities under application of Fourier's basic principle of the molecular heat transport and under the assumption that $\lambda = \lambda_x = \lambda_y = \lambda_z$ = const. it follows that:

$$\rho c \frac{\partial T}{\partial t} = \dot{Q}_i + \lambda \left(\frac{\partial^2 T}{\partial x^2} + \frac{\partial^2 T}{\partial y^2} + \frac{\partial^2 T}{\partial z^2} \right) = \dot{Q}_i + \text{div}(\lambda\, \text{grad}\, T) \tag{10}$$

6.1.2 *Convection*

In convection the heat transfer takes place between two thermodynamic systems moving relatively to each other, e.g. the flow-by of a fluid along a wall. The fluid's flow can be forced on by outer forces (pump) (forced convection) or arise due to a natural buoyancy of the warmer and therefore lighter fluid particles (free convection).

The convective change of the amount of heat per time unit within a volume element dV owing to a fluid flowing with seepage velocity v is:

$$\dot{q}_v = (\rho c)_f \left(\frac{\partial}{\partial x}(v_x T) + \frac{\partial}{\partial y}(v_y T) + \frac{\partial}{\partial z}(v_z T) \right) = (\rho c)_f\, \text{div}(vT) \tag{11}$$

where $\dot{q}_v$ = amount of heat per unit of time and volume [W/m^3], ρ_f = fluid's density [kg/m^3], c_f = fluid's specific heat capacity [J/(kg · K)], $v = -k\, \text{grad}\, h$ = seepage velocity [m/s], k = permeability coefficient [m/s], grad h = hydraulic gradient [–].

6.1.3 *Dispersion*
Dispersion describes the fanning out of the heat, transported with the carrier fluid as a function of time. The fanning out is caused by the differently, statistically distributed flow velocities of each water volume within the current space. The change of the amount of heat per element of time and volume owing to dispersion is:

$$\dot{q}_v = \operatorname{div}\left(D_\lambda \operatorname{grad} T\right) \tag{12}$$

with the empirical formulation for the heat dispersion coefficient D_λ:

$$D_\lambda = (\rho c)_f \delta_\lambda |v| \tag{13}$$

where D_λ = heat dispersion coefficient [W/(m · K)], δ_λ = heat dispersivity [m], c = specific heat capacity [J/(kg · K)], ρ = density [kg/m^3], v = seepage velocity [m/s].

6.1.4 *Thermal radiation*
In thermal radiation the heat transport takes place via electromagnetic oscillations. Thermal radiation is not bound to material, in a vacuum it is the only heat transferring mechanism.

6.1.5 *Heat transport equation*
Bear (1972) noticed that heat transport within the soil takes place mainly through heat conduction and convection. Sauty (1980) stated that dispersion has to be considered too. For the soil with its respective scope of temperature heat transport through radiation can be neglected (Brehm 1989).

Considering the heat transfer mechanisms heat conduction, convection and dispersion for an isotropic medium and under the assumption that there is local thermal equilibrium so that $T_s = T_f = T$, where T_s and T_f are the temperatures of the solid and fluid phases, respectively, and the assumption that heat conduction in the solid and fluid phases takes place in parallel so that there is no net heat transfer from one phase to the other, the differential equations for heat transport for the solid and the fluid phase result in:

$$\operatorname{div}\left((1-n)\lambda_s \operatorname{grad} T_s\right) + \dot{Q}_{i,s} = (1-n)(\rho c)_s \frac{\partial T_s}{\partial t} \tag{14}$$

$$\begin{aligned} &\operatorname{div}\left(n\lambda_f \operatorname{grad} T_f\right) + \\ &\operatorname{div}\left((\rho c)_f \delta_\lambda |v| \operatorname{grad} T_f\right) - \\ &(\rho c)_f \operatorname{div}\left(v T_f\right) + Q_{i,f} = n(\rho c)_f \frac{\partial T_f}{\partial t} \end{aligned} \tag{15}$$

where n = porosity [–], $(\rho c)_f$ = volumetric heat capacity of the fluid phase [J/(m^3 · K)], $(\rho c)_s$ = volumetric heat capacity of the solid phase [J/(m^3 · K)].

Adding Equation 14 and Equation 15 the differential equation for heat transport comes to:

$$\begin{aligned} &\operatorname{div}\left(\left(\lambda + (\rho c)_f \delta_\lambda |v|\right) \operatorname{grad} T\right) - \\ &(\rho c)_f \operatorname{div}(vT) + \dot{Q}_i = \rho c \frac{\partial T}{\partial t} \end{aligned} \tag{13}$$

where ρc = volumetric heat capacity of the subsoil [J/(m^3 · K)].

6.2 *Seasonal thermal storage*

Geothermal use of the subsoil is possible in two ways: On the one hand the subsoil can be used for the storage of energy – e.g. by a Seasonal Thermal Storage System. On the other hand the subsoil can be employed as energy source by using its natural energy potential, if a sufficient thermal regeneration is given.

The term "Seasonal Thermal Storage System" refers to a soil region equipped with energy exchanger elements in order to store heat seasonally. The different seasonal building tempering demands – heating in winter and cooling in summer – determine the respective energy extraction: During winter energy is extracted from the subsoil and conveyed to the structure to be tempered. The thereby induced decrease of temperature in the soil indicates a volumetric energy deficit – to be understood as storage of coldness. During summer the system works the other way around: Energy from the tempered structure, which is heated up by solar radiation or the normal use causing a heat overage, is extracted from the building and transferred to the soil. Thereby the building is cooled and the equipped soil region is heated up – the Seasonal Thermal Storage System is recharged just like a battery (Katzenbach et al. 2007).

To assure the annual functionality of the storage system the annual seasonal energy extraction and deposit should be balanced. Overall, the extraction and deposit must be well-balanced over the long term in order to maintain the function of the system. Thermal storage is possible in a regime of rather slow groundwater flow velocities. Generally, single use for energy extraction or deposit without seasonal storage operation does not benefit from slow groundwater velocities; in contrast in this case higher groundwater flow rates provide a better thermal regeneration of the subsoil.

The three self-contained system cycles for a Seasonal Thermal Storage are shown in Figure 8:

- system 1 consists of the energy piles or other geothermal systems in the subsoil,
- system 2a consists of the installation of the building services for heating and
- system 2b consists of the installation of the building services for cooling.

For the extraction of geothermal energy from subsoil different technologies are available. These can be categorized into so-called open and closed systems. This categorization refers to the circulated water. Open systems use the groundwater directly by pumping water from an extraction well to the tempered structure and re-injecting it after using through another well into the aquifer. Closed systems use a closed circuit for the energy transfer between the soil und the tempered structure. Common closed systems are borehole heat exchangers, collector mats and energy piles.

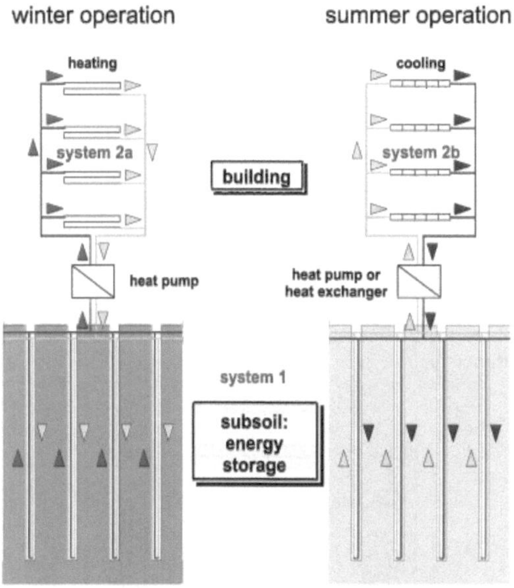

Figure 8. Seasonal Thermal Storage System (Ennigkeit & Katzenbach 2001).

Figure 9. Energy pile reinforcement cage (left) and installation of the reinforcement cage (right).

Energy piles have a double purpose: Besides their main purpose – load transmission into the foundation soil – they are also used as energy exchanging elements for the sustainable heating and cooling of buildings. To equip a foundation pile for his function as energy pile energy exchanger tubes of plastic are embedded. The tubes are attached to the reinforcement cage in large U-loops. An energy transfer fluid is circulating through the energy exchanger tubes. Figure 9 shows photos of an equipped reinforcing cage of an energy pile and the installation of the reinforcement cage in the borehole.

Basically, both cast-in-place piles as well as pre-cast driven piles are suitable to be used as energy piles. By the temperature difference between the medium and the surrounding foundation soil, energy is absorbed from the ground. Thus the whole pile acts as an energy exchanging element to transfer energy between the buildings' tempering system and the subsoil. Usually the energy is transferred to the technical building equipment by a heat pump or by heat exchangers. Cast-in-place piles are very flexible in terms of their geometry, i.e. length and diameter. Important is the protection of the energy exchanger tubes and connection elements during the construction process from mechanical damage to assure their functionality and therefore the mounting position of the pipes is inside the reinforcement cage. Pre-cast driven piles with full cross-section are completed in the production plant including the plastic pipes for the energy transfer medium. Here the main problem is to protect the tube connections from damage by the driving process. To guarantee the energy piles' functionality and quality in every step of the construction process appropriate protection and documentation measures are required. During transport, installation and concreting, the pipe system is to be kept under controlled and documented pressure and to be prevented from mechanical damage. The equipment of foundation piles and retaining walls as energy piles and their integration in the buildings' energy supply concept is state-of-the art and successfully implemented in various projects.

7 PROJECT EXAMPLES

7.1 *Monitoring of a deep excavation in Frankfurt am Main, Germany*

The continuous monitoring of deep excavation pits is not only a part of the observational method with the goal of an optimisation of the design but also an instrument to prevent or to minimise damage via adaptation of the construction process in terms of contingency actions to changing boundary conditions. This is shown in the following example that.

The upper layer of the subsoil is a 6 m thick filling. These strata a quaternary deposits, consisting of loam, gravel and sand, by about 7 m thickness follows. Underlain the quaternary deposits an overconsolidated tertiary deposits of the Frankfurt Clay exists.

To minimise deformations the up to 14 m deep excavation pit was supported by a 3 level anchored secant bored pile wall. Figure 10 shows the anchors of the retaining wall in the northern area of the excavation extend beneath the foundation raft of the Deutsche Bundesbank.

This Project was classified within the Geotechnical Category 3 in cause of the complexity of the construction works and the presence of settlement sensitive adjacent buildings and a comprehensive monitoring programme was established.

To control the excavation pit and the high-rise foundation a monitoring programme, composed of an architectural conservation of evidence, geodetic and geotechnical measurements, measurement of vibrations and the continuous measurement by using a motorised digital levelling system which also continuously monitored the adjacent buildings was installed (Figure 11). Thereby an automatic survey of defined points at a fixed time interval was possible.

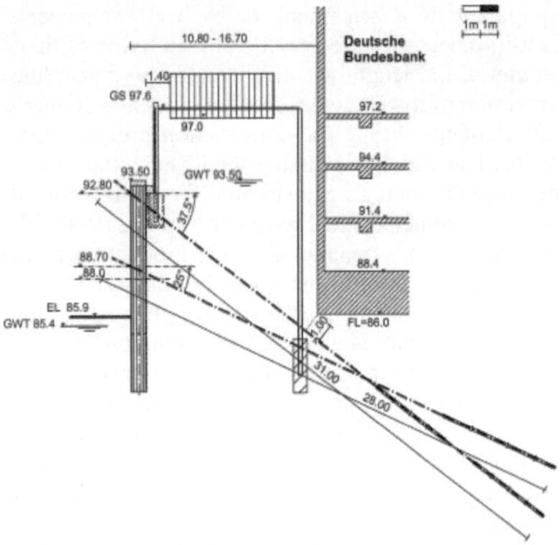

Figure 10. Section of retaining wall and adjacent building (Deutsche Bundesbank).

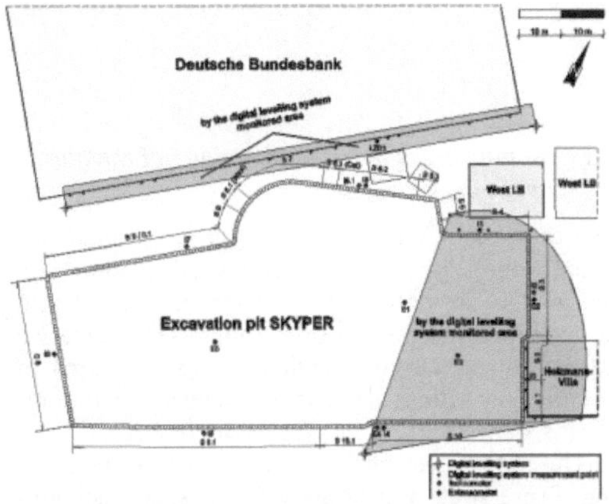

Figure 11. Location plan of the measurement devices.

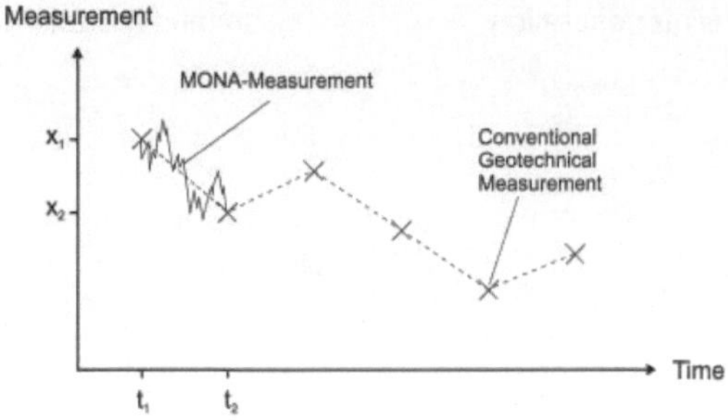

Figure 12. Influence of the measurement interval on the information.

Figure 13. "Messeturm", Frankfurt am Main, Germany.

The measured displacements indicated relative values with respect to the location of the digital levelling system and were used to observe the specified compliance of the limit deformation value which in the present case was set to 0.1 cm relative displacement (Katzenbach et al. 2010).

During the anchor construction it was possible to record the displacements with this system (MONA) in such detailedness that it was possible to correlate the obtained data with different stages in the anchor construction process. In Figure 12 the so called MONA-Measurement compared to conventional geodetic measurements is shown.

7.2 *"Messeturm", Frankfurt am Main, Germany*

The "Messeturm" in Frankfurt am Main is 256.5 m high. The foundation of the "Messeturm" is a CPRF, which is based in the Frankfurt clay (Figure 13). The foundation raft is 58.8 m × 58.8 m wide with a maximum thickness of 6 m in the centre and a thickness of 3 m at the edges. The base of the foundation raft is about 11 m to 14 m below the ground surface.

The raft is combined with 64 bored piles with a diameter of 1.3 m and a length 30.9 m in the centre ring and 26.9 m at the edges (Figure 14). The total load, including 30% of live loads, is about 1,855 MN. The progression of settlements was controlled in course of the observational method by measurement monitoring. The maximum settlements of about 13 cm were measured in December 1998 (Figure 15).

The CPRF was calculated with the Finite Element Method (FEM). Thereby a section of the foundation was modelled, using the symmetry of the plan view (Figure 16). The settlements of a pure raft foundation were calculated to 32.5 cm. The calculated settlements of the CPRF are nearly equal to the in-situ measured value of 13 cm (Figure 17). The pile raft coefficient is about $\alpha_{pr} = 0.43$ (Reul 2000).

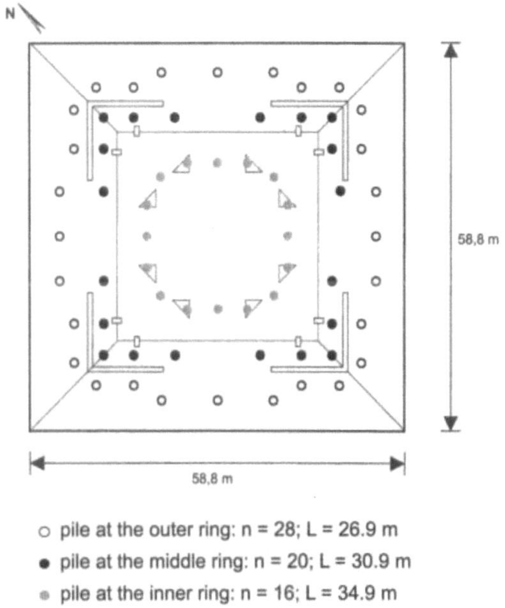

Figure 14. Foundation raft and pile alignment.

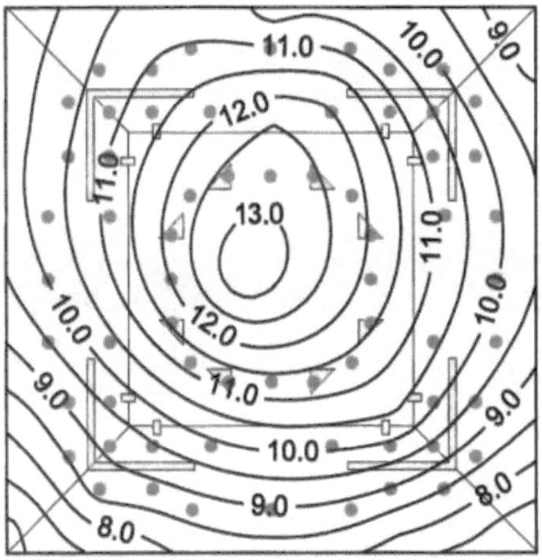

Figure 15. Settlements [cm] measured in Dec. 1998.

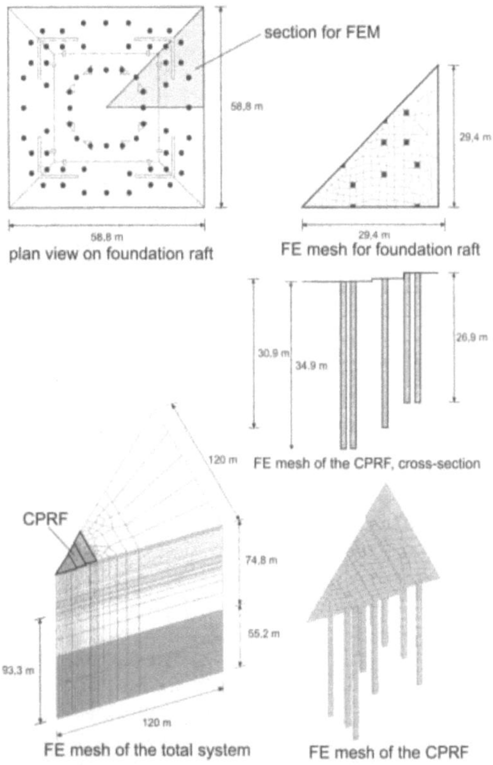

Figure 16. FE mesh of the foundation.

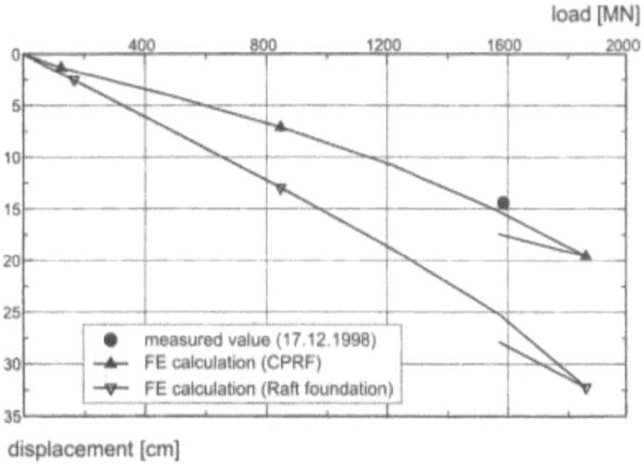

Figure 17. Measured and calculated settlements.

A pure pile foundation would have required 316 piles with 30 m length. In comparison to the executed CPRF with 64 piles and an average length of about 30 m a pure pile foundation would have required more resources e.g. concrete, energy and would have been approximately 3.9 Million Euro more expensive.

7.3 *"PalaisQuartier", Frankfurt am Main, Germany*

7.3.1 *Project description*

In the center of Frankfurt am Main on a construction site of 17,400 m^2 the high-rise building project "PalaisQuartier" has been realised (Figure 18). The construction was finished in 2010. It is located next to one of the most frequented shopping streets in Germany, the "Zeil" (Janke et al., 2010).

The complex consists of several structures with a total of 180,000 m^2 floor space, thereof 60,000 m^2 underground (Figure 18). The project includes the historic building "Thurn- und Taxis-Palais" whose façade has been preserved (Unit A). The office building (Unit B), which is the highest building of the project with a height of 136 m, has 34 floors each with a floor space

Figure 18. Project "PalaisQuartier", Frankfurt am Main, Germany.

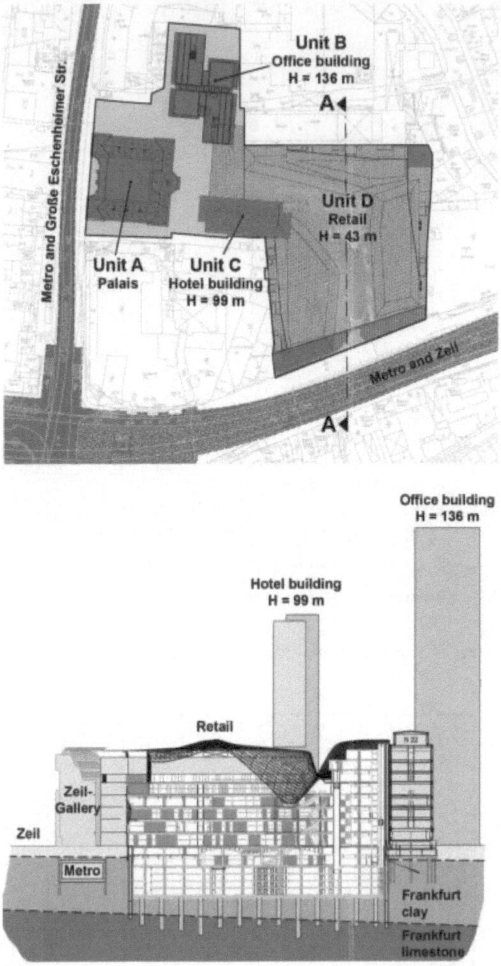

Figure 19. PalaisQuartier complex in Frankfurt am Main, Germany: plan view (top) and cross-section A-A (bottom).

of 1,340 m^2. The hotel building (Unit C) has a height of 99 m with 24 upper floors. The retail area (Unit D) runs along the total length of the eastern part of the site and consists of eight upper floors with a total height of 43 m. The underground parking garage with five floors spans across the complete project area. With an 8 m high first sublevel, partially with mezzanine floor, and four more sublevels the foundation depth results to 22 m below ground level. Thereby excavation bottom is at 80 m above sea level (msl). A total of 302 foundation piles (diameter up to 1.86 m, length up to 27 m) reach down to depths of 53.2 m to 70.1 m above sea level depending on the structural requirements. The pile head of the 543 retaining wall piles (diameter 1.5 m, length up to 38 m) were located between 94.1 m and 99.6 m above sea level, the pile base was between 59.8 m and 73.4 m above sea level depending on the structural requirements. As shown in the sectional view (Figure 19), the upper part of the piles is in the Frankfurt clay and the lower part of the piles is set in the rocky Frankfurt limestone layers.

7.3.2 *Pile tests*

Due to the large number of piles and the high pile loads a pile load test has been carried out within the "PalaisQuartier" project. Osterberg cells have been installed in two levels in order to assess the influence of pile shaft grouting measures on the limit skin friction of the pile in the Frankfurt limestone layer (Figure 20). The test pile with a total length of 12.9 m and a diameter of 1.68 m consists of three segments and has been installed in the Frankfurt limestone layer 31.7 m below ground. The upper pile segment above the upper cell level and the middle pile segment between the two cell levels can be tested independently. Pile shaft grouting measures have been carried out for the middle pile segment. In the first phase of the test the upper part was loaded by using the middle and the lower part as abutment. A limit load of 24 MN could be reached (Figure 21). The upper segment was lifted about 1.5 cm, the settlement of the middle and lower part was 1.0 cm. The mobilised shaft friction was about 830 kN/m^2.

Subsequently the upper pile segment was uncoupled by discharging the upper cell level. In the second test phase the middle pile segment was loaded by using the lower segment as abutment. The limit load of the middle segment with shaft grouting was 27.5 MN (Figure 21). The skin friction was 1,040 kN/m^2, this means 24% higher than without shaft grouting. Based on the results of the pile load tests using Osterberg cells the majority of the 290 foundation piles were made by applying shaft grouting measures. Due to the findings of the load tests the total pile length was reduced significantly.

7.3.3 *Geothermal concept*

Already in an early planning phase of the project, the possibility of using the large number of foundation piles and construction pit secant wall piles (845 units) as geothermal energy exchangers was discussed. The geothermal use of the subsoil as a regenerative energy source in Frankfurt was executed on numerous major projects, such as Gallileo's skyscrapers, the MainTower and Skyper (Ennigkeit 2002,

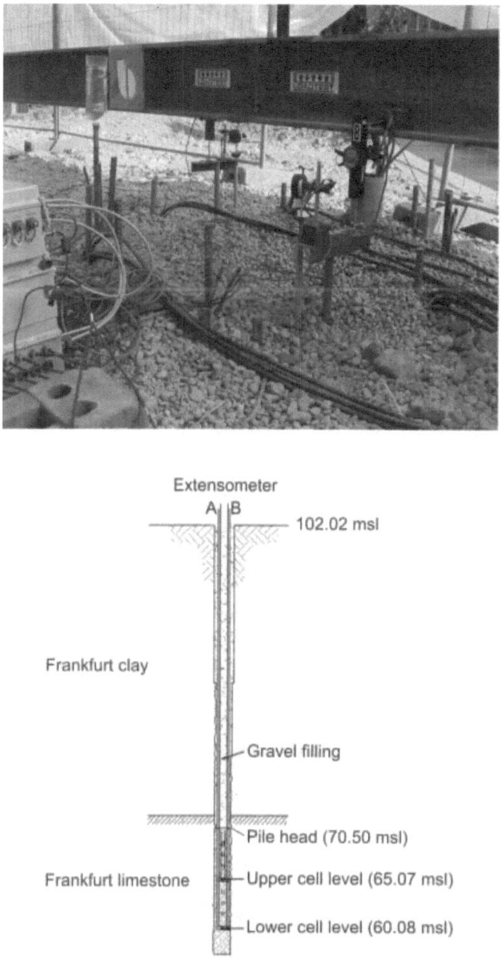

Figure 20. Pile load test setup.

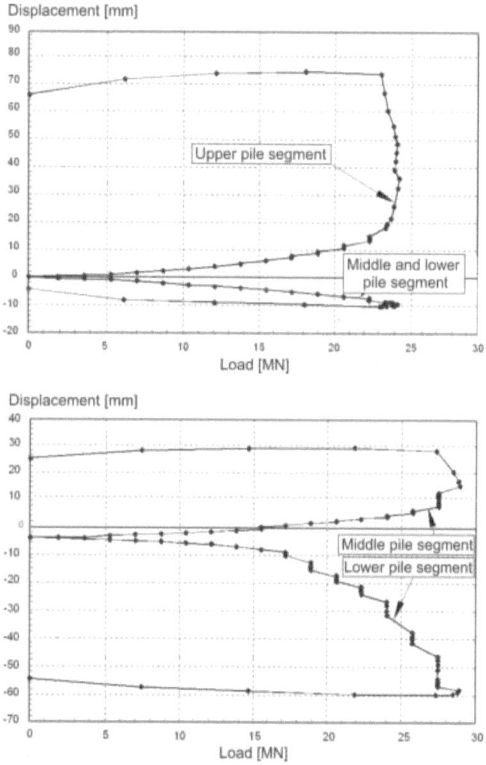

Figure 21. Load displacement curve of test phase 1 (top) and test phase 2 (bottom).

Figure 22. left: With heat exchanger pipes equipped reinforcement cage of a pile; right: Horizontal connection of energy piles within basement elevation.

von der Hude & Sauerwein 2007) and successfully integrated into the energetic building concept.

As a result of the comparably low additional costs of the geothermal installations and the high energy recovery potential of the energy pile system, the geothermal concept was realized in the project "PalaisQuartier". For thermal activation of the subsoil 262 of the total of 302 foundation piles and 130 of the total of 289 reinforced secant wall piles were equipped as energy piles (Figure 22). Hence, a total of 392 foundation piles and secant wall piles of the construction pit structure with a total capacity of 913 kW for geothermal exploitation of the building site were made available.

The energetic concept of the complex foresees a seasonal "in-out" service for the use of energy pile potential, which means that the building is used as a Seasonal Thermal Storage (Figure 8). In winter the circulating heat transfer medium (water) in the piles revokes energy out of the subsoil and supplies this to the building complex. If the subsoil during the summer months is not naturally thermally regenerated, it needs to be heated by external energy again. This avoids a long-term cooling of the building site. To prevent this, energy is removed from the heated building complex at summer and fed to the Seasonal Thermal Store (subsoil). Thus, the initial temperature level for winter operation is restored and the energy balance for the period of one year is in balance. The annual performance of the energy pile system was estimated during design for the heating period of approximately 2,350 MWh/yr. For the cooling mode an approximately 2,410 MWh/yr energy bill was estimated. The seasonal operation of the "in-out" system would therefore lead to a sustainable energy balance (Figure 23).

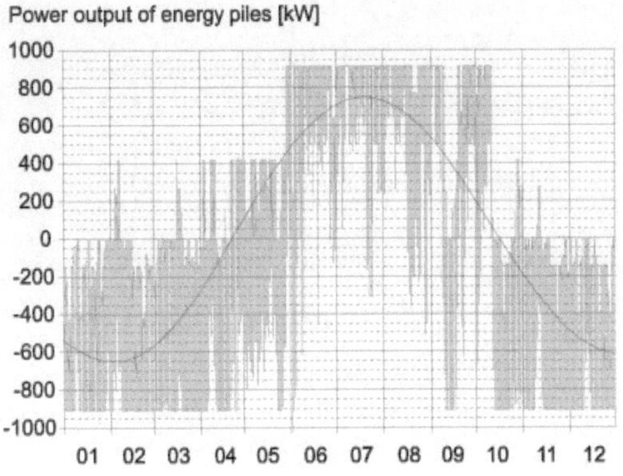

Figure 23. Calculated balanced yearly geothermal energy demand of the building complex "PalaisQuartier".

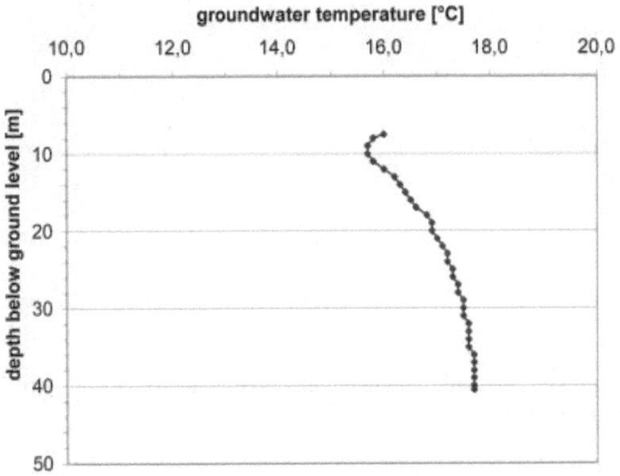

Figure 24. Groundwater temperature diagram measured on 18.05.2004.

7.3.4 *Unaffected ground temperature*

The groundwater temperature was recorded along the depth of the groundwater measuring points on the construction site, to identify the unaffected ground temperature on the construction site of the project PalaisQuartier (Figure 24). The measured groundwater temperature is relatively continuous over the depth between 15.8°C in 8 m below surface and 17.7°C in a depth of 36 m.

These relatively high temperatures are basically caused by anthropogenic influences as heat insertion from underground structures such as garages, metro and underground tunnels and the sewage tunnels. The measured temperatures verify previous recordings in the Frankfurt downtown area. In the period between 1997 and 2002 a total of 25 groundwater recording points and 3 inclinometers with hydrothermal readings were analyzed. Thereby the highest temperature with 20.7°C was measured right next to a metro tunnel. Within this test series the temperatures lay between 14.5°C and 19°C in a depth of 15 m below the top ground surface. In the areas with buildings firstly a decreasing of the temperature in depths of 30 m to 50 m below top ground surface was detected, while

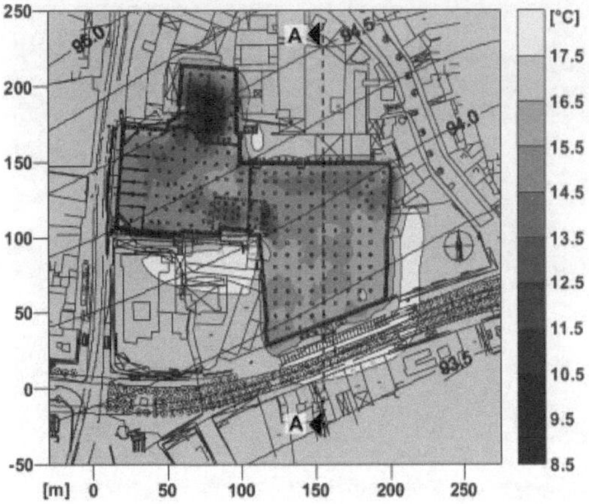

Figure 25. Calculated distribution of the isotherms in an horizontal section in the Frankfurt limestone with groundwater contour lines [mNN] (after winter operation).

in larger depths the temperature increased again. In areas without buildings the expected natural temperature trend was found. In this case the temperature continuously increased from a depth of approximately 10 m to 15 m below top ground surface on downward with the geothermal graded index of 3°K every 100 m. The 10 m to 15 m is the depth in which the temperature is not influenced noteworthy anymore by the solar radiation.

7.3.5 *Thermal influence on neighboring properties*

To analyze the possible thermal influences on neighboring properties caused by the operation of the geothermal Seasonal Thermal Storage of PalaisQuartier extensive calculations were made. These included three-dimensional numeric calculations with coupled groundwater flow models and heat energy transport models based on the Finite Element Method (FEM). Special attention was thereby paid to the special subsoil situation. All retaining wall and foundation piles lie with their upper part in the relatively impermeable Frankfurt clay and with their lower part in the Frankfurt limestone layers, which noticeably have higher water permeability.

The results of the FE-calculations show the expected highest thermal influence on the ground in the areas where the energy piles are set very close to each other. The core of the office building (building component B) is thermally stressed the most due to the existing density of piles (Figure 25). The highest expansion of the temperature field occurs towards the direction of the groundwater stream. Due to the higher permeability of the rocky Frankfurt's lime and the therefore higher groundwater streaming velocity, compared to the Frankfurt's clay, the expansion of the thermal influenced area is larger in the lime beds (Figures 26 and 27).

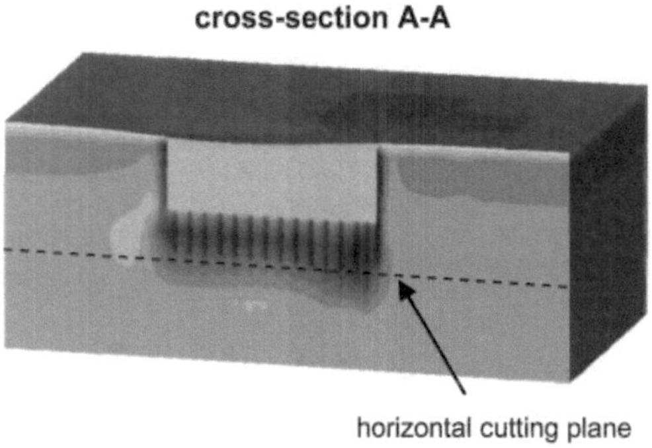

Figure 26. Vertical section A-A (exaggerated).

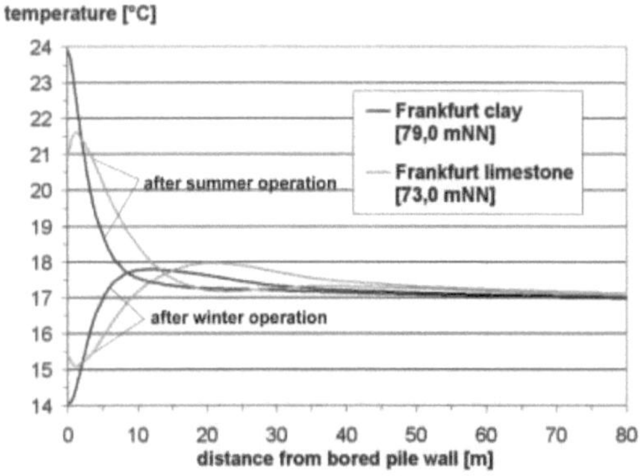

Figure 27. Calculated temperature distribution in groundwater flow direction after summer and winter operation.

8 CONCLUSIONS

The theoretical basics of deep foundation systems and retaining structures and the application in engineering practice have been explained on several examples.

The observational method in geotechnical engineering is state-of-the-art for the verification of the applied design and the applied models. Also, the observational method is a powerful tool for damage detection and prevention as well as quality control.

The possibility to incorporate deep foundation systems and retaining structures in a geothermal energy supply system is shown on a large construction project that is in its service time since two years.

REFERENCES

Bear, J. 1972. *Dynamics of fluids in porous media*. Environmental science series, American Elsevier, New York.

Brandl, H. 2006. Energy foundations and other thermo-active ground structures. *Géotechnique* 56 (2).

Briaud, J.-L., Ballouz, M., Nasr, G. 2000. Static Capacity Prediction by Dynamic Methods for Three Bored Piles. *Journal of Geotechnical and Geo-environmental Engineering*, Vol. 126, No.7, July 2000, ASCE, Reston, Virginia.

Brehm, D. R. 1989. Entwicklung, Validierung und Anwendung eines dreidimensionalen, strömungsgekoppelten Finite Differenzen Wärmetransportmodells. *Gießener Geologische Schriften*, Nr. 43, 1989

Duncan, J.M. & Bentler, D. J. 1998. Evaluation of deep excavation technology. *Darmstadt Geotechnics No. 4*, Vol. 1, 139–150.

Ennigkeit, A. 2002. Energiepfahlanlagen mit Saisonalem Thermospeicher. *Mitteilungen des Institutes und der Versuchsanstalt für Geotechnik der Technischen Universität Darmstadt*, Heft Nr. 60.

Ennigkeit, A. & Katzenbach, R. 2001. The double use of piles as foundation and heat exchanging elements. *Proceedings of XVth International Conference on Soil Mechanics and Geotechnical Engineering (ICSMGE)*, 27.-31.8.2001, Istanbul, Turkey, Vol. 2, 893–896.

Gudehus, G. 2004. Prognosen bei Beobachtungsmethoden. *Bautechnik* 81, Heft 1, Ernst&Sohn Verlag, Berlin.

Hanisch, J., Katzenbach, R., König, G. 2002. *Kombinierte Pfahl-Plattengründungen*. Ernst & Sohn Verlag, Berlin.

von der Hude, N. & Sauerwein, M. 2007. Practical application of energy piles. *Darmstadt Geotechnics* No. 15, 14th Darmstadt Geotechnical Conference 15 March 2007; pp. 111–127.

Janke, O., Zoll, V., Sommer, F. Waberseck, T. 2010. PalaisQuartier (FrankfurtHochVier) – Herausfordernde Deckelbauweise im Herzen der City. *Mitteilungen des Institutes und der Versuchsanstalt für Geotechnik der Technischen Universität Darmstadt*, Heft 86, 113–124.

Katzenbach, R., Schmitt, A., Turek, J. 1999. Co-operation between the geotechnical and structural engineers – experience from projects in Frankfurt. *Conference COST Action C7, Soil-Structure Interaction in urban civil engineering*, 01–02. October 1999, Thessaloniki.

Katzenbach, R. 2005. Optimised design of high-rise building foundations in settlement-sensitive soils. *International Geotechnical Conference of Soil-Structure Interaction*, 26.–28. May 2005, St. Petersburg.

Katzenbach, R., Bachmann, G., Gutberlet, C., Turek, J. 2005. The importance of measurements for evaluating numerical analyses of foundations of high-rise buildings. *11th Conference of the International Association for Computer Methods and Advances in Geomechanics*, 19.–24. June 2005, Turin.

Katzenbach, R., Bachmann, G., Leppla, S., Ramm, H. (2010). Chance and limitations of the observational method in geotechnical monitoring. *Proceedings of full papers of the XIVth Danube-European Conference on Geotechnical Engineering – From Research to Design in European Practice*, 02.–04.06.2010, Bratislava, Slovakia

Katzenbach, R., Clauss, F., Waberseck, T., Wagner, I. 2007. Geothermal Energy as a Sustainable and Efficient Energy Source and Storage Technology. *Jornada Sobre Aprovechamiento Geotérmico Superficial en la Edificación*, Madrid, Spain, 27.11.2007.

Moormann, C. & Katzenbach, R. 2002. Three dimensional effects of deep excavations with rectangular shape. *2nd International Conference on Soil Structure Interaction in Urban Civil Engineering*, 07.–08. March 2002, Zurich, Switzerland, Vol. 1, 135–142.

Nield, D.A. & Bejan, A. 1999. *Convection in Porous Media*. Springer, New York, Berlin, Heidelberg, 2nd edition

Nussbaumer, M.F. & Wolffersdorff, P.-A. 1997. Theme Lecture: Retaining structures and excavated slopes. *14th International Conference on Soil Mechanics and Foundation Engineering*, 06.–12. September 1997, Hamburg, Germany, Vol. 4, 2285–2300.

Peck, R.B. 1969. Deep excavations and tunneling in soft ground. *7th International Conference on Soil Mechanics and Geotechnical Engineering*, 1969, Mexico City, Mexico, Vol. 3, 147–151.

Poulos, H. G. 2001. Piled-raft foundation: design and applications. *Géotechnique* 51 (2).

Randolph, M. F., Clancy, P. 1993. Efficient design of piled rafts. *Proceedings Deep Foundations on Bored and Auger Piles*, Ghent.

Reul, O. 2000. In-situ-Messungen und numerische Studien zum Tragverhalten der Kombinierten Pfahl-Plattengründung. *Mitteilungen des Institutes und der Versuchsanstalt für Geotechnik der Technischen Universität Darmstadt*, Heft 53.

Rodatz, W., Gattermann, J., Bergs, T. 1999. Results of five monitoring networks to measure loads and deformations at different quay wall constructions in the port of Hamburg. *5th Symposium on Field Measurements in Geomechanics*, 1.–3. December 1999, Singapore.

Sauty, J. P. 1980. An analysis of hydrodispersive transfer in aquifers. *Water Resour. Res.* 18 (2), 253–265.

Baltic Piling – Mets & Raudsepp (Eds)
© 2013 Taylor & Francis Group, London, ISBN 978-0-415-64334-4

Experience of "BMGS" company in using of ground anchors for complicated engineering constructions

S. Polukoshko
Engineering Research Institute "VIRAC" of Ventspils University College, Ventspils, Latvia

A. Gaisin & A. Stasens
JSC BMGS, Riga, Latvia

J. Marnauza
Juras Projekts Ltd, Riga, Latvia

ABSTRACT: Ground anchors are widely used in industrial, civil, transport and hydraulic engineering: for fixing of cofferdam of foundation pits, the bottoms of docks and closed-end caissons, retaining walls and walls of underground structures, slide slopes and banks, foundations of high structures and others. In this paper using of ground anchors for construction of new berthing facilities and renovation of old one in the Baltic ports of Riga, Tallinn and Ventspils is considered on the base of the company BMGS long-term experience. Using of the soil anchors for retaining walls of construction pit near the railway embankment without stop the traffic is also presented. Extensive use of ground anchors is explained by positive effect of their application.

1 INTRODUCTION

1.1 *General*

JSC "BMGS" during 60 years is engaged in hydraulic engineering construction and civil engineering in Baltic Region. Company has the great experience of construction of harbor facilities, pile foundations of any type, railway structures, bridges and overpasses. "BMGS" company uses ground anchors for temporary and for permanent installations: for fixing retaining walls, support of excavation near the existing buildings, creating foundations of micropiles and others.

Last decade the dredging of ports aquatorium and reconstruction of the berths built in the 20th century became actual because of transport vessels deadweight and draft increasing. One of the most effective ways of the reinforcement of the piers and quays structures at deepen are ground anchors. In some cases at certain soil conditions quay structure with ground anchors are more effective than traditional ones for the new berth construction also.

Since "BMGS" company except building works also carries out geological survey and design work together with the "Juras projekts" Ltd, it provides the opportunities for the careful design, installation and testing of ground anchors. JSC "BMGS" follows the European Standard EN 1537 for execution of ground anchors, German Norm DIN 4125 for design, construction and testing ground anchors and, as reference, Russian Norm BCH 506-88 and USA recommendation RHWA-IF-99-015 for design and construction of ground anchors. For anchors forces estimation BMGS uses Plaxis and GGU softwares. Plaxis (Delft, the Netherland) is the finite element program for the analysis in geotechnical engineering, GGU (Gemany) – program for analysis and design of sheet pile walls, soldier pile walls and in-situ concrete walls.

1.2 *Installation of ground anchors*

For ground anchors installation drilling rig with a set of anchor rods, pump station for injecting of the grout, jacks with hydraulic pump unit for anchors tension and verification of anchor forces are used. JSC "BMGS" uses German drilling rig KLEMM KR-802 and pumping unit OBERMANN VS-100. Drill (anchor) rods of the two suppliers manufacturers Atlas Copso MAI (Austria) and ISCHEBECK TITAN (Germany). Each anchor consists of a set of 3-meter drill rods, interconnected by means of couplings, and at the lower end there is the drill bit.

1.3 *Anchor testing methods*

EN 1537 describes three methods of anchors testing:

Method 1: the anchor is loaded in incremental cycles from a datum load to a maximum test load; displacement of the anchor head is measured over a time period at maximum load in each incremental cycle;

Method 2: the anchor is loaded in incremental cycles from a datum load to a maximum test load or to failure; the loss of load at the anchor head is measured over a period of time at the lock-off load and at the maximum load in each incremental cycle;

Method 3: the anchor is loaded in incremental steps from a datum load to a maximum test load; displacement of the anchor head is measured under maintained load at each loading step.

Investigation, suitability, acceptance tests and creep displacement may be executed by means of each of testing method.

1.4 *Ground bearing capacity calculations methods*

In practice for estimation of the ground bearing capacity of soil anchors the designers use DIN 1054-2005 recommendation, taking into account shaft friction of soil q_{sk}. USA recommendation RHWA-IF-99-015 uses the same base, taking into account soil density and its aggregate gradation.

DIN 1054 formulas for ground bearing capacity estimation are:

$$F_d = \pi D l_b q_{sk}, \quad D = d_b k_d \tag{1}$$

where D = grout body diameter; d_b = bore bit diameter; k_d = grout body diameter increasing factor, for clay $k_d = 1.3$, for loam $k_d = 1.4$, for sand $k_d = 1.5$, for gravel $k_d = 2$; l_b = length of grout body; q_{sk} – shaft friction, for clay and loam $q_{sk} = 100$ kPa, for sand $q_{sk} = 150$ kPa, for gravel $q_{sk} = 200$ kPa.

Russian building codes BCH 506-88 proposes to calculate ground bearing capacity of anchor taking into account natural soil stress in embedded part area of anchor. BCH 506-88 formulas are:

$$F_d = \pi D l_b (1 + \sin\varphi_i)(\sigma_{od} \tan\varphi_i + c_i) K_p \gamma_c, \tag{2}$$

where D and l_b = the same as in formula (1); ϕ = average weighted on anchor bond length internal friction angle; c = average weighted on anchor bond length cohesion of soil; σ_{od} = averaged over the side surface of embedded anchor natural stress of soil, K_p – coefficient depending on bore hole diameter to grout body diameter ratio, soil natural stress, strength and deformative characteristics of soil, located in a zone of anchor embedment, γ_c = work condition factor, for clay condition $\gamma_c = 0.64$, for sand $\gamma_c = 0.72$;

σ_{od} is defined by the formula:

$$\sigma_{od} = 0.5(\gamma_i h_k + g)(\xi_0 + \sqrt{\cos^2\alpha + \xi_0^2 \sin^2\alpha})$$

where ξ_0 = side earth pressure coefficient, for clay $\xi_0 = 0.72$, for sand $\xi_0 = 0.43$; g – surface load reduced to the uniform in the center of grout body; α – anchor inclination angle to horizon; h_k – depth of the center of grout body; γ_i – weighted means of soil specific gravity on the depth h_k taking into account buoyancy effect.

K_p is defined by the formulas:

$$K_P = \left(\frac{1.01 - \left(\frac{d_b}{D}\right)^2}{1.01 - \frac{A_1^2}{1 + A_1^2}} \right)^{\theta}, \quad \theta = \frac{\sin\varphi_i}{1 + \sin\varphi_i},$$

$$A_1 = \frac{E_0}{(1 + \nu_0)(\sigma_{od} + c_i \cot\varphi_i)\sin\varphi_i}$$

where E_0, ν_0 = weighted mean value on the length of grout body of the deformation modulus and Poisson ratio respectively.

Russian Scientific, Investigation and Design Institute "Fundamentprojekt" proposes to calculate the ground bearing capacity of anchor taking into account cement grout excess pressure during injection. Fundamentprojekt formula is:

$$F_d = k m_p \pi d_b l_b p_b tg\varphi \tag{3}$$

where k = soil homogeneity factor, $k = 0.6$; m_p = coefficient. taking into account stress state of soil dependent on injection, for clay $m_p = 0.4 \div 0.2$; d_b, l_b = the same as in formula (1); p_b = excess pressure during injection, $p_b = 4000 \div 6000$ kPa; φ – internal friction angle of soil.

In this work for testing anchors calculation of bearing capacity of ground anchors was fulfilled in accordance with equations (1), (2), (3).

2 GROUND ANCHORS SYSTEM FOR MARINE STRUCTURES

2.1 *Oil jetty in free port of ventspils*

Jetty Nr. 3 for oil transportation in port of Ventspils, designed by Lenmorniiproekt, was built by "BMGS" company and was put into operation in 1974. Jetty with total length 407 m is divided into four sections: toe, working section, middle and head dolphins, which are interconnected with crossovers (Fig. 1). The toe section is designed as open jetty on shell – type concrete piles Ø1.2 m with precast reinforced concrete grillage of slabs with dimension 5.23 × 5.23 m. Working section of oil jetty with length of 148 m, 36 m width and 14m depth (with perspective deepening till 17.2 m) is designed as two anchored sheetpile bulkhead with front wall of sheet piles Larsen-5 and precast prestressed hollow concrete tube piles Ø1.6 m, driving with step 3.36 m (Fig. 2). Concrete tube piles 12 of two opposite rows are mutually anchored at CD +0.70 m with steel rods 13 with Ø95 mm, 30 m length and 3.36 m step; sheet pile wall 11 is anchored to wale of tube concrete piles at CD +0.70 m by means of

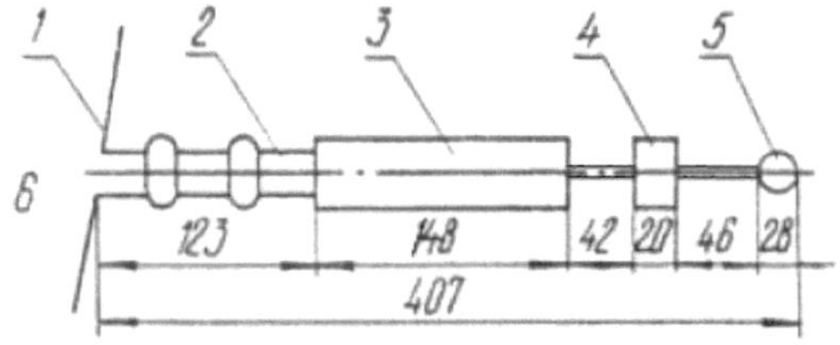

Figure 1. Lay-out of jetty No.3: 1 – bank protection, 2 – jetty toe, 3 – working section, 4 – intermediate dolphin, 5 – head dolphin, 6 – refueling area.

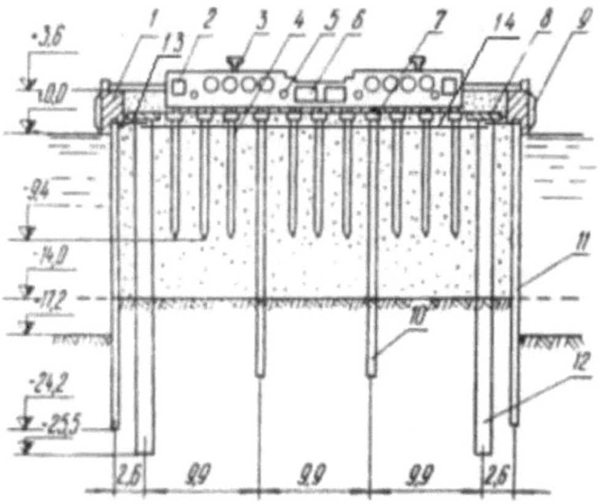

Figure 2. Cross-section of the working part of the old jetty: 1 – concrete wall capping; 2 – electrical cable gallery; 3 – bollard; 4 – concrete prismatic piles; 5 – process pipelines; 6 – feed pipe channel; 7 – girders and plates; 8 – relieving plate; 9 – fenders; 10 – steal casing supports; 11 – Larsen – 5 sheetpile wall; 12 – concrete pipe piles, 13 – anchor rod Ø 90 mm, 14 – anchor rod Ø 95 mm.

Figure 3. Construction of working section of jetty.

anchor rods 14 with Ø90 mm, length of 3.80 m and step 1.90 m. Filling is covered by the precast reinforced concrete slabs 7 on the piles foundation 4. The space between sheetpiles and tube piles are covered with reinforced concrete relieving slabs 8. In Figure 3 the construction of working section of jetty is shown.

In the upper structure of the jetty the industrial channels 5 and 6 are carried out, in which piping for loading oil tankers, cable networks, oil lines, steam

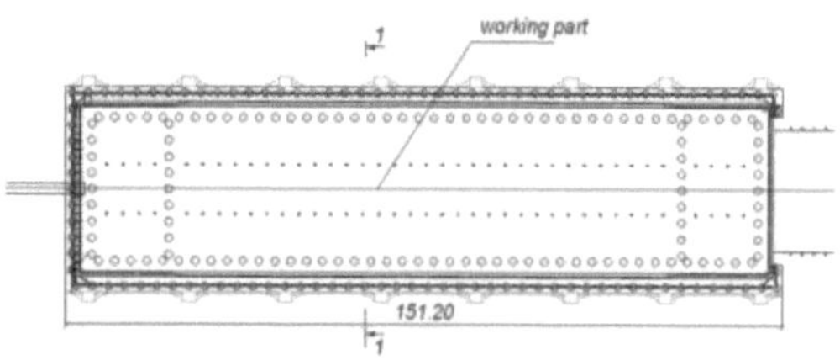

Figure 4. Scheme of working section after reconstruction.

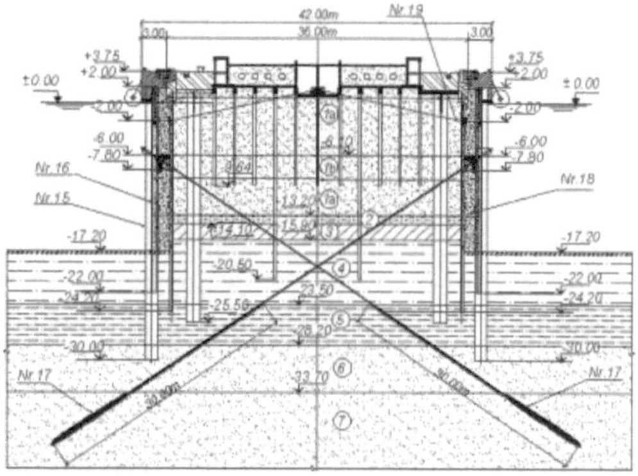

Figure 5. Cross-section 1-1 of jetty working part after reconstruction: 15 – tube pile d = 1520 × 19.1 mm, 16 – sheetpiles package L605+L755+L605; 17 – anchors TINAN 103/43; 18 – sandy-gravel fill; 19 – prestressed anchor of 5 tendons 5K7.

pipes, water, etc. are installed. Jetty is equipped with bollards 3 on the force of 100 tons and fenders 9. On the working part of jetty hose devices and remote control devices are located. Oil filling into vessels is performed by means of automatic equipment. Intermediate and head dolphins are identical in design with a working part.

In 1999 the state of the anchor has been examined and irreversible deformation was found because of yield point stress, some anchors were ruptured. For this reason in 2000 new additional anchor were installed by "GT Corporation" company: prestressed anchors of 5 tendons 5K7 with force 400 kN, at CD −2.00 m with step 3.36 m. This company since 2003 carries out the regular monitor of jetty structures deformations and displacements. At the same time the necessity of the deepen of the aquatory area to design CD −17.20 m was appear, but even with dredging up to CD −16.00 m near the jetty, deformation of sheetpile wall increased dramatically. Maximum displacement of the front wall has reached 500 mm. Therefore port authorities made a decision to reconstruct the working section of oil jetty in order to provide its safety operation. Design was elaborated by "BMGS" and "Juras projekts" companies and reconstruction was carried out by "BMGS". For jetty renovation the construction the combiwall round all working section with anchorage of king piles with ground anchor was proposed (Figure 4, 5). Combiwall consists of steel tube pile

Table 1. Characteristic values of soil parameters.

Soil layer	Unit weight γ_{unsat} kN/m³	Young's modulus E kN/m²	Poisson's ratio ν –	Cohesion c_{ref} kN/m²	Friction angle φ °
1a	18/20	20000	0.2	1	30
1b	18/20	40000	0.2	2	35
1c	18/20	20000	0.2	1	30
2	14/17	10000	0.3	1	27
3	16/20	10000	0.2	2	27
4	15/19	5500	0.2	10	22
5	18/21	10000	0.2	5	30
6	18/20	35000	0.3	1	35
7	18/21	45000	0.3	1	40

Figure 6. Reconstruction of jetty – combiwall sheetpiles driving.

Ø 1520 × 19.1 mm, driving to CD −30.00 m with step 3.36 m, and sheetpiles package L605+L755+L605 driving to CD −22.00 m. Ground anchors Ischebeck TITAN 103/43 with free length 28 m and length of grout body 30 m are installed on CD −6.00 m with inclination angle to horizon 34°.

Soil layers presented in table 1 are: 1a – loose sandfill; 1b – moderate sandfill; 2 – clayey sand; 3 – sandy loam; 4 – silty clay; 5 – sandy clay; 6 – moderate sand; 7 – dense sand.

Reconstruction of jetty is shown in Figures 6, 7.

The testing of two anchors was carried out by Department of Soil Mechanic and Foundation of Cottbus BTU under direction of Prof. Dr.-Ing. Wichter L. Properties of testing anchors are in table 2, testing program – in table 3.

Because of the deep anchors position under water the measurement of its deformation under reasonable forces is not possible, it was proposed to execute the anchors Ischebeck TITAN 103/43 testing in accordance with Test Method 2 of EN 1537. Type of test: suitability test with incremental application of the force, measurement of load loss at each load level during a certain time interval. The work load (design load) $P_W = 1566$ kN, maximum test load (proof load) $P_P = 1960$ kN, safety factor $\eta_p = 1.25$.

Figure 7. Reconstruction of jetty – anchor installation.

Table 2. Properties of testing anchors.

	Anchor No.15	Anchor No.19
Installation level, m(abs)	−6.0	−6.0
Inclination angle, deg	34	34
Free length, m	16.5	28.5
Anchor diameter, mm	103/43	103/43
Cross section area, mm²	6023	6023
Drill bit diameter, mm	175	175
Grout bogy length, m	30	30
Yield point, kN	3010	3010
Ultimate load, kN	3788	3788

Table 3. Testing load, observation time, allowable force loss.

Load stage	% P_P	Test load kN	Observ. time, min	Allowable force loss k, %
0 (initial load)	10	196	15	–
1	40	784	15	2
2	55	1076	15	2
3	70	1372	15	2
4	85	1666	15	2
5	100	1960	30	2.5

Tests results (Fig. 8, tab. 5) show the conformity of anchors. After testing the anchors are fixed, load by fixing $P_0 = 780$ kN; fixing the coefficient $\eta_0 = 0.50$.

Reconstructed jetty was put into operation in 2009.

2.2 *Terminal for containers in port of Muga*

New container berths Nr. 14 and 15 in Tallinn seaport of Muuga was built in built in 2003 ÷ 2005 with accordance to the project of "Estkonsult" company. Berths were dry-erected; mechanisms were placed on natural ground level, with subsequent dredging after the construction of piers (Fig. 9, 10).

Quay was constructed using ground anchors, which diameter (73/53 mm) and length (total 57 m) were

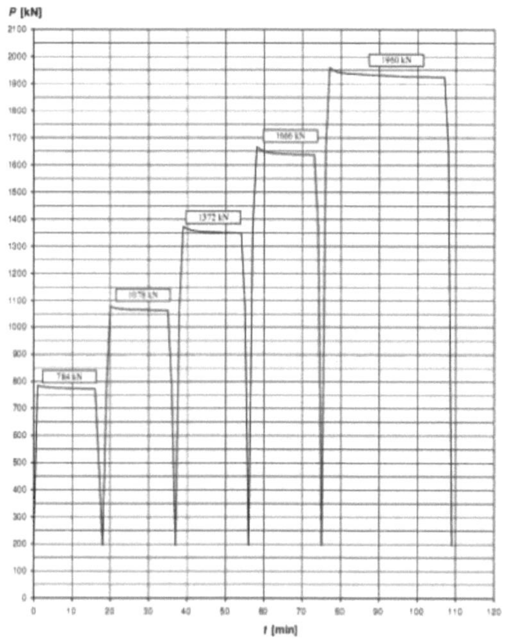

Figure 8. Results of anchor Nr. 15 testing.

Table 4. Test results.of anchors No. 15 and No. 19.

	Anchor No. 15	Anchor No. 19
Max. test load P_P, kN	1960	1960
Observation time, min	30	30
Measured force loss ki, %	1.8	1.6
Allowable force loss k, %	2.5	2.5

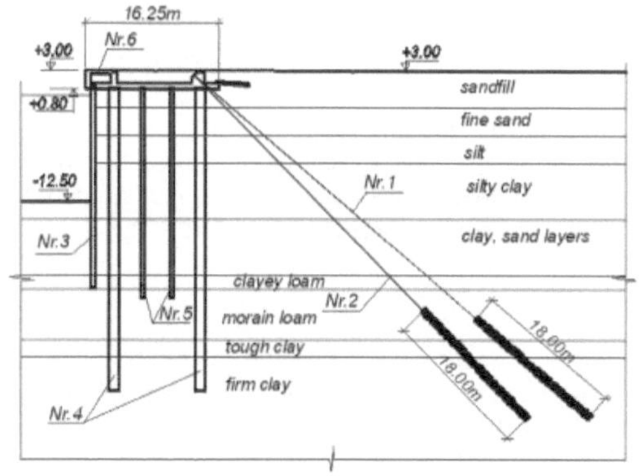

Figure 9. Cross-section of container terminal in Muuga: 1, 2 – soil anchors, 3 – sheet piles AZ-46, 4 – tube piles Ø 1220 × 12 mm, 5 – concrete precast piles, 6 – concrete slab.

determined empirically by drilling them into the load-bearing layers of stiff clay at an angle of 45° to horizon. The construction of piers is piled quay wall with front sheetpiles AZ-46 and two rows of pipe piles

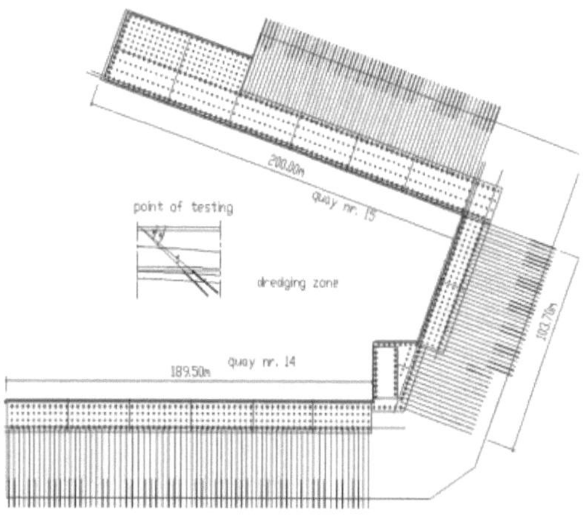

Figure 10. Lay-out of quay walls in Muuga.

Figure 11. Anchors TITAN 73/53 installed through concrete plate of berth in Muuga harbor.

Ø 1220 × 12 mm with distance of 10.5 m and step of 2.8 m. Between tubular piles two rows of precast concrete piles of prismatic cross-section 40 × 40 cm were placed. Tubular and concrete piles are the basis for foundation grill slab and crane beams.

Construction was carried out by the general contractor JSC "FKSM" involving JSC "BMGS" as the main subcontractor, who was engaged in the steel sheet piles, steel pipe and reinforced concrete piles driving and ground anchors construction, which were installed after grillage plate was concreted by the general contractor. At first, the piling works were carried out in the middle berth, including a ramp with loading platform and parts of the berth Nr. 15 (Fig. 10). Then, with some lag, grillage slab was concreted to provide working area for construction of ground anchors. Second, in the same sequence works on the berth Nr. 14 and then on the berth Nr. 15 was completed. Ground anchors were installed after casting of concrete grillage slab and gaining concrete strength not less than 70%. In the grillage slab there were pre-set sleeves from pipe of Ø 250 mm at an angle of 45°, through which the drilling performed (Fig. 11). Drill hole was cleaned

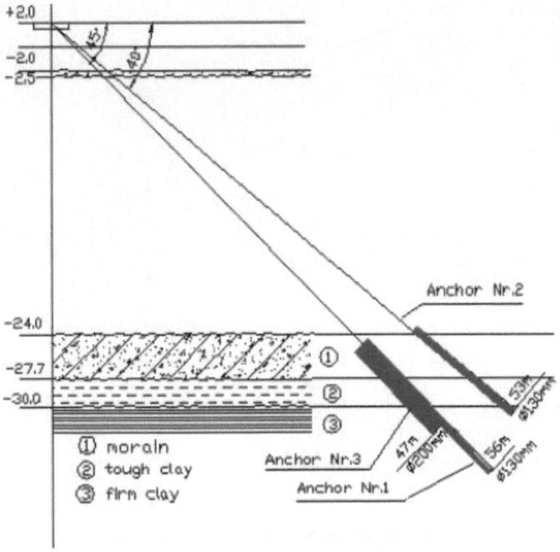

Figure 12. Trial anchors TITAN 73/53 in Muuga harbor.

Table 5. Properties of testing ground anchors.

	Anchor No. 1	Anchor No. 2	Anchor No. 3
Inclination angle, deg	45	40	45
Free length, m	38.0	40.0	38.0
Grout bogy length, m	18.0	13.0	9.0
Anchor diameter, mm	73.53	73/53	73/53
Cross section area, mm^2	1631	1631	1631
Ultimate load, kN	1160	1160	1160
Drilling bit diameter, m	130.0	130.0	200

out with cement grout with a water cement ratio of 1/1 during drilling. In the final part of the borehole water-cement ratio decreased to 0.4. Embedment of anchors in the dense layers of firm clay was carried out on the length from 13 to 18 m. As the investigation before construction show, this length is sufficient to obtain the design load 580 kN.

After cement grout hardening in the embedded part the tension of anchor is made up to design loads, taking into account the factor of safety. Tension and testing of each anchor was performed with ISCHEBECK RCH-603/1003 rig. Altogether 350 ground anchors Ø 73/53 with a maximum length of 57 m were installed in container berths.

Before ground anchors designing three anchors were tested to determine their ground bearing capacity, the optimal technology of it executing, required diameters and a length (Fig. 12). Properties of testing anchors are given in table 5. Tests were carried out on the construction site (Fig. 13).

Concrete slabs with dimensions 2.60 × 5.00 × 0.30 m and total weight about 35 tons were used as support.

Test is performed according to requirements of DIN 4125. The work load (design load) $P_W = 580$ kN, maximum test load (proof load) $P_P = 870$ kN, safety factor $\eta_p = 1.5$. The value of each stage corresponded to 25% of design value.

Figure 13. Trial anchors construction and testing.

Table 6. Testing load and observation time.

Load stage	% P_P	Test load, kN	Observation time, min
0 (initial load)	20	116	1
1	50	290	30
2	75	435	30
3	100	580	120
4	125	725	180
5	150	870	1440

Table 7. Trial anchor testing results.

	Anchor Nr. 1	Anchor Nr. 2	Anchor Nr. 3
Measured ΔSi, mm	0.81	1.64	2.09
Allowable ΔS, mm	2.00	2.00	2.00

Total anchor loading time was 1800 minute or 30 hours in the last stage of loading (1.5 F_W). Strain for each time of loading was fixed. Criterion of suitability of an anchor is condition at last loading stage:

$$\Delta S = S_{1440} - S_{120} \leq 2.00\,\text{mm}.$$

For each anchor the strain rates in the last stage of loading were recorded, difference between strains on 1440th and 120th minutes is presented in table 7.

Plot of anchor deformation depending on time on the last stage of loading is given in Figure 14.

Analysis of anchors testing results shows that their design and soil bearing capacity in general meets the criterion of suitability. In the process of anchorage construction the optimal mode of drilling was determined. Due to the fact that using of boring bit of Ø 200 mm increase cement consumption up to 30%, results showed the unreasonableness of its using. Based on the results obtained during the test, it was

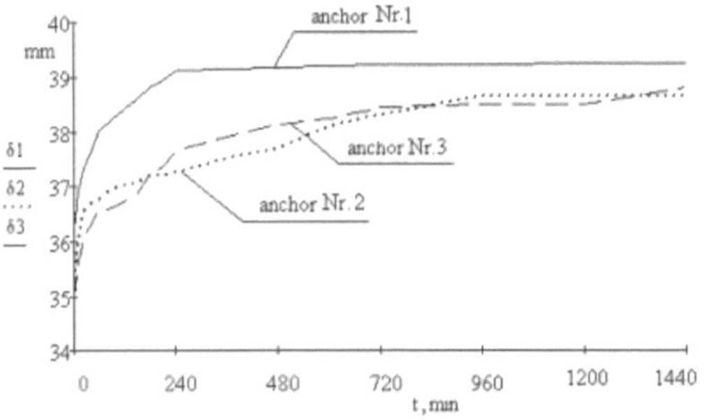

Figure 14. Testing of the ground anchors: deformation dependence on time at last stage of loading.

Table 8. Calculated bearing capacity of testing anchors, Kn.

In accordance with	Anchor No. 1	Anchor No. 2	Anchor No. 3
DIN 1054-2005	956	690	735
BCH 506-88	760	736	546
JSC Fundamentprojekt	945	682	727

recommended to accept: anchor type – TITAN 73/53, maximum length of 57 m (depending on superface of moraine clay stratum location the length of the anchor may be reduced), length of the grout injection of 13.0 ÷ 18.0 m, drilling diameter of 130 mm (Clay Bit-Cross Cut).

These recommendations were adopted in the design of anchorage, which were performed by JSC «BMGS» in the construction of berths No. 14 and No. 15 in port of Muuga.

Ground bearing capacity of each anchor, calculated by different methods is presented in table 8. At design value of anchor force 580 kN maximum safety factor has anchor No.1

2.3 *Coal terminal "STREK" in seaport of Riga reconstruction*

In Figure 15 example of the construction of ground anchors in the redeveloped coal berth Nr. 21 of "STREK" Company in the seaport of Riga is shown. Berth was built in 1993 as anchored piled guay wall with depth 10.5 m. Front wall of quay is combiwall, which consists of Larsen-5 boxes king-piles driving to CD −27.20 m with step 1.68 m, and sheetpiles Larsen-5 driving to CD −17.00 m. Front wall is connected with anchor wall of concrete piles by means of anchor rod Ø 60 mm with center to center distance of 1.50 m.

Steel tube piles Ø1020×10 mm of screen row and back row are installed with ctc distance 1.50 m to CD −27.60 m. Later in 2003 the necessity of the deepening of the berth to CD −13.50 m appeared and the quay wall was reinforced with additional anchors by the design of "Juras projekts". Ground anchors MAI T76N Ø76/53mm, with free length 37 m and bond length

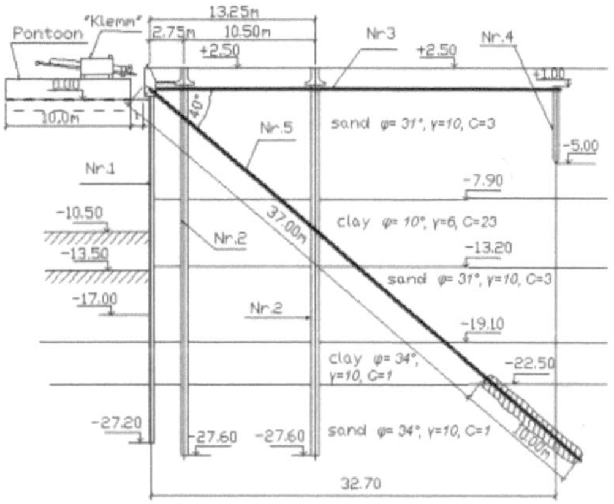

Figure 15. Reinforcement of "Strek" berth in port of Riga by means of grout anchors: 1 – combiwall of sheet piles Larsen-5, 2 – steel pipes Ø1020 × 10 mm; 3 – anchor rod Ø 60 mm, ctc = 1.50 m; 4 – reinforced concrete prismatic anchor piles 5 – ground anchor MAI T76N.

10 m were installed with step 3.00 m and inclination angle 40°.

Before anchor installation in concrete capping beam and steel sheet pilling the hole of Ø 150 mm was drilled, and after this the anchor at an angle of 40° to the horizon was drilled-in. In the process of drilling grout is pumped under pressure of 40–60 bars. After 7 days, the anchor is tested for load of 900 kN with a design force of 680 kN.

Calculated ground anchor resistance in accordance with DIN 1054-2005 – 919 kN, BCH 506-88 – 915 kN, JSC "Fundamentprojekt" – 2262 kN, safety factor is 1.35, 1.35 and 3.32 respectively.

Suitability and acceptance test were carried out in accordance with DIN 4125.

3 GROUND ANCHORS SYSTEM FOR RETAINING WALLS

The example of the application of ground anchor system in the retaining wall is construction of Stockmann Centre "Cinema Palace and Department Store" in Riga near the main railway station. For building placing the retaining wall should be installed by the railway embankment and the embankment should be removed in front of retaining wall (Fig. 16). Construction was carried out by the design of IP-Engineering Finland Ltd, JSC "BMGS" installed the retaining wall in 2002.

Retaining wall is installed of Larsen – 604 steel piles Larsen 604, L = 18 m, walling HEB 300 with three (Fig. 17a) and two (Fig. 17b) rows of ground anchors TITAN. Properties of ground anchors are given in table 10. During the building procedure the anchor may be cut, loosened or treated, therefore after concrete works the earth pressure can be taken by the basement wall. The work load of anchor No. 1 $P_W = 210$ kN, anchors No. 2, 3 $P_W = 530$ kN.

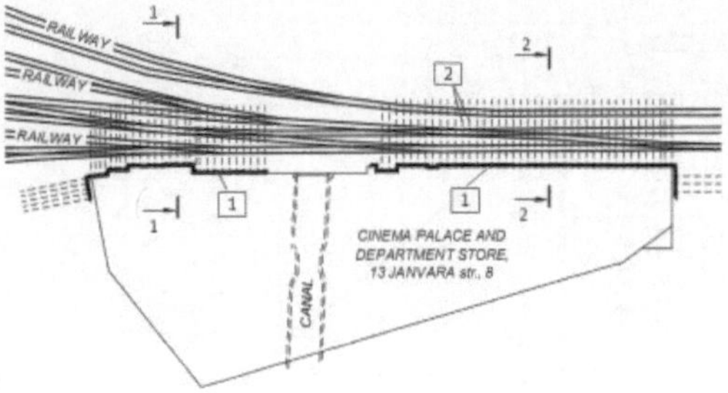

Figure 16. Situation plan of Stockmann Centre: 1 – sheet-piling retain wall, 2 – soil anchors.

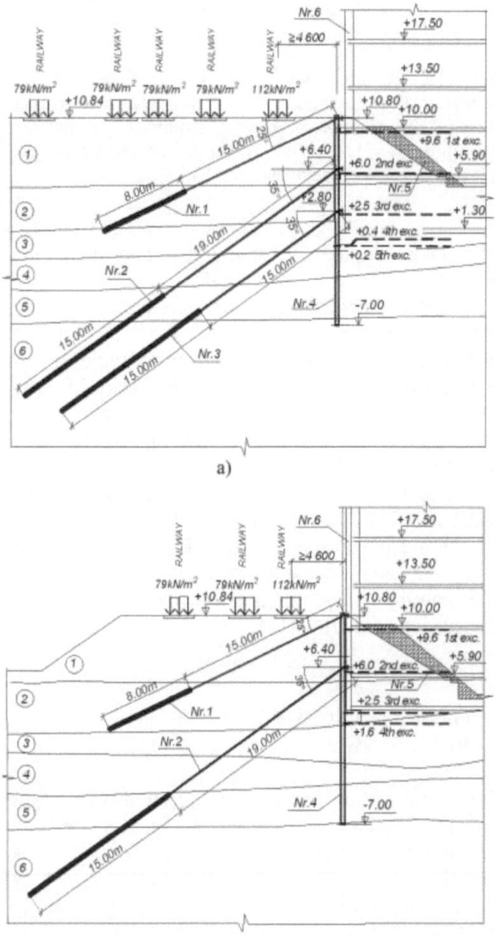

a)

b)

Figure 17. a) cross-section 1-1, b) cross-section 2-2 of the retaining wall and embankment: 1 – grouted anchors TITAN 40/20; 2, 3 – grouted anchors TITAN 73/53, 4 – sheetpiles wall Larsen 604, 5 – temporary working embankment, 6 – designed structures of Cinema Palace and Department Store.

The soil layers indicated in the table 9 are: 1 – sand, 2 – fill sand, gravel, organic, 3 – medium dense fine sand, 4 – silty sand and peat, 5 – medium dense sand, 6 – dense silty sand.

Table 9. Geotechnical soil parameters.

Soil layer	Unit weight γ_{unsat} kN/m^3	Young's modulus E kN/m^2	Poisson's ratio ν –	Cohesion c_{ref} kN/m^2	Friction angle φ °
1	18	28,000	0.2	2	32
2	16	10,000	0.2	4	26
3	18	13,000	0.2	3	32
4	17	13,000	0.2	4	15
5	19	23,000	0.2	3	23
6	20	30,000	0.3	3	30

Table 10. Properties of ground anchors.

	Anchor No. 1 upper row	Anchor No. 2 middle row	Anchor No. 3 bottom row
Installation level, m(abs)	10.80	6.40	2.80
Inclination angle, deg	25	35	35
Free length, m	15.0	19.0	15.0
Anchor diameter, mm	40/20	73/53	73/53
Cross section area, mm^2	726	1631	1631
Yield point, kN	430	970	970
Ultimate load, kN	539	1160	1160
Drilling bit diameter, m	0.090	0.130	0.130
Grout bogy length, m	8.0	15.0	15.0
C.t.c. distance for 1-1, m	3.60	2.00	1.30
C.t.c. distance for 2-2, m	3.60	1.40	–

Figure 18. Installation of the first raw of the anchors.

Sequence of the work: preparatory work; excavation and backfilling of the first working level; driving the sheet pile wall to CD −7.00 m; excavation to CD +9.60 m to the first anchoring level and installation of anchors including testing; excavation to +6.00 m to the second anchoring level and installation of anchors including testing; excavation to +2.50 m to the third and installation of anchors including testing, excavation to the fourth and fifth excavation level; finishing and rehabilitation work.

Table 11. Load stage and observation time for suitability test.

Load stage	% P_P	Test load, kN Anchor No. 1	Anchor No. 2, 3	Observation time, min
0 (initial load)	20	42	106	–
1	50	105	265	5
2	75	157	398	5
3	100	210	530	5
4	125	263	663	5
5	150	315	795	15

Table 12. Load stage and observation time for acceptance test.

Load stage	% P_P	Test load, kN Anchor No. 1	Anchor No. 2, 3	Observation time, min
0 (initial load)	20	42	106	–
1	50	105	265	5
2	75	158	398	5
4	100	210	530	5
5	125	263	663	5

Anchors and anchor works were high quality and met all requirements of standards. All works were journalized, cement grout consists of Portland cement CEM I 32.5 R and drinking water, water-cement ratio for rinsing injection in the free length section of anchor of W/C = 1, for injection in bearing part of anchor W/C = 0.4.

All anchors are tested not earlier than 7 days after their installation. Tests with static load were performed according to requirements of DIN 4125. Suitability test and acceptance tests were performed.

Three anchors were selected by random choice are tested in each row for suitability test performing. Stages of test loads and observation time intervals are given in table 11.

At last load stage anchor deformation is taken after 1, 2, 5, 10 and 15 minutes. Positive result criteria is considered deformation difference between 5th and 15th minute $\Delta S = S_{15} - S_5 \leq 0.50$ mm.

All anchors which did not undergo suitability test have to pass acceptance test.Stages of test loads and observation time intervals are diven in table 12.

At last load stage anchor deformation is read after 1, 2, and 5 minutes. At positive result criteria is considered deformation difference between 2nd and 5th minute $\Delta S = S_5 - S_2 \leq 0.20$ mm.

For measuring deformation is used camber meter with accuracy 0.01 mm.

After keeping of each stage of load the anchor is released up to initial load. If the positive test result is not reached, time for keeping of load is extended. If the positive test result is reached, anchor is fixed at load 0.80 F_w.

Table 13. Design bearing capacity of testing anchors, kN.

In accordance with	Anchor No. 1	Anchor No. 2	Anchor No. 3
DIN 1054-2005	509	1378	1378
BCH 506-88	296	1009	1140
JS Fundamentprojekt	1059	2860	2880

Ground bearing capacity of each anchor, calculated by different methods are presented in table 8; safety factor in accordance with DIN 1054 for each anchors is more than 2.

During excavation work the observation of the state of rail road and nearby buildings situated at the distance of 15 meters. Marks on the sleepers were drafted every 10 meters and measurements were taken once a day. In addition, the deformation of sheet pile wall was measured at the top and anchorage level. Displacement in vertical direction was measured with an accuracy of 1 mm, in horizontal direction – 2 mm. Unacceptable deformations were not observed.

Stockmann Centre "Cinema Palace and Department Store" in Riga was put into operation in 2004.

4 CONCLUSIONS

In this paper three different cases of using soil anchors in marine structures are shown: fastening the front wall of quay, fastening the back end of superstructure, fastening the front end of superstructure. In spite of difficulties of the construction of long-length ground anchors (accuracy of works, complexity of their testing because of significant elastic deformation) JSC "BMGS" proved the possibility of their usage in especially difficult conditions in which other solutions would be more expensive and less reliable.

In order to more exactly describe the formation of the grout body it is necessary to create of mathematical model of borehole cavity expansion under excessive pressure.

REFERENCES

DIN 4125. 1990. *Ground Anchorage. Design, construction and testing.* Berlin.

DIN 1054-2005. *Subsoil – Verification of the safety of the earth work and foundation.*

EN 1537:1999. *Execution of special geotechnical work – Ground anchors.* Brussels.

BCH 506-88. 1989. *Design and construction of ground anchors.* Moscow.

Plaxis b.v. 2007. Delft, the Netherland.

Sabatini, P.J. Pass, D.G., Bachus R.C. 1999. *Geotechnical Engineering Circular No. 4: Ground Anchors and Anchored Systems.* US Department ofTransportation, Federal High way Administration, Performing Organization: GeoSyntec Consultants, Office of Bridge Technology, Washington, http://www.fhwa.dot.gov/bridge/if99015.pdf.

Baltic Piling – Mets & Raudsepp (Eds)
© 2013 Taylor & Francis Group, London, ISBN 978-0-415-64334-4

The history of the foundations in Tartu

Vello Pallav & Mait Mets
Department of Rural Building, Estonian University of Life Sciences, Estonia

ABSTRACT: Tartu is situated on the slopes and marshy banks of the primeval river Emajõgi, where due to the complex geological conditions the foundations of buildings have been a problem throughout the history of the local human settlement development. This article focuses on constructions of the city of Tartu, and their dependence on foundation development.

1 INTRODUCTION

Based on various data the first traces of human settlement on the Estonian territory manifested more than ten thousand years back. In order to survive, man had to find a dwelling place near water bodies and on the most suitable natural elevations.

The history of Tartu settlement as a city is directly connected to its location on the banks of the river Emajõgi and it is probable that in the end of the first millennium the housing was foremost situated on the skirt of elevation, and the buildings were constructed on the places which remained dry during the floods period. At that time both the castle and the urban area probably formed a somewhat closed administrative unit (Metsallik 2007). The necessity to use waterways determined a demand for riverside facilities, but due to the unfavourable geological conditions it was difficult to build them. Consequently, the issue of foundations in Tartu has always been topical.

2 THE HISTORICAL DEVELOPMENT OF TARTU AND ITS CONNECTIONS TO THE BUILDING ACTIVITIES

An ancient castle was situated in Tartu already in the I millennium AD, and Russian Chronicles have recorded the capturing of Tartu in 1030 by the army of the Prince of Kiev, Jaroslav the Wise. This datum is considered to be the first mentioning of Tartu in written sources. The Estonians managed to regain their castle in 1061. This historical document is of high importance, because here Tartu is mentioned as one of the oldest city centers of Eastern Baltics (Pullat 1980).

During the war that has begun in the 13th century, the Teutonic Order finally conquers the city in 1224 and Tartu becomes a diocesan centre of the medieval vassal state governed by Hermann, Bishop of Tartu. Being situated on the banks of the actively used waterway which links the Baltic Sea and Russia, Tartu becomes a member of the Hanseatic League in 1280. Therefore, Tartu is developing into a flourishing trading city on the so-called trading road that leads to Pskov and Novgorod, and this fact has forced the construction of quays and ports in order to provide a place for downloading goods.

All these activities brought the settlement closer and closer to the river, although the primary dwelling place had been on an elevation, called Toomemägi nowadays, that was suitable for building a castle, but afterwards the housing descended to the primeaval valley of the river Emajõgi. The hydrogeological and geological conditions in the valley of the river Emajõgi are very specific for building. As it was extremely difficult to erect a house on a marshy or fragile soil on the riverbanks, people have searched for solutions how to fortify the bases or establish a strong foundation.

The travelogues from the first half of the 15th century describe Tartu as a well-fortified city with beautiful stone houses.

Due to its advantageous location Tartu was developed into a magnificent medieval city by the Middle Ages, but the beginning of the Livonian war had a destructive impact on its splendour. In 1556 Russian army occupied Tartu. Most of its inhabitants were deported to Russia and Tartu belonged to Russia till 1582, when Southern Estonian areas became a part of Poland due to conclusion of the treaty of peace between Russia and Poland. A notable fact from this period is founding a Jesuite Gymnasium by the polish authorities in Tartu in 1583. The beginning of the 17th century was marked by wars between Sweden and Poland, the city belonged in turn to one and the other and in 1625 remained under Swedish government. In 1630 a gymnasium was founded in Tartu and in 1632 a university, named *Academia Gustaviana* after its founder, the Swedish King Gustav II Adolf. The university provided education in Tartu, although with interruptions, till 1699 (Vahtre jt.1985).

A sequent campaign from Russia subjugated Tartu again in 1656 and it stayed under the Russian domination till 1661, when it became free, i.e. belonged to Sweden again for a while. 18th century treated harshly the housing of Tartu. As a result of the Great Northern War (1700–1721) Tartu was taken from the Swedish kingdom and submitted under the rule of tsarist Russia. In summer 1704 Peter I captured Tartu after days of cannonade. Tartu had been sieged from three sides and when it finished, the city was basically in ruins.

In 1708 Peter I considered it very probable that the Swedish army, after a successful campaign in Poland, was going to attack St. Petersburg from the east and ordered either to fortify the existent fortresses or destroy them to avoid the Swedish using them as their own strongholds.

Unfortunately the final decision of the Russian tsar determined the destruction of the medieval city of Tartu. The stone buildings were mined with loads of gun powder, the city was prepared to ignition and in July 1708 the planned action of destruction changed the city into smouldering ruins. The city walls and gates were destroyed, and so were former bishop's residence on Toome hill, churches of Maarja and Jaani (St. Mary and St. John), as well as quarters of stone houses in the down city. The medieval housing of Tartu ceased to exist. The occupation of whole Estonia and Livonia, including the war-torn city of Tartu, by the Russian power was fixed in Uusikaupunki (Nystad) peace treaty in 1721.

The citizens who returned to the city after the Northern War, were impoverished and could not afford more than erecting on the old ruins but one-storey wooden buildings, covered with straw or board roofs.

Moreover, at the time the imposing stone buildings were authorized to be erected only in the new capital city – St. Petersburg, in other cities it was forbidden. As such, Tartu was exposed to the danger of fire and thus the damage happened. In 1763 a large part of the city centre burnt down (Pullat 1980).

New hope for the city of Tartu arose after the Empress Catherine II's visit in 1764, after which there was decided building of Tartu in early classical style and already in 1764–1768 the Russian engineering commando prepared the first plans for rebuilding of Tartu. The works for building up the city began actively, and the burnt and destroyed buildings were replaced by new ones.

Between 1763–1767 Tartu was constructed as a fortified city and once again it acquired an importance as a vital point of junction in St. Petersburg – Koenigsberg post road. A new Town Hall of Tartu and the legendary Stone Bridge were constructed in 1770–1780.

However, in 1775 Tartu was hit by the biggest fire accident in its history. The fire started from the Rüütli (Knight's) Street, near the St John's Church, and due to the strong gusts of wind the fire spread rapidly and the mainly wooden buildings took fire easily. In a short time most of the city centre was in flames and burning the wooden bridges, it spread to the other side of the river. Only the northern part of the old city – Lai Street area – remained intact. 290 buildings, including 43 stone buildings perished in the fire. In order to prevent the fire spreading, 18 buildings were pulled down later, so that there were left only 160 houses in Tartu, and 120 of them were situated in the outskirts. After the fire there were established first building regulations of the city, which consisted primarily of fire safety regulations. As a result, erection of wooden houses, outbuildings and fencing was banned in urban areas (Palamets 2003). The city walls destroyed during the North war were not repaired any more, and most of them ceased to exist in the course of time (Pullat 1980).

Even nowadays we can see a part of the old city wall such as it has been preserved in Lai Street. There can be seen the old city walls and the so-called rampart passageways on the city plans from 1760, but after the fire in 1775 the city plans from 1777 and 1780 depict the network of streets and buildings mainly such as we see it today. The strict rules of building established by the Tsarist state made the officials to adhere to the rules of construction and the city of Tartu was built up as a complex classicist city centre, and the active construction activities continued throughout the 19th century.

In 1802 the University of Tartu was re-opened as a German speaking imperial university, and belonging into the Russian Empire, Tartu becomes an important research centre whose reputation extended far beyond the borders of the empire. Thanks to the university a world-class observatory was built on Toomemägi (Toome Hill), the same period also witnesses building the university clinics, libraries, and a cutting edge botanical garden at the time.

Tartu becomes a second city of the Livonian province of the Russian Empire and the success can also be observed in building activities.

The I half of the 19th century was a flourishing period of classicism in Tartu, and the 2nd half witnessed new architectural ideas and trends. The housing landscape of Tartu was enriched with Historicism, Art Nouveau and national romanticism. Between 1843 and 1914 more than 2000 new houses were built in Tartu, and the existing ones were rebuilt extensively. If in 1843 there were counted 973 living houses in Tartu, then by 1910 their number had risen up to 3125. In the early 20th century the number of houses might have been *ca* 4000, including the suburban areas. Thus, most of the preserved historical buildings date from this period (Siilivask 2006).

Today the Old City of Tartu is a complex area in classicist style and it is under national protection as a complex national architectural monument. Its registration number is 6889 and in the old city there are more buildings and facilities that are under national protection.

The construction of buildings in the primeval valley area of the river Emajõgi has not been easy at all – the soil needed reinforcement using suitable fillers or wood piles and rafts, the latter being the only possibility at the time, as the archaeological excavations have exposed.

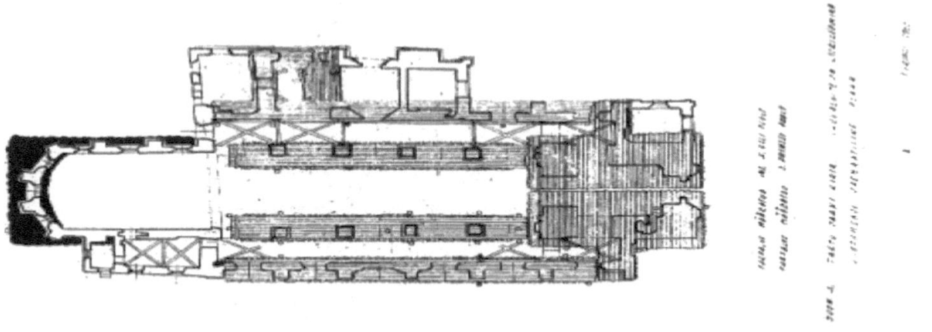

Figure 1. Schematic drawing of raft base under Jaani (St. John) Church. Drawn by O. Prints in 1962 (Oll 1984).

In the VII square of the Old City there were found traces of a wooden fence, which coincided with the latter boundary of the land plot (Aun 1990). There was also found a row of wood piles driven into the ground in the yard at Munga Street 8, probably dating from an earlier era (Trummal 1974). Some parts have been preserved from the pre-conquest buildings in a land plot situated at the corner of Lai and Rüütli Street (Tiirmaa 1990), and Poe Street 8, (Tiirmaa 1980) and in some other places. Worth mentioning are also fragments of clay insulation, found from cultural layer in many places (Metsallik 2007).

The archaeological excavations in Lossi St and surveillance in the area of Rüütli and Lai Street expose that the local wood pavement in its most ancient form must have been built already some time before German occupation (Trummal 1989). In 1987 experts studied an earlier medieval wood pavement in the northern side of Lossi Street, Street name signed as Väike-Maarja (Little Mary) on a reconstruction plan by R. Otto. The oldest pavement had initially been made of wood sticks and branches, of which only a flimsy stripe was preserved in the cultural layer (Metsallik 1992, 2007).

The city wall was probably finished during the 14th century as well. In 1981 the excavations on Barklay Square exposed an East-West directional row of thick wood piles, which seems to be closely connected with the fortification zone (Metsallik 1992, 119). The results of excavation demonstrate that it was not everywhere expedient to found the foundations of the houses on wooden basis. There are often found traces of foundations of medieval stone buildings being built on boulder filled (Metsallik 2007).

As the above-mentioned information shows, we have no comprehensive overview of the state of foundations of the old city and we can only assume that in 1763, when the Russian Empress Catherine II gave precise orders for rebuilding the city centre, people tried to make maximum use of the earlier foundations of buildings. In 1775, after the great fire, which destroyed most of the buildings, there were no longer taken into account the earlier streets, city wall and land plots, and it seems that then the civil engineers skilfully made use of the remnants of the demolished houses, as their toil and moil has been successful and the buildings have remained in a relatively well state even nowadays.

Jaani (St. John) church that has survived the fire is according to the existing data the oldest wood based building in Tartu, and after having been in ruins for decades, submitted to years long research and restoration it was re-opened in 2005. Based on investigations in 1954–1968 the plan of a raft of this building was drafted by O. Prince. The raft base has two layers: the top layer of logs runs along the wall, the lower part – crosswise the wall. It is believed that under the raft foundation there are also piles, but there is no data about their arrangement. The church building, tower and part of Lübeck chapel are founded on a raft (Oll 1984). The research has been done by Nikolai Oll, architect Udo Tiirmaa and art historian Kaur Alttoa.

City archivist MPhil E. Tender in his article treats the building of the city hall in 1688–1693, when Tartu belonged to Sweden, and later. The plan was prepared by an engineer-colonel-lieutenant Paul von Essen, and the budget was made by one of the local master builder. The original plan is lost, but the budget and pay bills have been preserved. E. Tender said that in order to build a permanent basis, the five-fathom (ca 10 m) pine logs were to be driven into the ground and on them was to be placed a lattice of six- and four fathom (respectively 12 and 8m) logs. The logs were probably driven into the ground by statute labour, because the invoice of City Hall contains no comment for payment of wages. Here it should be noted that the budget maker has greatly exaggerated. There was no need for the logs of that length to be driven into the ground at that place nor was it possible at that time. Such budgetary excesses also occur below (Oll 1984).

After the fire, during the 18th–19th century, the complex classicist centre was built in Tartu, and it has been declared to be a National Heritage of the Old City area.

As the City Hall ceased to exist after the fire in 1775, a new administrative building was built according to Master J. H. B. Walter's draft, which added colour to the old city centre of Tartu. The new City Hall was built between 1778 and 1784, and the project documentation contains also the foundation plan. According to the proposed solution for the building the pile foundation

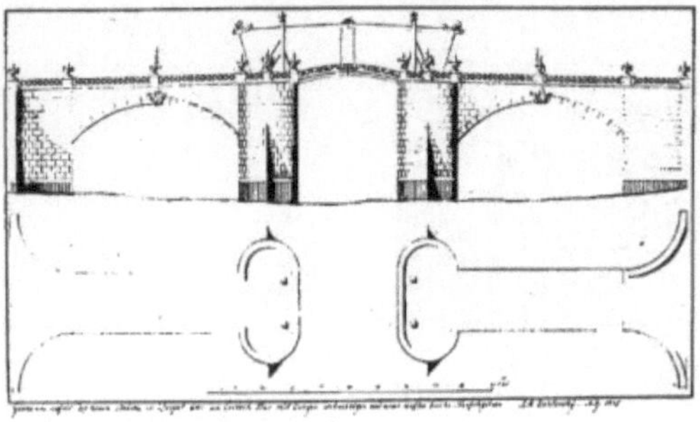

Figure 2. Stone Bridge project by Zaklowsky, version from 1775 (Oll 1984).

has four rows of piles under exterior walls, three rows under inside bearing walls and two rows under the partitions of the cellar. The distance between the rows from axis to axis is 0.6 . . . 0.7 m and spacing of piles in a row is 1.25 m. On top of the piles there is a grillage of longitudinal and transverse logs. The building was designed by master builder of Tartu, J. H. B. Walter. (Oll 1984).

The Stone Bridge of the city of Tartu was built between 1775 and 1784 and it became a peculiar symbol for the city. The bridge construction works were carried out according to a project from 1781, drawn by the local wall master Johann Zaklowsky. Figure 2 shows the version from 1775 by J. Zaklowsky, which was later dramatically altered, especially concerning the top part.

The lower part is unchanged and the figure shows that the bridge columns rest on closely side by side driven piles. Such construction method was already proposed in 1699 by Nicolai Goldman in his book about constructing bridges. The Stone Bridge construction budget drawn up in 1776 mentions the logs for piles, scaffolding and bank fortification (Oll 1984). Unfortunately, the Tartu Stone Bridge was destroyed during the Second World War.

In the Old City is located well-preserved Uspensky Church, built between 1776 and 1782. For its building there were drawn up two budgets. Both budgets are drawn up in 1769, first by the masters of Tartu, the second in Riga. The first budget included 800 logs for piles of 5 and 6 feet, and 80 logs for the grillage. Probably the short piles for tightening were meant to be used here. The second budget included 350 five fathoms long logs for piles and grillage. There is nothing said about using 5 fathom long piles, as it can be read from the article by E. Tender, who wrote about the Town Hall dating from the Swedish time (Oll 1984).

The development of Tartu befitted from re-opening the university in 1802 and its development into an important educational centre in Tsarist Russia. At that time the contemporarily modern complex of buildings was built by design of J.W. Krause. It includes clinics, bridges, university library in the ruins of the Cathedral, the old anatomical theatre and the Tartu Observatory designed and built on Toome Hill. Tartu Toome Hill, as a close elevation to the river Emajõgi, is also a great

Figure 3. Cross section of the main building of the University of Tartu (Oll 1984).

Figure 4. Section of foundations and geology of the University of Tartu drawn by N. Oll in 1966–1971 (Oll 1984).

place for the construction of the point of view of foundation. The main building of the University of Tartu as the central representative building was intended by J. W. Krause to belong into the city-centre complex, and when starting the building it was necessary to get rid of the remnants of the St. Mary church that had been demolished in 1704 and remained for more than 100 years in ruins, because the sites of the two buildings overlapped. The old piles pulled out from the trench were in perfectly good condition, and there was still fully preserved spruce wood plank floor. These data suggest that St. Mary's Church had been built on wooden base. Whether it was a pile foundation, however, it cannot be confirmed because J. Krause, when presenting each month a report to the university government about the status of the building process has mentioned that under the old Swedish (i.e. St. Mary) church's heavy basements there was a raft base found in perfect order (Oll 1984).

The year 1804 marked the beginning of driving the piles into ground for the pile foundation of the University of Tartu. By August 21, the same year, 548 piles of length from 9 to 21 feet had been driven into ground. The work was completed the following year. It is noted that it took 5000 logs to build the pile foundation. The main building of the University of Tartu was completed in 1808.

The new St. Mary's Church was built in Tartu between 1837 and 1841 in classicist style by G. F. W. Geist's project.

Figure 5. Maarja (St. Mary) church before the World War II.

We find the data about construction of buildings on the raft bases from budgets drawn up by G. F. W. Geist, master builder in I half of the 19th century. For example, for a raft base (liegende Rost) of a hospital there were used logs of 8 m length and 0.25 m diameter, but the gendarmes' barracks budget includes the logs of the same thickness which are 7 m long. Budgets do not include, however, plans for a raft foundation (Oll 1984).

In the 19th century the city continued rapid development, supported by existence of the University to Tartu and the railway that reached Tartu in 1875. In 1876 was completed the building of Tartu railway station, which became one of the pearls from the construction point of view. The building foundations are built of cobblestones as spread foundation, and regardless intensive vibrations caused by rail traffic it is still in good condition. Tartu station restoration works were completed and the station was re-opened on 25th July 2012.

Tartu is characterized significantly by wood architecture in its suburbs, often destroyed by wars or fire, but always restored and expanding, as we have nowadays quarters as Supilinn, Tähtvere, Karlovo and later Tammelinn. The most difficult conditions for foundations are in Supilinn, where the peat soil, floods and high level of the groundwater cause still problems. There is no good solution for these problems and the foundations there range from historical wood rafts, and pile foundations to the modern r/b piles.

The effectiveness of foundations can be observed in constructional stability of the buildings during longer period, and mistakes or miscalculations cause many problems. There is an illustrious example of buildings built together where are located Tartu Art Museum (City Hall Square 18 (1793)) and Tartu Oskar Luts City Library (Company 3/5 (1787–1803)). For rational reasons most probably the riverside wall was built on former city wall and for founding the west side wood piles were used, as well as remnants of ruins for filling and wood rafts (Oll 1957, 1984). Using various foundation types in buildings on swampy ground may cause different settling, as it can clearly be observed

Figure 6. The Pool of Tartu 1929.

by the Tartu Art Museum house, which western side has strongly settled. This process has been stopped by using different engineering solutions, but the house has got the nickname "Oblique House".

The 20th century will continue building momentum in Tartu, and a number of architecturally notable buildings are built, provided with new foundation solutions. The impact of the university on construction activities continues at the beginning of the century when representative student corporation houses and school buildings are built. There are many beautiful examples, as the Estonian Student Society building from 1902, the German Corporation house in Kastani Street 1 from 1904, Korp! Sakala house in Veski Street 69 from 1911, and university academic building in Vanemuine Street 46 built in 1914 (UT 2012).

There are three important buildings built in the same period as architectural, cultural and educational buildings: the building Vana Vanemuine, completed in 1918, its project's author is the city architect Arved Eichhorn, present Miina Härma Gymnaasium building, being projected by engineer Georg Hellat, it was completed in 1915; a symbol of Tartu, St. Paul's Church, its project was drawn by the Finnish architect Eliel Saarinen, and its construction works lasted from 1914 to 1919. The quality of these works can be seen in foundations that have been resistant in a course of time, war damages and reconstructions made on the basement floor. Renovations of the church continue until 2015 (Siilivask, Kimmel 2009).

After proclamation of the Republic of Estonia in 1918 the on-going civil war ended after the treaty of the Tartu Peace in February 2, 1920, but the destructions of war needed to be put in order and freedom activated construction works in Tartu. In this period between the two wars a number of remarkable buildings were built and the construction policy was directed for a long time by architect Arnold Matteus, according to whose projects were built the pool of Tartu in 1929, Tartu Teacher's College building from 1936, Ugala corporation house in 1938 and several other buildings and private residences.

Regarding foundations there is constructed a noteworthy house for the Bank of Tartu, designed by architect Arnold Matteus in 1935–1936, which foundation is formed by a system of rib bearing beams,

Figure 7. Ugala Corp. House 1936.

Figure 8. Tartu Market House 2011 (Pallav 2011).

Figure 9. Completed Tartu elevator in 1941.

and such a construction is fully justified on swampy soil.

One of the most remarkable buildings in Tartu has been designed by architect Voldemar Tippel, and Nikolai Oll designed load-bearing roof and foundations, using wood piles. In 1936 the test drivings into ground were started and according to Nikolai Oll's project solutions one thousand six-meter long pine piles were driven into the ground (Jaaniso 2000). The building was completed in 1939, it remained intact in war and has maintained its representativeness also nowadays.

For developing the water supply systems the architects V. Tippel and engineer V. Onton designed the

Figure 10. . . . and after blowing up in 1944.

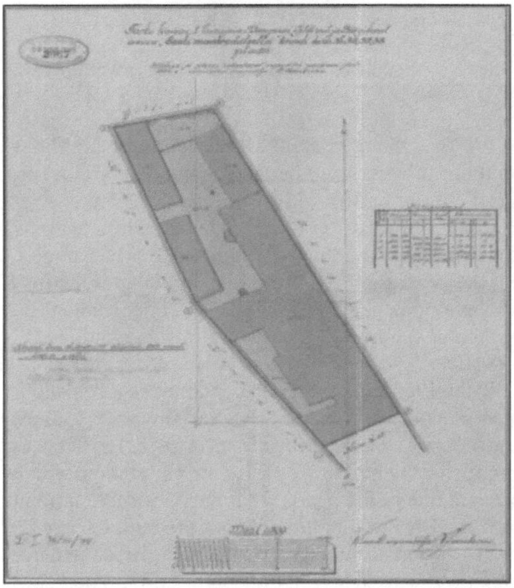

Figure 11. Land plot plan at Kompanii 3/5 in 1934, before reconstructions in 1936 (ERA).

water tower construction project in 1938 and the building was completed in 1939.

The Tartu elevator was completed in collaboration with the famous engineer August Komendant (1906–1992) in 1941, and for first-rate foundation solution speaks the fact that the German forces did not succeed in blowing it up in August 1944 (Kaljuvee 2006).

3 RESEARCH AND SOLUTIONS OF TARTU FOUNDATIONS

Before 1940 the soil survey data about Tartu geotechnical solutions is available only for some of the larger buildings, but there were no laboratory studies about soil characteristics, yet there were field experiments carried out in collaboration with engineers – the researches A. Matteus, N. Oll, I. Lenzius, A. Matteus, H. Tari, N. Kuzmin, K. Orviku, H. Sööt, A. Komendant, L. Saukas, E. Leppik (Jaanisoo 2000).

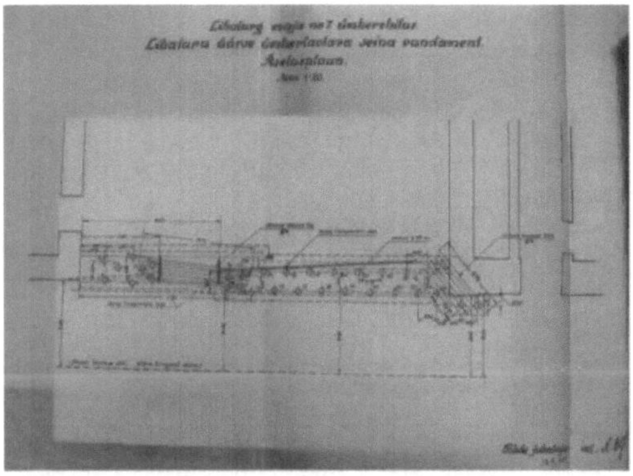

Figure 12. Plan of the foundation of the riverside wall on wood piles (ERA).

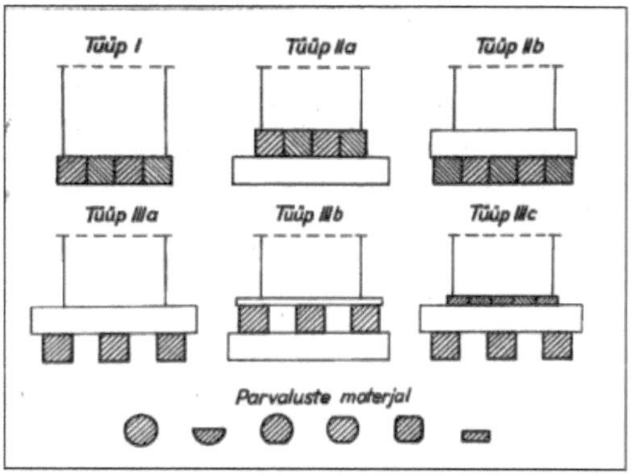

Figure 14. Raft foundation types in Tartu in 1871–1918 (Oll 1984).

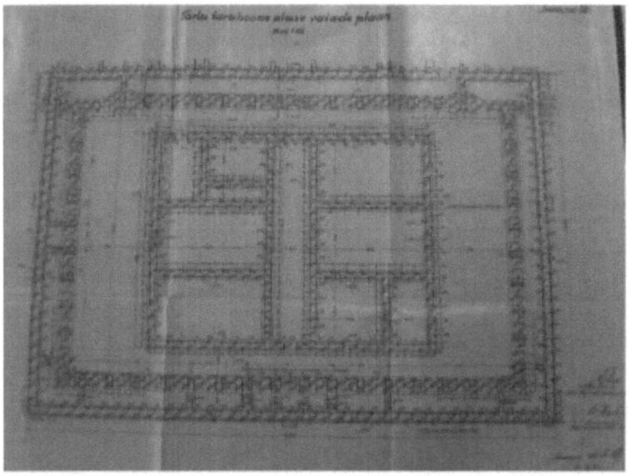

Figure 13. Plan of piles for Tartu Market House by Nikolai Oll (ERA).

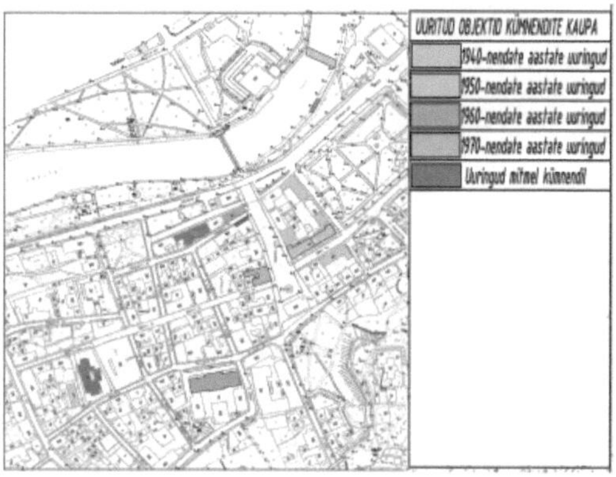

Figure 15. The location of researched raft bases and timber pile foundations in the old city of Tartu (Oll 1984) The author of the plan is I. Salumets (Salumets 2008).

The wood bases of Tartu old city have been thoroughly researched by Associate Professor Nikolai Oll, who has collected a huge specimens collection since 1935, when he began working in the Tartu City Government as a young engineer, being invited to be the head of construction in Kompanii Street 3/5 in 1935–1936 and (Lihaturg 7) reconstructions. He also has designed a reconstruction project of the foundations using wood piles. When the building was completed, he continued as a construction inspector and a head of construction inspection till 1948.

In 1945 Nikolai Oll started to work as an assistant at the Faculty of Constructions of the University of Tartu and dedicated to academic and scientific work, writing his dissertation about "Using Wood in Foundations on Weak and Highly Compressed Soils." In 1957 he focused on wood bases in the old city of Tartu.

The first field researches took place in St. John's Church in 1947–48, conducted by engineer Nikolai Oll. These were technical by nature and focused mainly on explaining how the foundations were founded. At that time there were made 27 smaller excavations in St. John's church (Oll 1948, Metsallik 2007).

N. Oll continued to study the wooden raft and pile base foundations throughout his life, resulting in a bulky six-volume study: "Research of Wood Preservation in Buildings on Raft and Pile Foundations in Tartu Old City" Tartu 1984.

The specimens have been taken for these studies during various excavation and construction operations since 1935; larger samples are taken during the post war period and are shown on the plans, see figures 11–14. N. Oll conducted these works.

N. Olli works present the types and occurrence of the rafts used between 1871 and 1918.

4 SUMMARY

The geotechnical conditions of the primeval river valley, near which is located the city of Tartu, are highly variable and complex, from the marshy riverbanks to the elevations suitable for construction. Thus, the choice of building site and foundation problems are always hot spot issues that require regular observations and on-going studies.

An invaluable contribution to the studies of Tartu geotechnical conditions has been made by Associate Professor Nikolai Oll, a young engineer who started in 1935 systematic observations and studies, and has devoted his entire life to this work. He has published numerous scientific studies and articles which have not yet been published. Also, in 1953 he started regular measuring of settle-down of the houses in Tartu, which regrettably has been interrupted since 2004 (Jaaniso 2000).

As a result of nowadays renewed private capital market-oriented engineering there has evolved an application of rapid and modern foundation solutions, but it does not promote organization of complete and complex research, which might prevent introducing sustainable and environmental friendly solutions in the future.

REFERENCES

ARC Projekt OÜ, 2008. Kompanii 3/5 ajalooline õiend. Tartu, veebruar, 2008.

Jaaniso, V. 2000. Nikolai Oll kui Tartu geotehniliste probleemide lahendaja. Tartu 2000.

Kirjavahetus linna maja Lihaturg nr. 7 ümberehituse asjus koos projektidega ja kalkulatsioonidega 1935–1940 aastatel. ERA, f. 2966.

Lihaturg maja nr.7 ümberehituse projekt – Töövõtjale väljaantud detailjoonistuste koopiad. ERA, f. 2966.

Metsallik, R. 2007. Tartu arheoloogiast ja vanemast ehitusloost. Tartu veebileht.

Nortim OÜ, 2003. Ekspertiisiakt E – 0302 Tartu Linna Keskraamatukogu tehnilise seisundi kohta. Tartu, 22. jaanuar, 2003.

Oll, N. 1957. Puidu kasutamisest hoonete rajamisel nõrkadele ja tugevasti kokku surutavatele pinnastele: Dissertatsioon. Eesti Põllumajanduse Akadeemia (käsikiri). Tartu. (Käsikiri Eesti Maaülikooli maaehituse osakonnas).

Oll, N. 1984. Puidu säilivuse uurimine Tartu vanalinna hoonete parvalustes ja vaivundamentides. Tartu 1984. I-IV köide.

Palamets, H. 2003. Lugusid toonasest Tartust. Tartu: Kleio (Võru: Võru Täht).

Pullat, R. 1980. Tartu ajalugu. Tartu: Eesti Raamat.

Saar, J. 2007. Uus raamatukogu kipub parklasse. Tartu Postimees. 22. oktoober, 2007, lk. 1.

Salumets, I. 2008. Magistritöö. Tartu vanalinna puitvundamentide uurimine Nikolai Olli tööde põhjal. Tartu 2008.

Siilivask, M. 2006. Tartu arhitektuur 1830–1918. Tartu 2006. Rahvusarhiiv.

Siilivask, M., Kimmel, T. 2009. Jalutaja teejuht. Tartu I. Südalinn ja Toometagune. Tallinn 2009. Solnessi Arhitektuurikirjastus.

Tartu Ülikool 2012 (http://www.ut.ee/et).

Trummal, V. 1974. Eesti ala vanem asustus ja rahva etniline kujunemine. Tartu, TRÜ Kirjastus, 1974.

Vahtre, S., Piirimäe, H., Siilivask, K. 1985. Tartu Ülikooli ajalugu 1632–1982. Tallinn, Perioodika, 1985.

Vanad asjad 2012 (http://vanadasjad.wordpress.com/2011/04/03/tartu-elevaator/).

Baltic Piling – Mets & Raudsepp (Eds)
© 2013 Taylor & Francis Group, London, ISBN 978-0-415-64334-4

Use of piles in complex reconstruction of cities

V.M. Ulitsky
State Transport University, Saint Petersburg, Russia

A.G. Shashkin & M.B. Lisyuk
Georeconstruction Engineering Co, Saint Petersburg, Russia

ABSTRACT: The paper presents examples of use of piles for reconstruction of historical monuments in St. Petersburg. Two important historical monuments are considered – Konstantinovsky Palace and Kamennoostrovsky theatre.

1 INTRODUCTION

In reconstruction of historical cities piles and pile foundations are used very extensively. We could categorize the use of pile foundations during reconstruction for the following purposes:

1. Strengthening of subsoil and foundation of existing buildings (underpinning of historical buildings using piles).
2. Extension of the existing buildings and structures.
3. Construction of protection walls of underground structures.
4. Reducing pressures on the existing collectors and communications.

One of the brightest examples of the pile foundation application in the 19-th century is the largest cathedral in St. Petersburg – St. Isaac's Cathedral, which is in fact the heaviest building in the city and the largest East Orthodox temple in the world (Fig. 1).

It was constructed in place of the cathedral designed by architect Rinaldi in 1768. August Montferrand, the creator of the new edifice, left foundations of the altar parts and dome pillars of the old building, which were based on 13.000 piles of 10.5 m under the pillars and 8.4 m under the walls. To make a new foundation 5 meter reduced excavation had been provided and from the bottom thereof 24,000 26 cm thick piles were driven being and 8.4 m long under the pillars and 6.3 m under the walls (Dashko, Alexandrova, 2003). In between the piles at the level of 35 cm quarry stone rubble was compacted into subsoil and topped with lime-sand mix.

Underlying the bearing structures and the corners the foundation contains granite slabs. Underneath the porticoes layers of quarry stone masonry courses alternated with those of granite slabs, and throughout the rest of the building with those of limestone. Masonry walls elevated by 2.0 m above the ground level forming a massive 7.5 m thick raft within which galleries were furnished. Expert opinion has it that the maximum pressure on subsoil therein totals 0.32 MPa. Immediately under the slab pilecap there are saturated silty sands and loams with piles having their toes embedded in flowing loamy strata. Currently the overall settlement of the building by various assessments has reached 0.8 . . . 1.0 m with the highest settlement values registered in the heaviest dome section of the cathedral.

Figure 1. St. Isaac's Cathedral.

In this case the relatively short piles serve as an additional reinforcement of the subsoil. Considering massive (up to 8 m) pile raft and relatively short piles,

Figure 2. A photograph of Constantinovsky Palace taken in 1910.

according to modern assumptions this structure is built on pile-raft foundation. Undoubtedly, uneven loading locally generates excessive strain in the ground. State-of-the-art diagnostic methods and numerical modeling techniques applied to such a complicated structure highlight most critical locations suitable for subsequent strengthening.

Examples of using piles during reconstruction of two important historical monuments in St. Petersburg are given below.

2 USE OF PILES DURING RECONSTRUCTION OF KONSTANTINOVSKY PALACE IN STRELNA

2.1 *Historical background*

Strelna Palace, more widely known as Konstantinovsky Palace, is a large palace located in the nearest suburb of St. Petersburg on the shore of the Gulf of Finland (Fig. 2). Peter the Great was fond of that location where he resolved to construct masonry chambers. Imperial residences in the suburbs of St. Petersburg would commonly begin as modest buildings and at a later date be expanded and reconstructed, graduating to more luxurious and pretentious edifices. Strelna Palace, however, had from its inception been conceived as a stately and imposing structure. It involved leading European and Russian architects of the time, such as Jean Batiste Leblon, Nicolo Micetti, and Francesco-Bartolomeo Rastrelli.

This palace was situated on the top of the 8-m high slope of the historical Baltic coast. The slope was reinforced with a retaining structure fashioned into a series of loggias (half-sphere niches with the open space towards the park).

Subsequently, the great Russian reformer lost interest in Strelna and shifted his attention to the town of Peterhof as the place for establishing the official suburban residence town. Therefore the palace, having been constructed up to the roof level, remained uncompleted. It was only following the accession of Empress Catherine that Rastrelli was commissioned to complete its construction. However, the court never moved into the new residence. The luxurious palace was again forgotten for 50 years and, as the case usually is with abandoned buildings, it was decaying quickly due to lack of maintenance and heating.

In 1802, the new owner, Emperor Paul presented the palace to his son Constantine, whereupon it became known as Konstantinovsky Palace. Refinishing of the palace was completed in 1 year. It was designed and supervised by A. Voronikhin. The sumptuous abode of the Grand Duke stood open to welcome its new owner. Fate, however, had no remorse as an enormous fire broke out on December 28, 1803, destroying the entire artistic decorum of the hapless building. Everything was to be renovated by architect L. Ruska. Architect A. Voronikhin designed a series of grottoes. The roof of the grottoes served as a spacious terraced square facing the palace. The structure of the terrace at the same time functioned as a retainer for the palace, conditioning stability of the entire palatial complex.

2.2 *Palace condition survey and site investigation*

Strelna Palace is a 3-story building on a high basement floor (socle floor). It was constructed on linear rubblework foundations on the crest of a natural slope which descends into Lower Park by three tiers. The natural slope in front of the palace was fashioned into a horizontal area 23 m wide in the middle and 17.3 m wide on the edges. The absolute level of the terrace surface is at 12.7 m Baltic Datum (BD) (Fig. 3, 4). The vertical terrace ramp (8.0 m high) is retained by a complex system of masonry structures forming grottoes and lateral loggias on the front elevation (Fig. 3), as well as the suite of wine cellars between the grottoes and the palace. The grotto is divided throughout its length into 9 equal bays, each approximately 4.75 m in length.

Symmetrically on each part of the central grotto there are 3 loggias. The gable wall for both the grottoes and the loggias is the actual retaining wall. The loggias retaining wall contains half-sphere niches forming the volume of every loggia. In these locations the retaining wall is especially thin (around 1.5 m), but gradually increases up to 3.2 m elsewhere.

The retaining structure of Konstantinovsky Palace serves not only as a podium for the palace on the Lower Park side, forming a spacious terrace in front of its north elevation, but also as a structure ensuring the building's stability on the brink of an 8-m slope at the historical coast of the Baltic Sea. Stability of the entire palace depends on the technical condition of its retaining structures.

Figure 3. A photograph of Constantinovsky Palace taken in 2000 before reconstruction.

Figure 4. Dilapidated grottoes of the palace.

The authors were commissioned by Governmental Monument Preservation Committee to provide a pertinent condition survey of this monument or, more precisely, of the structure's areas of critical dilapidation (Fig. 3, 4). The surveyors saw an abandoned palace gracing a high slope, strengthened by a retaining structure fashioned into a series of grottoes and loggias. The principal bearing wall, withholding the ground on the slope and the palace on top of it, was considerably damaged in a number of locations. Water had found its way inside, penetrating through fall-outs over piles of brick rubble. Later, as cold weather set in, the water was transformed into ice. Ice stalactites hung on the precipitation.

The structural layout and condition of all foundations, were studied and described. To accomplish this, 28 trial pits were excavated, 35 boreholes were drilled through foundation masonry courses, 2 large trenches were excavated on-site, the rigidity characteristics of brickwork were established, moisture conditions of the walls were studied, and the length of timber piles underneath rubblework foundations were defined.

As attested by site investigation in the park, underneath 1 m of fill there is a 3-m layer of soft varved clay loam, underlain by medium stiff moraine clay loam (Fig. 5). At the level of around 14.0 m

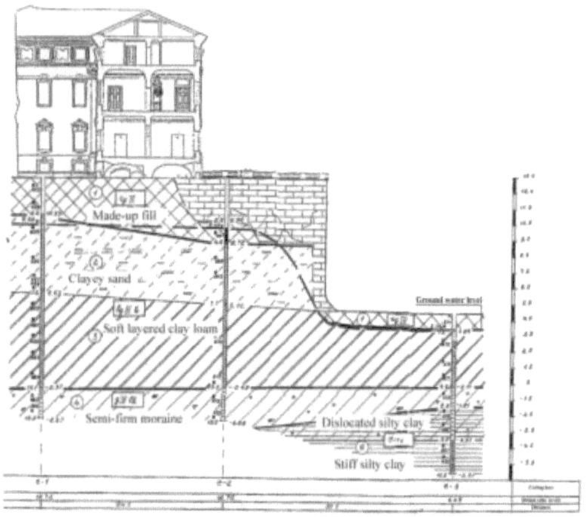

Figure 5. Cross-section of the palace and subsoil profile: 1. Made-up fill, 2. Clayey sand, 3. Soft varved clay loam, 4. Medium-stiff moraine clay loam, 5. Dislocated stiff Cambrian silty clay, 6. Stiff silty Cambrian clay.

(absolute – 1.3 m), there are medium stiff and stiff deep Cambrian clays.

The slope incorporating the terrace is compounded by lacustrine-glacial clayey sands followed by silty lacustrine-glacial loams. Straight upon the terrace there is a stratum of man-made ground of sand with admixtures of lime cement, above which there is 2 m of brickwork (in the section between the palace and the cellars) serving as a base for the terrace. Above the clayey sand there is a layer of man-made ground about 3 m thick.

Hydro-geological conditions are characterized by the presence of groundwater associated with man-made strata and silty sand inclusions in lacustrine-glacial clay loams which together with clayey sands act as a confining bed. Groundwater is discharged into the Lower Park canal. The groundwater table generally follows surface geometry.

2.3 *Condition survey results*

The condition survey results were as follows (see Fig. 4–7):

1. Foundations of the dilapidated retaining walls were constructed of bricks. The foundations were no longer capable of being classified as a structure. There was imminent danger of crushed brickwork movement with formation of local bulges.
2. The entire brickwork structure was soaked in water, causing dilapidation through cycles of freezing and thawing.
3. There were no foundations of the transverse walls of the cellars. Footing was level with the cellar's floor. Decomposed timber pile heads supported the transverse walls.
4. Dilapidated terrace gutters had caused weakening of some retaining wall sections adjacent to niches of the loggias and grottoes.

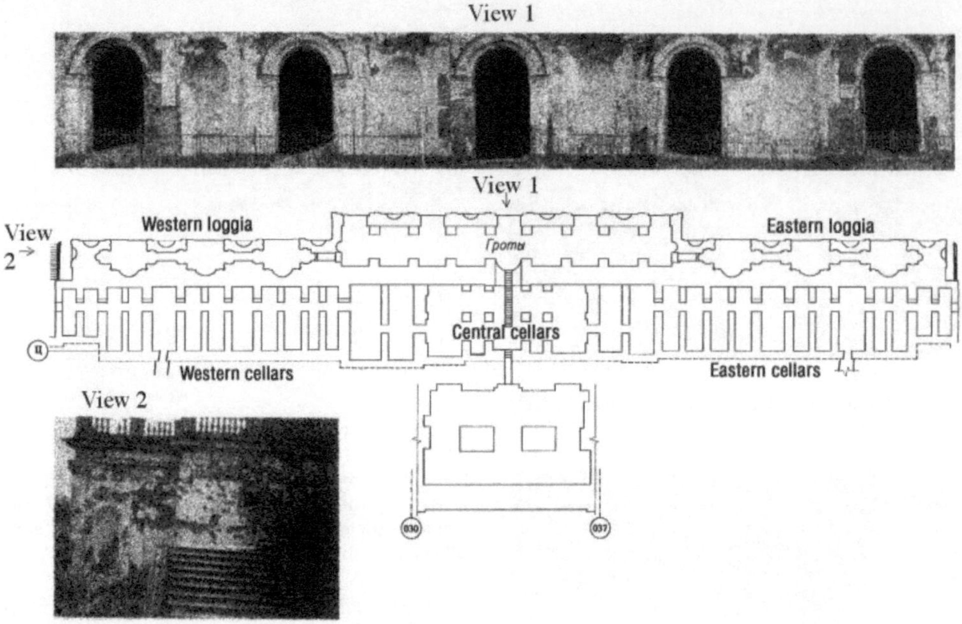

Figure 6. Dilapidated cellars, grottoes, and loggias of Konstantinovsky Palace.

Figure 7. Collapsing cellar wall behind east loggias.

5. The precipitation sewer consisted of three straight courses underneath the retaining structures, designed to divert precipitation and ground water from the palace. There was water flow through the ground underneath the cellar walls, and through dilapidated retaining wall sections.
6. Most structural damage (fallouts) was associated with the destroyed drainage sections underlying the retaining structures.

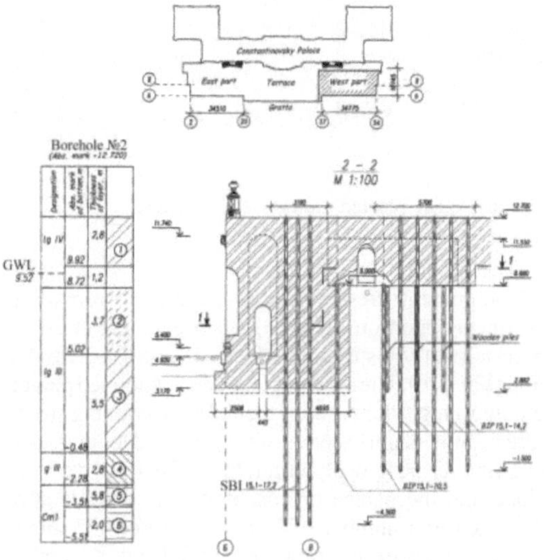

Figure 8. Underpinning of the retaining structure. See Fig. 5 for soil stratification.

2.4 *Design project of palace reconstruction*

The Palace condition demanded immediate rendering of complex strengthening works.

The design project envisaged pressure grouting and strengthening of brickwork, with underpinning of all retaining structures with piles embedded into stiff stratum (Fig. 8, 9).

Grouting of the brickwork was necessary in order to restore its strength and stiffness. Strengthening was required to properly allow the transfer of loads from the entire structure. Finally, underpinning piles had to be constructed in order to transfer the structure loads onto the incompressible subsoil stratum. It needs to be noted that conventional underpinning piles installed at an angle from the level of the lower terrace (around 4.0 m Baltic Datum) would prove ineffective as the dilapidated foundation brickwork was incapable of accommodating the heads of the underpinning piles, and the retaining wall itself being practically unavailable for underpinning.

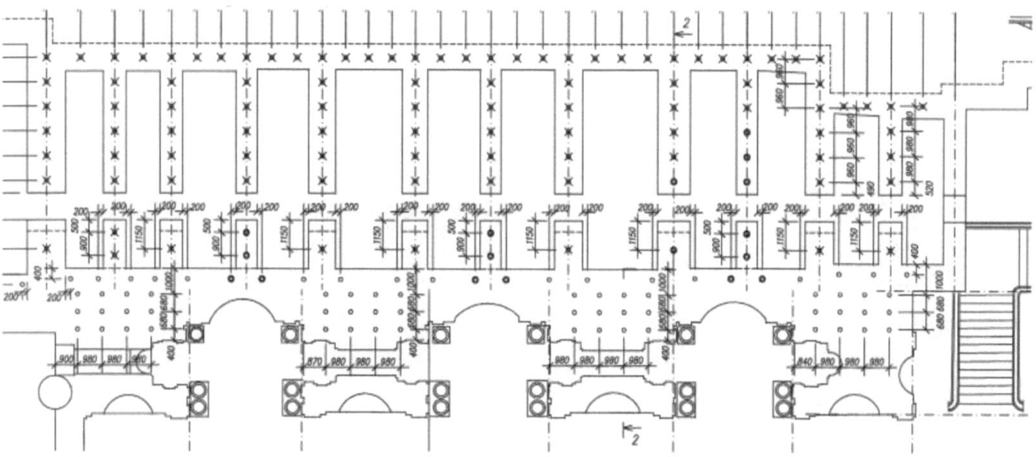

Figure 9. Location of underpinning piles in plan.

Based on the above, the foundations underpinning of the retaining structures was carried out in the following sequence.

Stage One

1. Strengthening of the critically dilapidated structures (3 left and 3 right loggias).
2. Provision of temporary propping scaffolding in cellar chambers installed on wedges in the cellar floors. Wedging of the scaffolding was regularly inspected. The unsupported spans in locations of the brickwork fallouts were propped.
3. Drilling of 42 mm vertical bores above the partitions of the retaining wall from the terrace in front of the palace down to the brick-wall footing level. Subsequently, the brickwork was grouted by intervals with packing lime mortar until completely permeated.
4. Redrilling of the bores by 151-mm core bores down to the top of the firm Cambrian stratum following 70% setting of the mortar. Drilling below foundation footing was either carried out using thixotropic grout or was casing protected. Cement grout with added plasticizing and shrink-proofing agents was pumped into the subsoil and brickwork at 0.2 and 0.1 MPa respectively, followed by a stain-proof reinforcement casing tube being oscillated into the grout mix. The resulting pile was embedded into the Cambrian stratum reinforcing and underpinning the retaining brickwork section. The tube was required to ensure both longer pile life and subsequent possible of deepening of the cellars.
5. Drilling of 42 mm vertical bores paced at approximately 1.0 m from the terrace in front of the palace along each transverse wall down to the brick-wall footing level (absolute level 8.9 m Baltic Datum) in order to reinforce the transverse walls and rear longitudinal wall of the cellars. This was followed by interval grouting and subsequent redrilling of the bores by 151-mm augers, used in the bored piles construction, down to absolute level of 1.5 m Baltic Datum. Those piles were reinforced through their entire length.
6. Construction of pile heads in the dilapidated areas of the transverse walls at the level of the brickwork footing (absolute level 8.8 m Baltic Datum), with subsequent construction of the pile caps and masonry courses within the original scope.

Stage Two

1. Completion of the retaining structure strengthening works.
2. Provision of works described in Stage 1 above for unreinforced sections of the walls.
3. Removal of terrace surface material and construction of a reinforced concrete wall connecting pile heads above the brick vaults.

Stage Three

Provision of the terrace surfacing incorporating drainage and snow melting systems, finished by tiling.

The provided strengthening should serve to ensure the reliability and long life of the retaining structures, preserving their appearance and historic materials almost completely unscathed by any patchwork or replacement of brickwork. Such approach proved most appropriate in relation to the significant architectural monument.

The constructed strengthening option was successful even when faced with an unexpected challenge. The architects suddenly decided to provide front access to the palace from the Lower Park and furnish a vestibule underneath the terrace. To do this, all cellars had to be deepened by 1.0–1.5 m and the transverse brick walls were temporarily suspended on the thin underpinning piles (Fig. 10, 11). Quality of the strengthening was attested by the fact that not one section of the brickwork was in any way displaced. It was therefore ascertained that the strengthening was successful.

It took about 1.5 years to completely reconstruct Konstantinovsky Palace in Strelna). The palace officially opened as the Congress Palace in 2003 during

Figure 10. Deepening of the basements by 1.0–1.5 m.

Figure 11. Bored pile with tube reinforcement viewed from underneath foundation.

Figure 12. Cellars of Konstantinovsky Palace, 2004 (at the location of reconstructed retaining cellar wall, see Fig. 10).

Figure 13. Konstantinovsky Palace, south façade. View after a final reconstruction stage (July 2003).

Figure 14. Kamennoostrovsky Theatre during reconstruction.

the tercentenary celebrations of Saint Petersburg (Fig. 12–13).

3 USE OF PILES DURING RECONSTRUCTION OF KAMENNOOSTROVSKY THEATRE

3.1 *Historical background*

Kamennoostrovsky (Stone Island) Timber Theatre was constructed in 1828, designed by architect S. Shustov to provide a temporary floor for the Imperial Opera and Ballet Troupe whilst reconstruction of the Large Masonry Theatre was ongoing. The critics noted rare elegance of the structure and it was due to that elegance that following a decade and a half a decision was taken to preserve it by means of adding rubblework foundations (prior to that the lowest courses of the building rested directly on timber piles). The architect Alberto Cavos believed he had managed to prolong the building's life by about 50 years. However the building was able to withstand the passage of time for as long as 180 years. Until the 1930s the Stone Island Theatre was used as a warehouse, when it was renovated and converted into a television theatre and later to a dance studio. The Stone Island Theatre being a world heritage site protected by the UNESCO (Fig. 14).

3.2 *Development of plan to renovate the theatre*

The Stone Island Theatre was revived in 2006 when the Russian President decreed for it to serve as Second House of Tovstonogov Academic Drama Theatre to commemorate the 80th anniversary of a famous Russian actor Cyril Lavrov.

Over the two centuries, however, thespian field has developed more sophisticated requirements regarding comforts of the audience. Indeed, these days there are codes and standards envisaging roomy foyers and cafes, convenient wardrobes and lavatories. A contemporary theatre is like an iceberg, and its bigger part is not visible to the audience. That invisible space houses stagecraft facilities – upper and lower stage mechanics (e.g. stage terracing, orchestra pit, lighting arrangements, backdrop hoists, etc).

The General Designer was given a highly complicated task featuring what at the time seemed as two mutually exclusive requirements – to convert an old timber building into a contemporary theatre with all the necessary technical sophistication without altering the original historic appearance in any way. The only way to make it happen was to locate all new addenda and enhancements required by a modern state-of-the-art theatre underground, immediately underneath the historic superstructure.

The geotechnical challenge at the Stone Island Theatre was serious due to a high ground water level. The absolute levels at Stone Island are rather low and the area is subject to flooding. Additionally, the subsoil is compounded largely by soft clay. To make matters worse there is a listed building in the vicinity – the Summer Mansion of Baron Kleinmichel.

The area of the theatre's location was levelled with made up ground, the absolute levels being in the order of 1.93 . . . 2.27 m BS (Baltic Datum). Made up ground is 0.8 . . . 4.5 m thick. Underlain by lacustrine and marine deposits, it is compounded by silty sand, medium grained sand, light silty loams, and silty sand clay. The underside of this stratum is at 5.5 . . . 11.0 m. Underneath down to 10.8 . . . 17.0 m the site is compounded by lacustrine and glacial deposits of the Baltic Glacial Lake – silty loams with consistency ranging from liquid to liquid-plastic. Directly underneath there are glacial deposits of the Luga Moraine, compounded by silty sand-clay and loams with gravel and pebbles, as well as boulders which are encountered down to the level of 21.0 . . . 26.5 m. Underlying the latter lacustrine and glacial deposits are identified – varved silty loams, stiff-plastic and liquid-plastic in consistency. Quaternary deposits extend down to 24.2 . . . 30.3 m, corresponding to absolute levels of −22.27 . . . −28.27 m BS, and are underlain by Wendian stratum, being light firm silty clay.

The area has an aquifer located around the boundary of quaternary deposits. Site investigation in October and November 2007 identified ground water at 1.3 . . . 1.8 m from the surface, corresponding to the absolute levels of 0.75 . . . 0.30 m BS.

Reconstruction implies provision of underground facilities under the entire footprint of the building and extending beyond its perimeter by 6.8 to 25 m (the size of the underground space is 80 × 40 m). The relative level of the underground structure is at minus 6 m. In order to construct the underground area, excavation was necessary down to relative level −6.7 m. Therefore, an excavation down to 5.6 m from the surface was necessary to accommodate the underground structure.

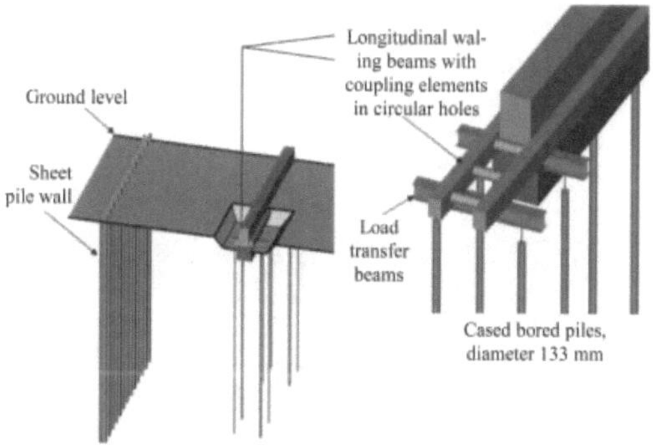

Figure 15. Construction of sheet piles, bored piles, and the unit transferring the load from the walls to the piles.

4 STAGES OF THE THEATRE RECONSTRUCTION

The sequence of works adopted for the project is shown in Figures 15–20. The works can be subdivided into 5 stages:

1. Drilling holes in existing walls to accommodate transverse beams of the unit responsible for temporary walls-to-piles loads redistribution.
2. Construction of bored piles, introduction of the transverse steel beams, preparation of rebar cages and concreting for reinforced concrete waling beams to facilitate load transfer (Fig. 15). Simultaneous construction of the sheet pile cofferdam. Bored piles were of 2-step construction "Titan" type: initiated with oscillation of the casing down to the level of the underground structure subsequently followed with pile construction as such. The casing was necessary to improve bending strength of the top half of the pile, which was needed during excavation for the underground structure.
3. Application of jacking devices on the transverse beams, tightening retaining nuts on bars of the underpinning bored piles thereby engaging the piles to assume loads. Following this the lower sections of the building's rubblework foundations could be removed (Fig. 16). Following removal of the lower parts of the rubblework foundations a slab is to be cast directly underneath, forming a rigid disc at the level of the foundations (Fig. 17). Stiffness strips are cast under the existing foundations to redistribute loads onto the slab. Bored piles also are embedded into the rigid disk slab. Following this step, the units described in 2 above are no longer required as the loads from the walls of the building

Figure 16. All loads have been redistributed. Old foundations removed.

Figure 17. Construction of the top slab.

are now transferred directly onto piles through the medium of the rigid disk.

4. Excavation of the pit for the underground structure down to the designed level. To provide for stability of the sheet pile cofferdam a waling beam and a shoring system are provided; those are to be supported by the rigid disk and mounted on special embedded elements in the slab (Fig. 18). Following excavation of the perimeter strips light-duty machinery is introduced into the pit and excavation begins directly underneath the theatre building. As the soil is excavated the piles are propped by struts to form through-section columns and enhance stability.
5. Construction of bottom slab for the underground structure. Forming embedment slots for the bored piles in the bottom slab. Concreting for the exterior walls, interior walls and columns of the underground structure (Fig. 19). After concreting for the exterior wall of the underground structure and construction of the intermediate slabs at the perimeter sections the spaces between the sheet piles and the underground structure are backfilled. Following completion of the underground structure and backfilling the struts are taken down (Fig. 20). As soon as supports are introduced under the rigid disk

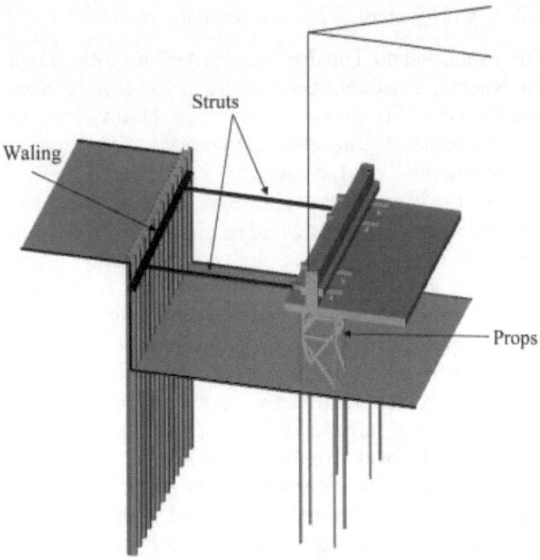

Figure 18. Excavation down to design level with struts.

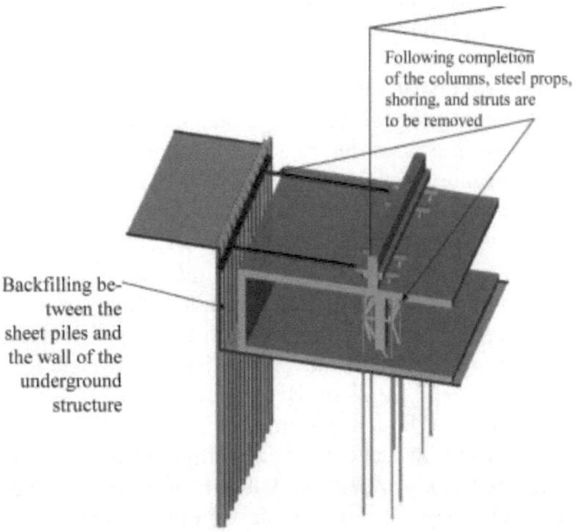

Figure 19. Concreting for the walls and columns of the underground structure.

Figure 20. Constructing the underground structure.

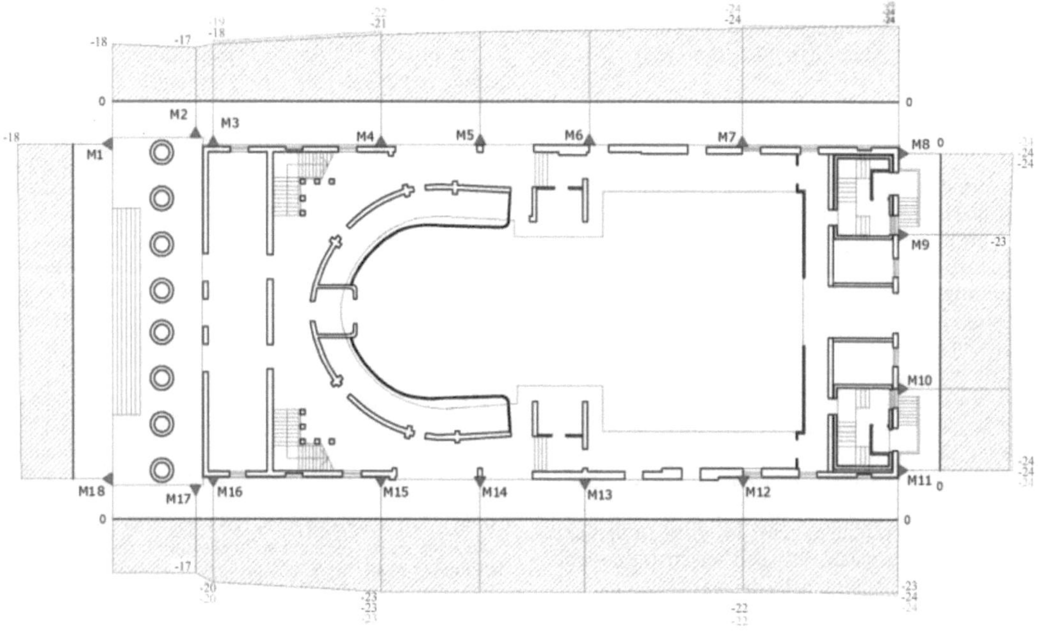

Figure 21. Settlements of Stone Island Theatre (mm).

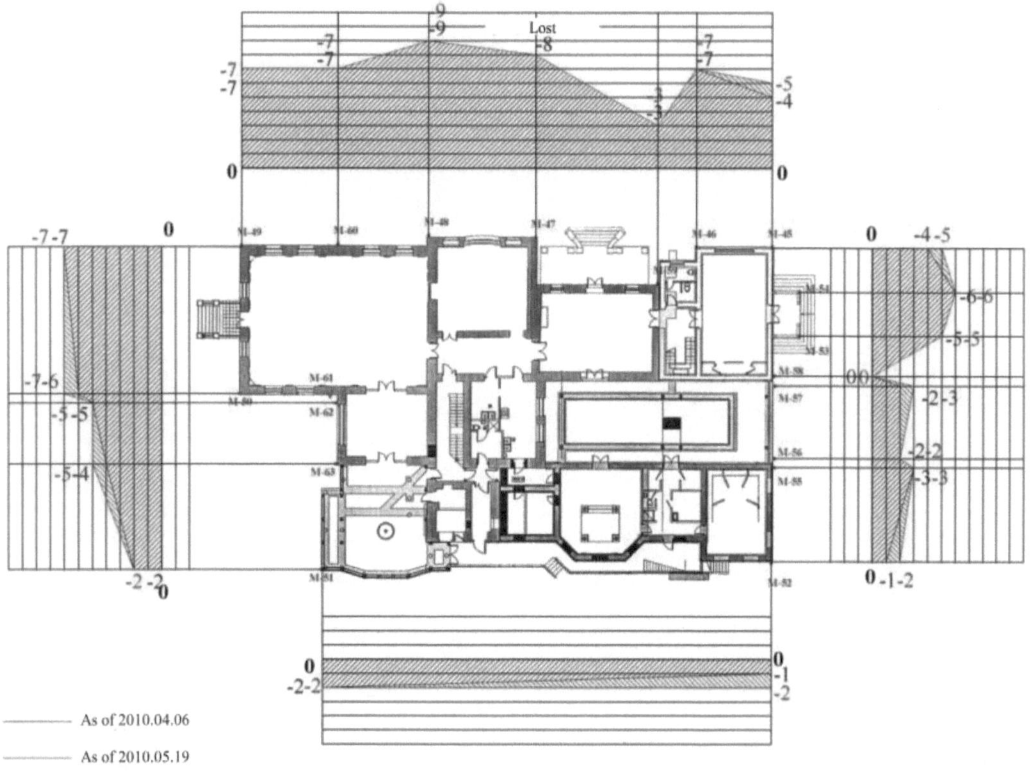

Figure 22. Settlement of Kleinmichel's Summer Mansion (mm).

slab, temporary through-section columns of bored piles are dismantled at which point the loads come to be transferred through the bottom slab.

5 MONITORING OF THE THEATRE

Monitoring in situ allowed to assess workability of the suggested design methodology as well as efficiency of the entire concept of geotechnical supervision.

It is worthy of notice that owing to geotechnical monitoring which was ongoing throughout the entire period of construction works the project team managed to rule out any influence of exceeding the permissible level of vibration acceleration (0.15 m/s^2). Cofferdam had been constructed with a very high quality and made it possible to keep the ground water at its natural level. Maximum settlement of the theatre building reached 24 mm, whereat no dangerous differential was ever registered (Fig. 21). Following redistribution of loads from the building onto the piles settlements became fully manageable as it had been made possible to adjust the entire building by means of retaining nuts on bars of the underpinning bored piles. Owing to a high culture of works implementation and a reliable design this was never implemented.

Horizontal displacement of the sheet piles reached 25 . . . 30 mm, and settlement of Kleinmichel's Summer Mansion were never in excess of 9 mm (Fig. 22). It is interesting to point out that the biggest danger to the Dacha was not the nearby excavation for the theatre project but seasonal fluctuations of the foundations related to frost heave in winter followed by spring thaw.

Therefore design, of the underground structure based on two groups of limit states for the existing buildings ensured not only safety of historic structures, but also safety of their interiors, whereas geotechnical monitoring ensured safety of soil in terms of remoulding during all types of construction works.

REFERENCES

Dashko R.E., Alexandrova O.Y. 2003. Causes for deformations of St. Isaac's Cathedral: technical, geological and environmental analyses. *Recontruction of cities and geotechnical engineering*. ASV Publishers, Saint Petersburg. Volume 5. pp. 57–65.

Ulitsky, V.M, Shashkin, A.G., Shashkin, K.G. 2010. Main regularities of soft clays soil behavior during deep excavations. *Proc. of International Geotechnical Conference 'Geotechnical Challenges in Megacities'.* Moscow, Volume 1, 2010 (in Russian).

Shashkin, A.G., Shashkin, K.G. 2005. Elasto-viscoplastic model of structurally unstable soils. 2005. *Reconstruction of Cities and Geotechnical Engineering*. Saint Petersburg, Volume 9, pp. 221–228 (in Russian).

Baltic Piling – Mets & Raudsepp (Eds)
© 2013 Taylor & Francis Group, London, ISBN 978-0-415-64334-4

Dynamic pile testing in the Baltic Countries

T. Riihimäki
Tampere University of Technology, Tampere, Finland

ABSTRACT: Before the year 2000 in the Baltic Countries the most common way to determine pile capacity was and still is static load testing. But after the year 2000 Tampere University has been in several projects in the Baltic Countries and has made several dynamic tests for piles. This is summary from these dynamic measurements from years 2003–2012. In this summary all of the projects are harbour projects and used pile types are driven large diameter steel pipe piles. As a result of this review hammer-, pile type and capacity combinations can roughly be estimated. It is possible to get reliable results from dynamic tests if right parameters and methods are used for testing.

1 BACKGROUND

1.1 *History of testing*

Before year 2000 the dynamic testing in Baltic Countries was very unusual. The most common way to determine pile capacity was by means of static load test. Static load tests are still very common but also dynamic tests have become general. This differs a bit from the Nordic Countries, for example in Finland the static load tests are very uncommon and the pile capacity with is almost always determined with dynamic test. The main reason for this is the differences in soil conditions. The use of friction piles in the Baltic Countries is more common because the hard soil layer is quite deep and it has been a common practice to test these friction piles with static load test.

2 CASES

2.1 *Special cases*

In this study three particular cases are studied more accurately. In these three cases the soil conditions made the pile driving quite challenging or the capacity of the pile was problem.

2.2 *Hammer and pile study*

This case history is outlined to driven large diameter open ended steel pipe piles in the marine constructions. This case consists of about 100 different piles driven in 11 different sites with different hammer types.

3 SPECIAL CASE 1

3.1 *General*

This case was Located in Muuga harbour, Estonia. Piling operator in this case was Fort Ehitus OU and the used hammers in this case were Junttan HHK 5A, Junttan HHK 10A and Junttan HHK 18S. Pile type in this case was open ended steel pipe pile, diameter 1064 and wall thickness of 16 mm and the length of the pile approximately 12 m. In Fig. 1 is presented the Junttan HHK 18S in the site.

Figure 1. Junttan HHK 18S in the site.

3.2 *Pile driving and dynamic measurements*

Pile driving in this case started in this case with Junttan HHK 10A. The piles were driven to refusal, but the pile toes were not in the designed level. The soil layer was so stiff that piles could not penetrate in to that layer. The driving continued with Junttan HHK 18S hammer, at this point the dynamic test was made and the capacity of the pile was 12 000 kN. The driving continued with full stroke with several hundred blows and the driving resistance was about 20 mm per 10 blows. At the end of the test about 0.8 m deeper the driving resistance was about 10 mm per 10 blows and capacity of the pile was about 14 000 kN. The pile was still above the design level.

Dynamic test was made also with sheet pile with Junttan HHK 5A to determine the stresses in the pile during driving. The capacity of the sheet pile with full blow was around 3000 kN and compression stress about 300 MPa.

3.3 *Conclusion*

The soil layer in the designed toe level was too stiff and pile could not penetrate in, even thought the hammer was enough efficient for this pile type. This problem reduced when pile driving continued away from the shore.

4 SPECIAL CASE 2

4.1 *General*

This case was Located in Saarenmaa harbour, Estonia. Piling operator in this case was Fort Ehitus OU and the used hammers in this case were Junttan HHK 5A and Delmag D46-32. Pile type in this case was open ended steel pipe pile, diameter 660 mm and wall thickness of 13 mm. In Fig. 2 is presented the Junttan HHK 5A in the site.

4.2 *Pile driving and dynamic measurements*

Pile driving in this case started with test pile and hammer was Junttan HHK 5A. The test pile was used to determine right pile length in this case. The target capacity of the testpile was around 4000 kN.

The pile was first tested when the pile toe was about 20 m below sealevel. The driving resistance at this point was about 40 mm per 10 blows and the capacity of the pile was about 3200 kN. The driving continued for next two meters and driving resistance was 50 mm per 10 blows and capacity was 2800 kN. At this point the pile had to be lengthened with another 12 m element. The driving resistance did not get any better during driving more at the end about depth of 33 m, driving resistance was 70 mm per 10 blows and the capacity of the pile was 1700 kN. There was no use to go any further, piles did not plug and hard soil layer is in very deep level.

At this point the piling stopped and something had to be done. The solution in this case with "basic engineering" was to use with the vertical piles a bottom plate and with inclined piles the plate in the middle of the pile. The 70 mm thick plate with hole in the middle was located 10 m above pile toe.

Figure 2. Junttan HHK 4A in the site.

This solution was tested by driving piles later and dynamic test was made for 3 vertical piles and capacity of the piles was 4000–7000 kN (pile lengths 15–19 m). Test was also made for 5 inclined piles and capacity was 4000–4300 kN (pile lengths 17–24 m). In this case the driving and dynamic test was made with Delmag D46-32 diesel hammer.

4.3 *Conclusion*

The solution to use bottom plate and plate in the plate in the middle of the pile worked very well in this case. Piles did not had to drive so deep and the capacity of the piles were enough compared to design.

5 SPECIAL CASE 3

5.1 *General*

This case was Located in Klaipeda harbour, Lithuania. Piling operator in this case was Josef Möbius Bau GmbH and the used hammer in this case was IHC S-90. Pile type in this case was open ended steel pipe pile, diameter 1220 mm and wall thickness of 12.5 mm. Soil conditions in the site consists of different kind of sandlayers. In Fig. 3 is presented the IHC S-90 in the site.

5.2 *Pile driving and dynamic measurements*

Pile driving in this case started with 18 m open ended steel pipe piles and hammer was IHC S-90. The piles

Figure 3. IHC S-90 in the site.

Table 1. Pile capacities from different test dates.

	Test date				
	L = 18 m 26.5.2011	L = 18 m 10.6.2011	L = 21 m 30.9.2011	L = 21 m 3.11.2011	L = 21 m 14.12.2011
1	6000	7500			
2	3850	5100		6900	
3		5100	8600		
4		4800	4400		7300
5			5300		8000
6				5000	8000

were measured after installation and capacity of the piles was 4000–6000 kN and the target was around 7000 kN. These same piles were measured again three weeks later and the capacity was about 20–30% better. Also few other piles were measured. Still some piles were below target capacity and at this point piles were lengthened about 3 m. Piles were measured again after 3 months, some of the piles were again below the target and this lead to another measurement and so on. In the Table 1 are collected some interesting piles from measurements.

From the Table 1 can easily catch that pile lengthening does not necessarily give the advance immediately (compare third and fourth pile).

5.3 *Conclusion*

The sand layers in the site were challenging. The soil is quite disturbed after driving and driving resistance is quite low, but when the soil settles down driving resistance and capacity of the pile gets higher, but in this case time for this process took months. In this case the right pile length together with waiting time led to successful result. Perhaps this was not the most economical way, but eventually the goal was reached.

6 HAMMER AND PILE STUDY

6.1 *Typical hammers used in projects*

In these projects were used variety of hammers, but basically there are three kinds of hammers: diesel hammers, hydraulic hammers and hydro hammers.

Diesel hammers with these cases were open ended and the ram mass from 4 to 6 tons. The maximum drop height of the ram depends on soil conditions and fuel supply, the maximum is usually about 3.5 m. Typical efficiency (measured energy from pile/theoretical energy) with diesel hammers is 0.5–0.7. The small fuel consumption makes diesel hammer economical to use in case of heavy driving (lots of blows per pile and lots of piles).

Hydraulic hammers with these cases were both accelerated and non-accelerated. The mass of the rams varied from 4 to 18 tons. Maximum drop height is limited by lift cylinder and it usually 1.2 to 1.5 m. Acceleration to ram is made by hydraulics and usually the real value of the acceleration is about 1 g (to win the friction effects). Typical efficiency with these hammer types are 0.8–1.0.

Ram mass of the hydro hammers used in these cases were 4.5 tons. The acceleration with hydro hammers is made by gas and the acceleration is about 2 g (to make the hammer more shallow and reduce drop height). The efficiency with these cases were lower than typical 0.6–0.8.

6.2 *Typical piles in this project*

In these projects was used huge variety of pile types. The diameter varied from 660 mm to 1220 mm and wall thicknesses from 10 mm to 16 mm. Steel grades with piles varied from unknown (used pipes) to high strength steel X-70 (yield strength 482 MPa).

6.3 *Pile capacities with different hammer and pile types*

In this case history is collected dynamic pile tests from about hundred piles from 11 different sites. The test results are collected in Table 2. In the table are presented on each row results from one site (hammer and pile type can change also in the same site).

With this table it is possible to estimate and select a suitable hammer for pile type in "typical" soil conditions in the Baltic Countries.

Table 2. Mean results from dynamic tests with different pile and hammer types.

Hammer model	Piletype D/t	Mean length of the piles m	Mean capacity of the piles kN
BSP 357	1020/14	30	6200
Delmag D42-32	660/13	20	4600
Delmag D42-32	1220/12	30	8000
Delmag D46-22	1220/12	42	8500
HC S-90	813/12	17	4200
IHC S-90	1220/12.5	20	6600
Junttan HHK 4S	711/12.5	28	3500
Junttan HHK 5A	711/12.5	20	5200
Junttan HHK 7S	1020/12.5	18	6500
Junttan HHK 10A	1020/10	10	8700
Junttan HHK 12S	1067/13.5	20	5200
Junttan HHK 16S	1067/13.5	25	7500
Junttan HHK 18S	1064/16	15	14000
Junttan HHK 18S	1220/12	35	8200
Menck 3-7	813/11	20	4400

6.4 *The influence of case damping factor to calculated capacity*

In most cases one or more capwap-analyses were made to determine the suitable case damping factor for site. The measurements are always done before the capwap-analysis and after signal matching the right case damping factor is obtained and capacity values have to be re-calculated. The PDA-measurement equipment manufacturer has recommendations to damping factor values in different soil types (PDI). If pile is mostly end bearing pile the capacity is not so sensitive to selection of damping factor (end bearing pile recommendation in Finnish national piling code is 0.5). With these cases the used case damping factor varied from 0.5–0.7.

7 CONCLUSION

7.1 *Pile driving*

In the Baltic Countries soil conditions can be challenging especially to determine the right pile length and pile type versus pile load. Sometimes the soil can also be so stiff that it is hard to penetrate to design level. With Table 2 it is possible to estimate roughly suitable hammer type to the certain pile type.

7.2 *Capacity of the piles with dynamic test*

With these cases the capacity of the piles varies quite a lot. Usually the capacity gets better if there is time to wait between pile driving and dynamic test.

The recommendation for case damping factor is 0.6–0.7 and the damping factor should be checked at least from one pile in the site. With this method it is possible to get reliable records from dynamic tests.

REFERENCES

PDI Pile Dynamics Inc. 2006. CAPWAP Case Pile Wave Analysis Program. Cleveland, Ohio, USA, Pile Dynamics Inc.

PDI Pile Dynamics Inc. 2000. PDA-W Manual. Cleveland, Ohio, USA, Pile Dynamics Inc.

RIL 254-2011 Paalutusohje 2011 (National code for piling), RIL, Helsinki, 2011.

Riihimäki T. Dynamic pile testing reports, Department of Civil engineering, Tampere University of Technology, 2003–2012.

Baltic Piling – Mets & Raudsepp (Eds)
© 2013 Taylor & Francis Group, London, ISBN 978-0-415-64334-4

Soft rock of perm as the base of pile foundations soils

A. Ponomarev, E. Sytchkina & D. Sursanov
Perm National Research Polytechnic University, Perm, Russia

ABSTRACT: Building and designing on soft rocks is problematic for many countries in the world. In particular, in Russia the problems of building on soft rocks are not given clearly enough in national construction norms and standards. In that paper laboratory research methodology and some results of experimental analysis of soft rocks are presented.

1 INTRODUCTION

Today high-rise buildings is actively constructed in the central districts of Perm. It leads to the foundation depth and load increase on the soils the as well. Under these conditions foundations interact with geological Lower Permian deposits are the most effective and sometimes the only possible type of foundations. Therefore in the conditions of building in Russia it is important to know the real properties of Lower Permian soft rocks for the forecast of the settlement and bearing capacity of the foundations based on them (Bartolomey 2007, Ponomarev 2010).

2 LABORATORY RESEARCH METHODOLOGY

2.1 *Initial information*

The study of physical and mechanical properties of soft rocks was carried out for following samples: sandstone (depth of burial – 19.5–23.0 meters) and argillite (depth of burial – 25.0–30.0 meters).

2.2 *Physical properties determination methodology*

In the course of study following physical properties of argillite samples were determinated: density of soil ρ, density of soil particles ρ_s, water content ω, plasticity index I_P, liquidity index I_L, according to GOST 5180-84. Moreover, next calculated parameters were determinated: unit weight γ, density of dry soil ρ_d, porosity n, void ratio e, degree of saturation S_r. Statistical treatment of tests results was carried out according to GOST 12248-96.

In the course of study following physical properties of sandstone samples were determinated: density of soil ρ, density of soil particles ρ_s, water content ω, according to GOST 5180-84, and granulometric composition, according to GOST 25100-95.

2.3 *Deformation properties determination methodology*

In the course of study following deformation properties of argillite and sandstone samples were determinated: oedometric modulus E_{oed}, deformation modulus E. Laboratory tests were carried out by means of oedometer. The device is allowed to create a vertical pressure 1.5 MPa to the sample. The height of samples was 20 mm, and the diameter of samples was 71.5 mm. Loading of samples was done by device with load steps equal 50 kPa each, from 25 kPa till 1000 kPa, according to GOST 12248-96. Treatment of test results was carried out by means of software.

2.4 *Strength properties determination methodology*

In the course of study following strength properties of argillite and sandstone samples were determinated: angle of internal friction φ; effective cohesion intercept c. Laboratory tests were carried out by means of shear box. The height of samples was 40 mm, and the diameter of samples was 71.5 mm. Study was carried out by consolidated-drained scheme, without water saturation. Normal pressure took following values 400 kPa, 500 kPa and 600 kPa. Before shear test began, each sample was compressed in the device of preliminary compression, by appropriate normal pressure. Time periods of preliminary compression load steps corresponded to requirements of GOST 12248-96. In course of tests tangential load increased continuously, and velocity of slice took following values: for sandstone – 0.5 mm per minute, for argillite – 0.02 mm per minute. The final criterion of shear test included one condition: the common slice deformation could be more than 5 mm. Treatment of shear test results was carried out by means of software.

Table 1. Normative values of argillite samples physical properties.

	Depth of burial, m			
Properties	13.0	16.0	20.0	27.0
ρ, g/cm^3	1.97	2.10	2.12	2.10
γ, kN/m^3	19.31	20.58	20.78	20.58
ρ_s, g/cm^3	2.76	2.75	2.77	2.76
ρ_d, g/cm^3	1.64	1.78	1.80	1.81
ω,	0.20	0.18	0.18	0.16
I_P,	0.19	0.15	0.20	0.16
I_L	−0.08	0.09	0.10	0.01
e	0.68	0.55	0.54	0.52
n, %	40.52	35.29	31.15	34.41
S_r	0.81	0.91	0.92	0.84

Table 2. Normative values of oedometric modulus E_{oed} for argillite samples with different water saturation.

Sample number	S_r	$E_{int\ 0.5\text{–}0.6}$ MPa*	Average value $E_{int\ 0.5\text{–}0.6}$ MPa
70	0.89	6.16	
72	0.89	7.00	
76	0.89	5.03	6.16
200	0.89	5.60	
207	0.89	6.23	
212	0.89	6.93	
120	1.00	6.94	
175	1.00	5.14	
176	1.00	5.84	5.53
194	1.00	3.76	
195	1.00	6.86	
196	1.00	5.02	

*Deformation modulus in interval E_{int} 0.5–0.6 MPa.

3 RECEIVED RESULTS

3.1 *Physical properties of argillite samples*

Physical properties of argillite samples are presented in the Table 1.

According to plasticity index I_P and liquidity index I_L argillite samples were classified as hard loam.

3.2 *Deformation properties of argillite samples*

Oedometric modulus Eoed in interval 0.5–0.6 MPa of argillite samples with natural water content changed in following limits 5.03–7.00 MPa.

Oedometric modulus E_{oed} calculation results of argillite samples from depth of burial 25.0–30.0 m are presented in the Table 2.

Compression curves for argillite samples are presented at Figure 1 and Figure 2.

Therefore, in the scheme without water saturation the oedometric modulus E_{oed} average value of argillite samples in interval 0.5–0.6 MPa was 6.16 MPa, and in the scheme with full water saturation the oedometric modulus E_{oed} average value of argillite samples in interval 0.5–0.6 MPa was 5.53 MPa. So, decrease of oedometric modulus E_{oed} value in the scheme of full water saturation was 10.2%.

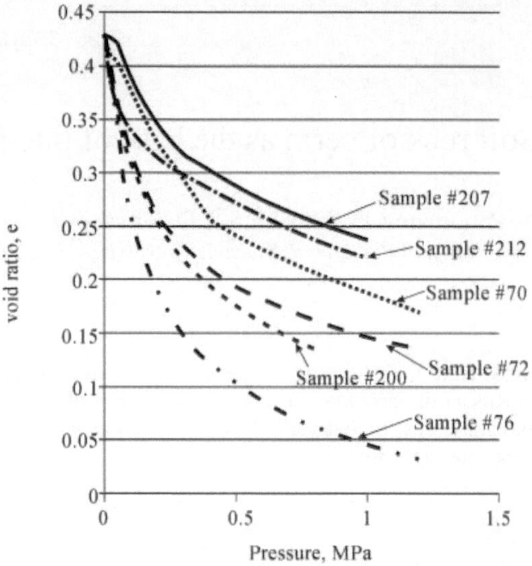

Figure 1. Compression curves for argillite samples (without water saturation).

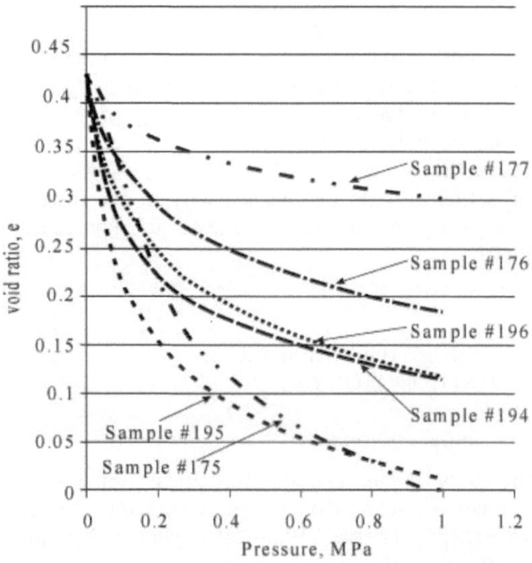

Figure 2. Compression curves for argillite samples (with full water saturation).

3.3 *Strength properties of argillite samples*

Strength properties of argillite samples, from depth of burial 25.0–26.0 m, are presented in the Table 3, strength straight lines are presented at Figure 3.

Table 3. Strength properties of argillite samples, with depth of burial 25.0–26.0 m.

Strength properties	φ °	c MPa
Effective	29.23	0.043
Residual	30.17	0.031

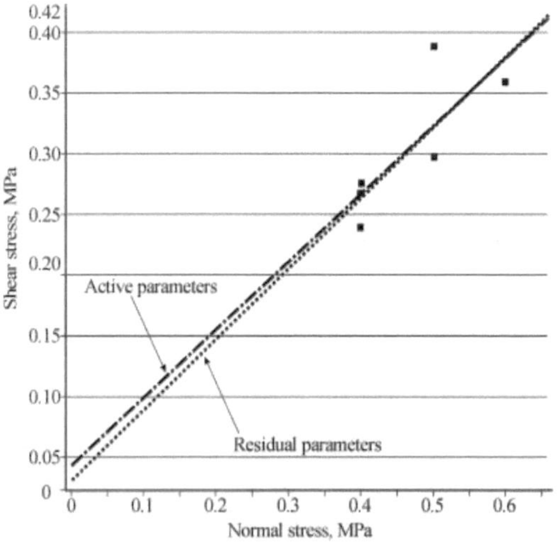

Figure 3. Argillite samples strength straight lines.

4 CONCLUSIONS

As the result of laboratory tests, properties of soft rocks were investigated. Also was proved that strength properties of soft rock samples depend on the rate of water saturation. Therefore strength properties of the samples decreased on 10.2%. A number of laboratory and especially field experiments are required in order to find an integrated solution of the problem under discussion.

REFERENCES

Bartolomey, L. & Bartolomey I. 2007. Soft rocks as the basement for high-rise buildings in Perm: *2nd academic Bartolomey lectures; Proc. intern. conf., Perm, 3–5 September 2007.*

Ponomarev, A., Zaharov, A., Sursanov, D. 2010. Some results of field tests of piles static pressing loading: *International Scientific and Technical Conference, devoted Dalmatov 100-anniversary of the birth; Proc. Intern. Conf., St. Petersburg, 2010.*

GOST 5180-84. Soils. Laboratory methods for determination of physical characteristics, Moscow, 1985.

GOST 12248-96. Soils. Laboratory methods for determining the strength and strain characteristics, Moscow, 1996.

GOST 25100-95. Soils. Classification. Moscow, 1996.

Baltic Piling – Mets & Raudsepp (Eds)
© 2013 Taylor & Francis Group, London, ISBN 978-0-415-64334-4

Hard proterozoic deposits as subsoils of deep foundations under high-rise buildings in St. Petersburg

V.M. Ulitsky
State Transport University, Saint Petersburg, Russia

V.A. Vasenin, A.G. Shashkin & C.G. Shashkin
Georeconstruction Engineering Co, Saint Petersburg, Russia

ABSTRACT: Due to increasing height of buildings and development of the underground space hard proterozoic deposits now are used quite often as subsoils of structures in Saint Petersburg. For high-rise buildings embedding pile ends into hard soils often is the only possible option of foundation design, since the depth of rocks in central Saint Petersburg is more than 200 m. One of the main tasks of pile foundation design subjected to big loads is estimation of settlements of such foundation. Meanwhile, deformation properties of hard soils are not studied well yet.

Research of these deposits was conducted by the authors on several sites in Saint Petersburg. It consisted of laboratory testing in triaxial cells, and in-situ tests. Results of investigations on one test site in the centre of Saint Petersburg are presented in the paper. Some results of soil-structure interaction modeling of a high-rise building are also presented.

1 GEOLOGICAL CONDITIONS OF THE SITE

The ground conditions of the site considered in the present paper are characterized by presence of quaternary deposits of variable origin and bedrock of Wendian strata of the Kotlin horizon. The quaternary strata within the boundaries of test areas for testing piles and barrettes extend down to 43–45 m. Following those from 45.0 m to 53.0 m stratification consists of bluish-grey laminated clays with thin inclusions of dislocated sandstone (max. 1 mm in thickness). Bluish-grey laminated clays continue to follow from 53.0 m down to 127.5 m. Within the boundaries of test areas under consideration, SPT tests on subsoil were conducted — predominantly on quaternary deposits (up to the upper boundary of dislocated Wendian strata, i.e. going down to 40–43 m).

2 RESULTS OF LABORATORY TESTS OF HARD CLAYS

The design for the high-rise building envisages toe levels of deep supports (barrettes) to be located in stiff Wendian clays. Significant attention during laboratory testing was given to strength and strain (deformation) properties of the Wendian clays. The tests were conducted by a professional testing facility according to three testing schemes to comply with requirements expressed in Russian and British Standards. The analysed the test results statistically. Fig. 1 shows strength dependency of hard Wendian clays on depth. Within the interval from 40 to 150 m, shear strength of samples tends to increase. There was, however, no significant difference between the executed testing schemes. Apparently, the difference in strength for hard clays consists only in degree of micro-cracks closing (the cracks formed in the process of coring and reconstituting samples) during loading at hydrostatic pressure stage, as well as in strain rate at the stage of action along compression paths.

Fig. 2 shows the corresponding change with depth of vertical relative strain value at failure, which reaches about 15% in the area of glacial dislocations at the level of 45 m and reduces with depth to 2–3%.

Fig. 3 shows a dependency of change in water content of Wendian clays samples tested according to various schemes. water content of the tested samples decreases with depth. At depths of around 50 m average water content amounts to approximately 15%, whereas at depths of approximately 150 m average water content reaches roughly 11.5%. Based on triaxial tests results, dependency of strength on natural water content of samples becomes distinctly visible (Fig. 4) for all hard clay strata. The division into various testing schemes (i.e. consolidated-drained, consolidated-undrained, and unconsolidated-undrained schemes) fails to distinguish any other regularity. This may be explained by practical absence of free water in hard clay voids. All pore water in such clays is bound, and it is because of that that the process of the so called "consolidation" in such soil is practically absent and is reduced in all test types to micro-cracks closing.

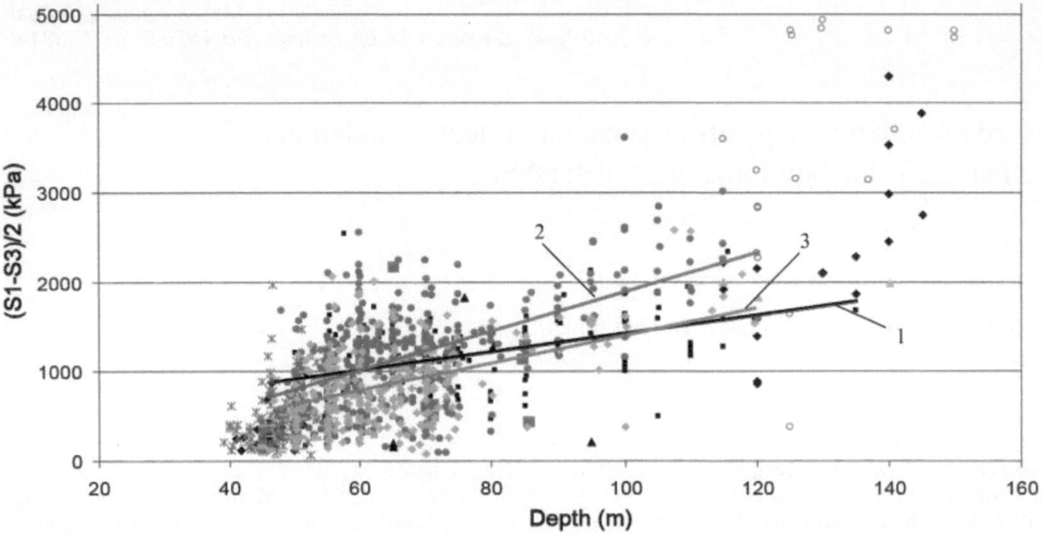

Figure 1. Change of shear strength (kPa) of Wendian clays with depth. According to various test schemes (dots represent soil samples): 1 – Unconsolidated-undrained (UU), 2 – Consolidated-undrained (CU), 3 – Consolidated-drained (CD) tests.

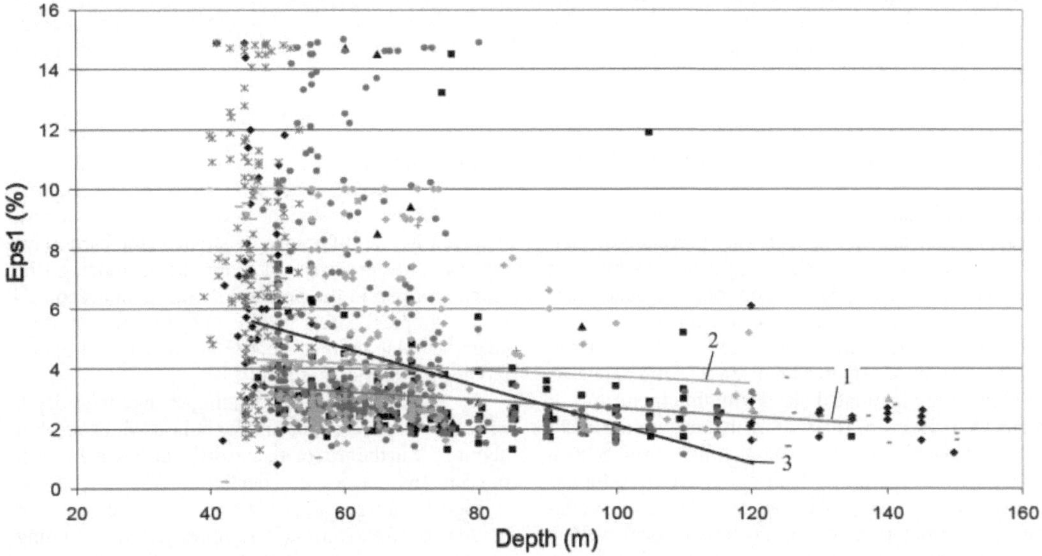

Figure 2. Change of vertical relative strain of Wendian clay samples (kPa) at failure with depth. According to various test schemes (dots represent soil samples): 1 – Unconsolidated-undrained (UU), 2 – Consolidated-undrained (CU), 3 – Consolidated-drained (CD) tests.

Fig. 5 shows comparison of dependencies of reached axial strain of samples on water content based on triaxial tests at various schemes of hard clay behaviour. There was no significant difference in strength or deformability of hard clay samples in various testing schemes. Thereat, however, a considerable scatter of both strength, and deformability is observed, which is related, first and foremost, to the scatter of natural physical soil properties.

Absence of significant differences during hard clay tests according to different schemes presumably means that the volumetric strain component is immaterial. Thus, the behaviour of hard clay samples can be described based on various conditions (of both drained and undrained testing modes).

3 SOIL-STRUCTURE INTERACTION MODELLING OF A HIGH-RISE BUILDING

The designed high-rise building (having about 80 floors) is of core and frame type. Its overall stability is ensured by joint action of the central monolithic reinforced concrete core and steel columns along the

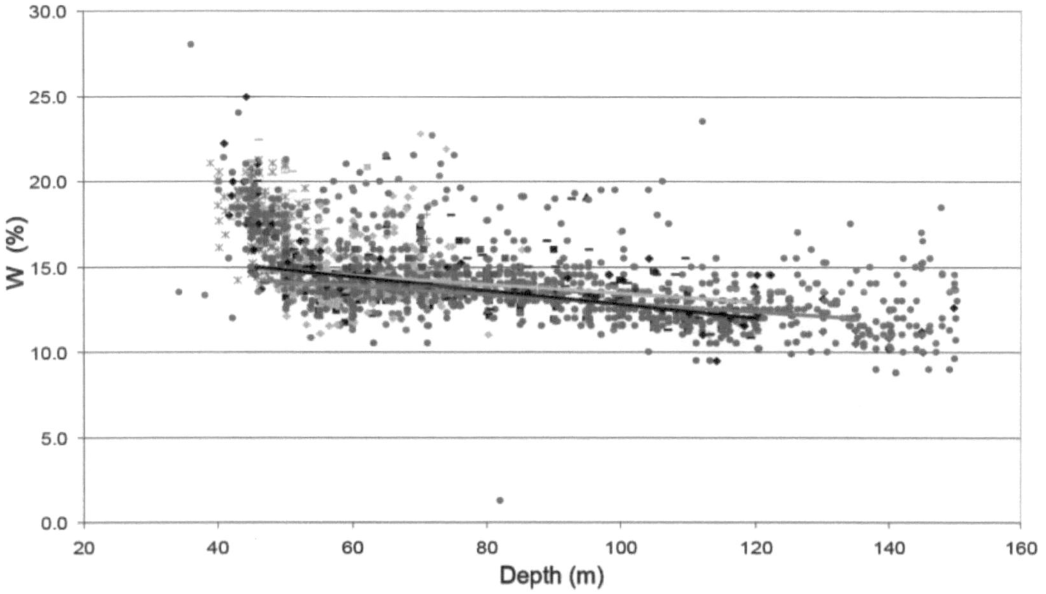

Figure 3. Change of degree of saturation of Wendian clay samples with depth. Variously shaped points correspond to various soil elements and boreholes.

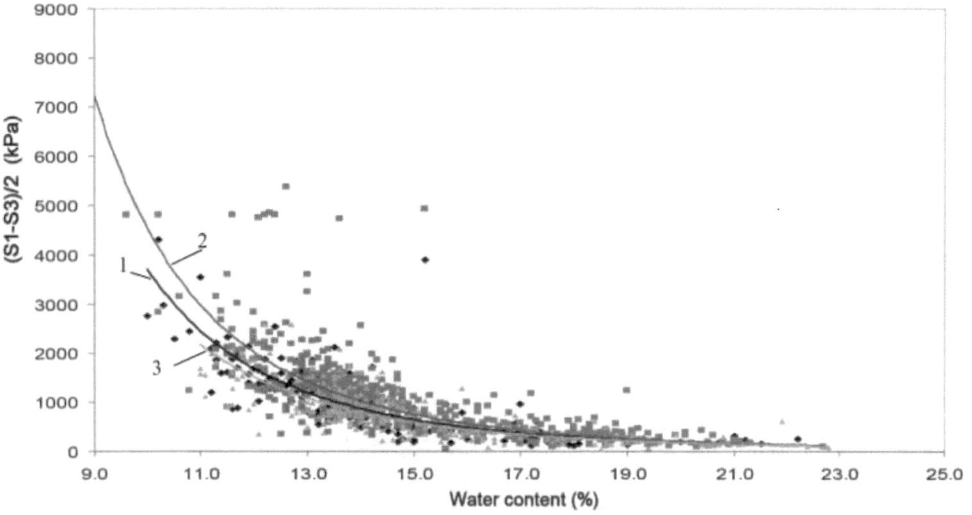

Figure 4. Dependency of strength on saturation in triaxial tests according to various test procedures: 1 – Unconsolidated-undrained (UU). $y = 7E{+}07x^{-4.3017}$, $R^2 = 0.7152$; 2 – Consolidated-undrained (CU), $y = E{+}08x^{-4.431}$, $R^2 = 0.7225$; 3 – Consolidated-drained (CD) tests, $y = 5E{+}\ 07x^{-4.2041}$, $R^2 = 0.6611$.

building perimeter, the columns being joined into a single system by means of steel beam cells, along which intermediate floor discs are placed; additionally there are special outrigger beams in three levels of maintenance floors. In such a layout the main element responsible for assuming the vertical and horizontal components of the loads is the core. Square area of the central reinforced concrete core decreases with height together with the square area of individual floors, which likewise decreases as the building becomes higher.

As the purpose of our analyses was the consideration of interaction of the designed superstructure and the subsoil, the necessity arose to adequately represent the building superstructure, wherewith to ensure the required calculation accuracy (Ulitsky et al, 2005, Shashkin, 2006). Modelling the action of subsoil and barrettes implied obtaining accurate loads distribution onto barrettes, as well as an account of their interaction with the surrounding soil, which made it necessary to model the barrettes with spatial finite elements. Pressure is transferred onto barrettes through

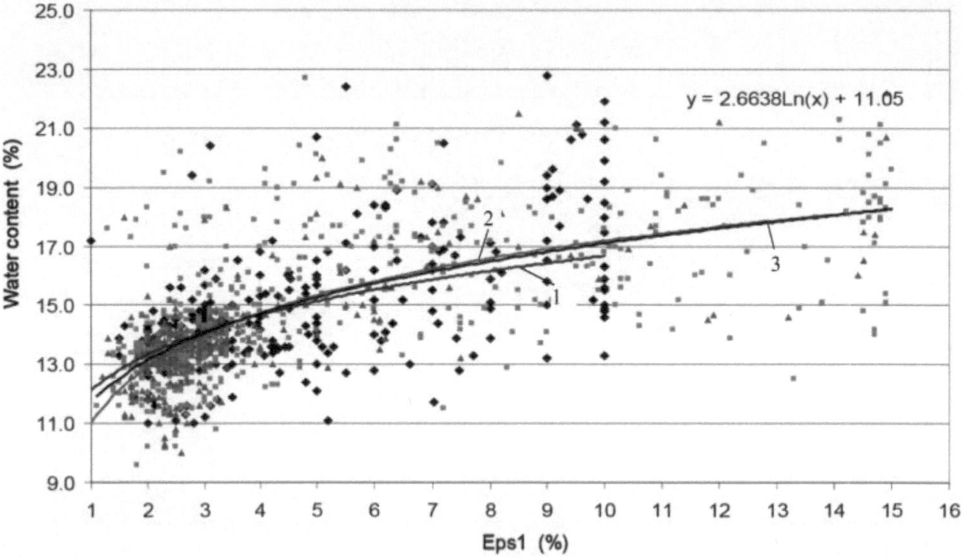

Figure 5. Dependency of max. vertical strain on degree of sample saturation in Wendian clays according to various testing procedures: 1 – Unconsolidated-undrained (UU), 2 – Consolidated-undrained (CU), 3 – Consolidated-drained (CD) tests.

the foundation mat, and considering the thickness of the slab as being 3.6 m, for a reliable assessment of loads distribution onto barrettes, again, it was necessary to model the pilecap slab with spatial finite elements. The major portion of loads onto subsoil from the designed high-rise building is generated by concentrated loads from its core. In the underground part of the designed building the thickness of the external core wall is 2 m. Moreover, in places where the core width changes, the width of the core walls reaches 3.5 m.

In this case while constructing a 3-D scheme for the building to perform soil-structure interaction analysis it was decided to build up the finite element profile according to the following sequence: modelling behaviour of subsoil, barrettes, pilecap slab and external core walls with spatial finite elements; modelling the internal core walls and intermediate floors with plate finite elements; modelling the columns and beam cells with spatial rod elements.

General outlook for the superstructure calculation profile is presented in Fig. 6(a).

Soil-structure interaction analysis of the designed high-rise building and its subsoil was performed according to several scenarios of subsoil behaviour:

1. Calculating structures of the building on elastic subsoil, as per requirements contained in currently applicable codes and in agreement with assumptions adopted on limiting depth of compressible stratum in the subsoil of the designed building.
2. Calculations with account of subsoil behaviour based on non-linear models, which permit automatic limitation of the depth of compressible stratum.

A considerably important issue in assessing settlements of the high-rise building under consideration will be the proportion of subsoil deformation brought about by volumetric and shear strains of hard clays. With this consideration in mind, two conflicting hypotheses in respect of volumetric

(a) (b)

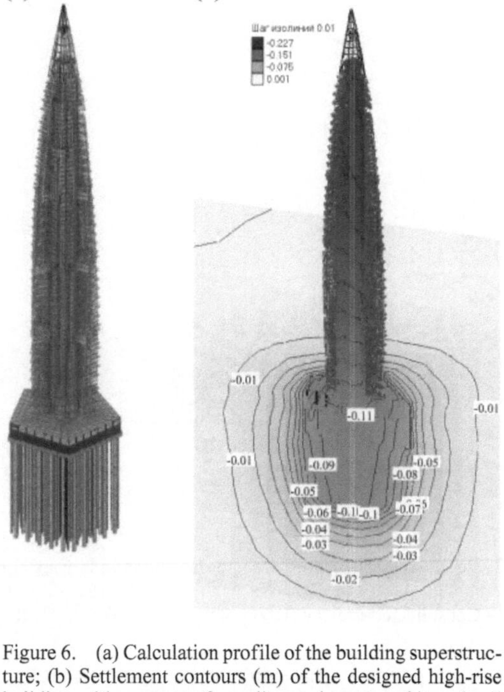

Figure 6. (a) Calculation profile of the building superstructure; (b) Settlement contours (m) of the designed high-rise building with account of non-linear character of its subsoil behaviour according to consolidated-drained scheme with account of dividing pressure into effective and hydrostatic – according to Scheme 2 – Option with barrette toe level at 105 m from the ground surface. (Contours spaced at 0.01 m.)

compressibility of hard clays were used in our analyses:

1. there is no volumetric compressibility at all;
2. volumetric compressibility corresponds to the obtained compression curves within the interval of stresses from natural pressures to acting stresses in subsoil.

The first hypothesis corresponds to an assumption of there being no considerable voids filled with air or free pore water in natural soil. The assumption of overconsolidated state of hard clays can lead to a similar conclusion as well. At overconsolidation ratio OCR = 2 or more, additional loading with the building's own weight will not exceed preconsolidation pressure, and, correspondingly, volumetric compressibility will be at its minimum and will be defined by the unloading-reloading modulus.

In respect of strength parameters (as well as of shear deformability) three various hypotheses were put forward:

1. strength depends, largely, on physical properties (natural degree of saturation) of samples and only weakly depends on total pressure; thereat strength is defined directly based on triaxial tests (same as in undrained conditions);
2. strength depends on total confining pressure (i.e. increases with depth), in calculations an account is made of angle of internal friction, effective stresses in soil skeleton (considering also the water column in total absence of a confining layer);
3. strength depends on total confining pressure (i.e. increases with depth), in calculations an account is made of angle of internal friction, stresses in soil skeleton equal total stresses (not accounting for water).

The first assumption, as far as its physical sense is concerned, corresponds to the hypothesis of there being no considerable volumetric compressibility of soil. Indeed, there being a dependency of clay strength on pressure is connected not so much with friction as such, as with soil compaction under load. An increase in density leads to a corresponding increase in quantity and quality of contacts between clay particles and, correspondingly, to an increase in strength. If there is no considerable volumetric compressibility, growth of strength at increasing loads is also hardly probable. In this case strength properties will be defined by natural density of soil, formed at this or that depth by acting natural stresses.

Prior to calculating superstructure of a building on non-linear foundation-subsoil we selected parameters of soil models, with which intention we modelled triaxial tests of soil samples.

Results of calculations performed according to various schemes of subsoil behaviour are contained the following table.

As can be seen from Table 1, various models of non-linear foundation-subsoil produce provisionally close values of settlement from 7 to 11 cm. Contours of vertical movements in subsoil and superstructure of the designed building, with account of non-linear character of its subsoil behaviour according to consolidated-drained scheme, are contained in Fig. 6(b) and Fig. 7. Differences in settlement values calculated based on possible schemes of subsoil behaviour are conditioned largely by accounting for or disregarding the volumetric component of strain. Here it is necessary to point out that the settlement input contributed by the volumetric portion of strain in hard clays will take a rather long time to develop. As this component of deformations will be entirely defined by consolidation process in hard clays, then, bearing in mind low values of permeability coefficient (approximately 10^{-6} m/day) and long permeability paths, this settlement component will take a long time to develop, which makes it of secondary importance for consideration over the building lifetime.

Table 1. Settlement and relative settlement differentials for various foundation options.

Subsoil behaviour scheme	Option with barrette toe level at 105 m from ground surface/Settlement		
	Max., cm	Min., cm	Relative differential
Elastic calculation through layer-by-layer summation method (compressible stratum equals ½ of subsoil width)	12.5	9	0.001
1 Unconsolidated-undrained (Cu-lowered)	6.5	5	0.0006
2 Consolidated-drained with account of effective stress	11	9.6	0.00055
3 Consolidated-drained effective stress equals total	6.5	5	0.0006

The most important for design is distribution of loads unto barrettes and stresses in the superstructure. Based on performed calculations, values of loads in future building structures were chosen as the least favourable for the considered schemes of subsoil behaviour.

One of the important moments in foundation design for high-rise buildings is the criterion of permissible values of absolute subsoil settlement. In this case the ultimate value criterion for foundation settlement was limited to the value of 10 cm. To observe this criterion of ultimate settlement, the barrette toe level in subsoil of the high-rise building must be approximately 100–105 m. If the criterion of ultimate settlement development of the high-rise building is lowered, the length of barrettes can be considerably reduced.

Another not less significant issue defining reliability of foundation-subsoil design is the guarantee of adequate bearing capacity of the barrettes. This issue for the site in question was studied separately and the results of this study are published in a separate paper.

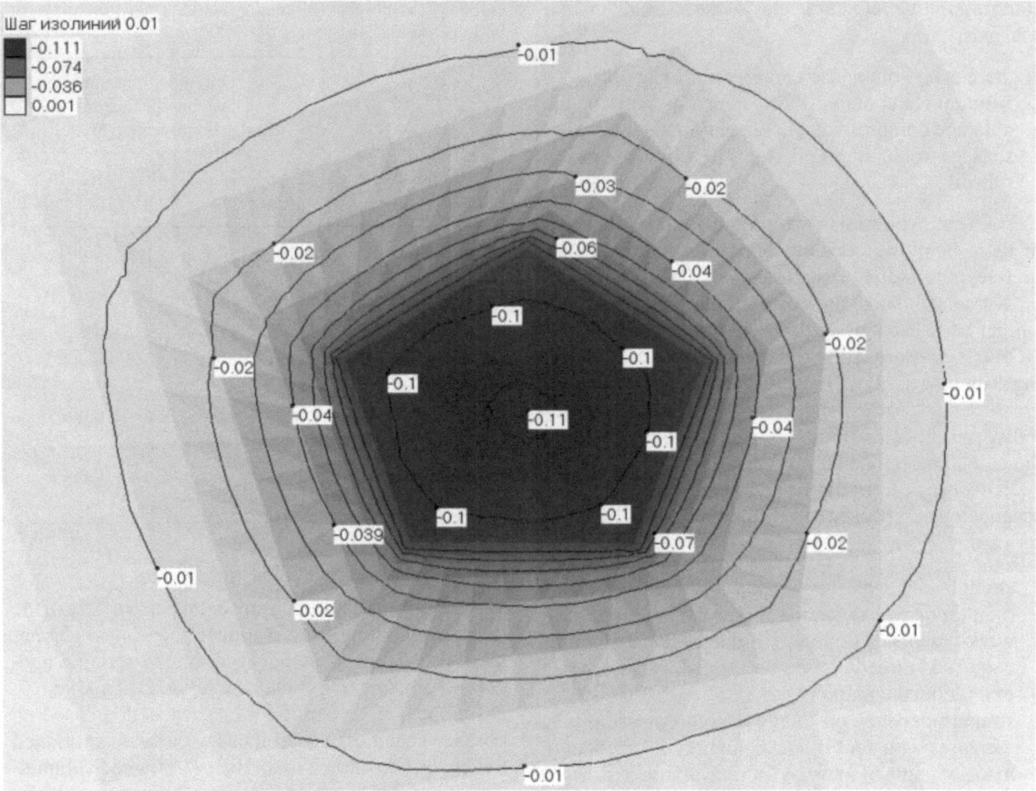

Figure 7. Contours of barrette settlement (m) when calculating building on non-linear subsoil according to consolidated-drained scheme with account of dividing pressure into effective and hydrostatic according to Scheme 2 – Option with barrette toe level at 105 m from the ground surface (contours spaced at 0.01 m).

Thus, analyses of various models of subsoil-foundation behaviour provided the possibility to evaluate the possible range of absolute settlements of the high-rise building, as well as the range of relative settlement differential of its individual structures. This in turn allowed definition of loads distribution range onto barrettes and the corresponding loads in the structures of the high-rise building.

4 CONCLUSIVE REMARKS

Test results of hard Wendian clays and soil-structure interaction analysis of a high-rise building are presented in the paper

The soils have been tested with different various testing schemes – consolidated-drained, consolidated-undrained, and unconsolidated-undrained. Graphs of dependency of strength on samples natural water content. Absence of significant differences during hard clay tests according to different schemes presumably means that the volumetric strain component could be very small. Thus, the behaviour of hard clay samples can be described based on various conditions (of both drained and undrained testing modes).

Soil-structure interaction analysis of the designed high-rise building and its subsoil was performed according to several scenarios of subsoil behaviour: with elastic subsoil, with assumptions adopted on limiting depth of compressible stratum in the subsoil of the designed building: and with account of subsoil behaviour based on non-linear models, which permit automatic limitation of the depth of compressible stratum. The computations have made it possible to estimate the settlements of the building with the various foundation options.

REFERENCES

Shashkin, C. 2006. Basic regularities of soil-structure interaction. *Proceedings of the XIII Danube European Conference on Geotechnical Engineering.* Ljubljana. Vol. 1, pp. 179–190.

Ulitsky V.M. 2005. The basics of soil-structure interaction. *Proceedings of the International Conference "Soil-structure Interaction: Calculation methods and Engineering Practice".* St. Petersburg, Vol. 1, pp. 3–10.

Baltic Piling – Mets & Raudsepp (Eds)
© 2013 Taylor & Francis Group, London, ISBN 978-0-415-64334-4

New drilling tools for hard rock

Gordian Ulrich
BAUER Maschinen, Germany

ABSTRACT: Nowadays the Kelly rotary drilling method is the most common way of installing bored piles around the world. When it comes to drill into rock people face difficulties regarding drilling performance and wear and tear. The challenge is to develop a drilling tool which can be combined with the Kelly rotary method.

For common geology such as sand, gravel or clay layers there are plenty of different rotary drilling tools available on the market. Equipped with either flat teeth or round shank chisels these tools work perfectly even for rock strengths up to 100 MPa. For harder rock conditions or boulders one solution could be to drop a chisel and destroy the rock with a huge impact. This technique might be still used but tool manufacturers have to fulfill their obligation to improve tools and technique constantly.

With the latest developments it is now possible still using the rotary drilling method and by just changing the drilling tool, rock strengths up to 300 MPa can be loosened and transported. This technical progress now enables one to use the rotary drilling rig in Kelly mode for penetrating through the overburden and also for cutting across very hard rock conditions.

For rock socketing purposes roller bit tools with air flushing are usually used. One of the latest developments is a tool which can cope with extremely hard rock, sloping rock surfaces or even boulders of any size. The New Bauer MHD System employs; Proven, Rotary Air Percussion Technology to advance the Rock Socket quickly and economically.

This paper will give a short overview of commonly used drilling tools before the latest developments, will be introduced. First projects in Scandinavian countries as well as different reference projects will complete the explanations.

Keywords: Rock drilling, Roller bit tools, Down the hole hammer, Multi Hammer Drill

1 CONVENTIONAL TOOLS

One of the most commonly used conventional drilling tools is probably the so called core barrel. This tool is equipped with round shank chisels and drills under high crowd pressure an annulus into the rock so the core is then released from the lateral support. Sometimes, especially in fractured rock, the friction between core and tool increases to an extend that the core shears off horizontally and can be taken up in one piece.

In case the core stays in the borehole it can be destroyed with a progressive rock auger in dry conditions or with a Cross Cutter in wet conditions. This working step requires to change the tool first.

Conventional drilling tools are equipped with round shank chisel. They are positioned on the drilling tool under certain angles and cut a groove into the rock while the tool is continously rotated. Following teeth cut the remaining rock. The ripper tooth tends towards coming up from the groove, therefore it is essential to have enough crowd pressure (between 1 to 2 tons per tooth) on the tool. The usage of drilling rigs with high torque (500 kNm), the ability to push the tool under high crowd pressure against the bottom of the borehole (500 kNm) plus the development of very sustainable ripper teeth, make it possible to drill rock with up to 100 MPa unconfined compressive strength.

Figure 1. Core barrel with round shank chisel.

Figures 2 and 3. Creating annulus – Core recovery.

Figures 4 and 5. Progressive rock auger – Round chank chisel.

Figure 6. Cross cutter.

1.1 *Case study with conventional tools*

For a hydro-power station in Germany existing concrete walls (mostly without reinforcement), with a strength of up to 60 MPa and 15 m depth, had to be drilled out. The holes were filled up with gravel after drilling. 1500 mm diameter casings were used to drill secant holes down to 22 m fully cased. Prior to the start of the project questions regarding suitable tools for the "rock" drilling was raised. A major concern was how to make sure the casing can be screwed in down to the desired depth economically?

Due to the fact that the boreholes were wet, augers couldn't be used. A test with different tools showed the best drilling progress with acceptable wear and tear can be achieved using two tools (Cross Cutter and Rock drilling bucket). More difficult was to get the casing to final depth even though a casing oscillator was used. The time for oscillating the casings and the wear and tear of the cutting teeth on the casing shoe was inacceptable at the beginning. The advantage of an oscillator is it's high torque (four to five times higher compared to drilling rigs) but the weakness is, the casings are only oscillating and not rotating 360° continuously. Consequently the "cutting effect" is less and that leads to higher drilling times which are linked with more wear and tear on tooling.

On the way for looking for reducing wear and tear measures, a so called under reamer tool was tried out. This tool is basically a single cut rock auger where on the opposite side of the cutting edge, there is a foldable cutting wing. When lifting the tool this foldable wing is hanging downwards and reduces hereby the total tool diameter. As soon as the tool touches the surface, the wing extends gradually while the auger is rotating. When the wing is fully extended the total drilling

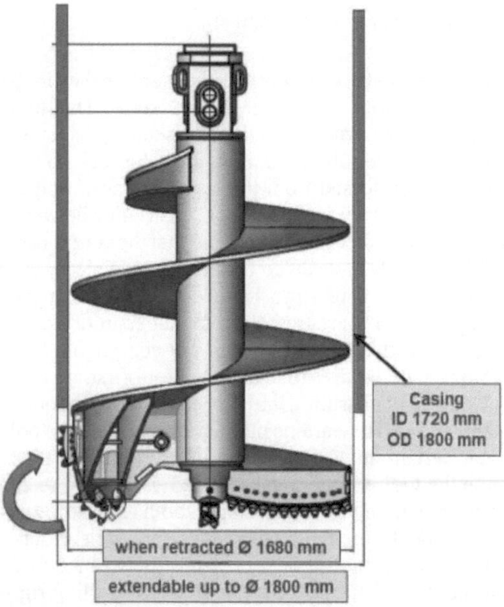

Figure 7. Under reamer tool.

diameter is then the same as the outer casing diameter. The auger is then cutting the rock below the casing, so there is much less effort necessary for the oscillator to advance the casings. This under reamer tool was the key to reduce the drilling times noticeable and consequently also the wear and tear.

2 TOOLS WITH ROLLER BITS AND AIR FLUSHING

For rock strengths >100 MPa tools are equipped with roller bits. These rollers fitted with carbide bits are rolled over the rock surface under very high crowd pressure. While drilling there is such a high punctual pressure that the rock is pulverized under the carbide bits so they can penetrate into the rock. Simultaneously, a surrounding stress condition occurs around the carbide bit which splits the rock laterally in chippings. Depending on the rock strength, the crowd pressure is between 3 to 5 tons per 100 mm roller bit width. This is one of the reasons why the roller bit tools are usually not bigger than 2000 mm in diameter. The crowd pressure on drilling rigs is limited unless drill collars are installed in the drill string. One of the difficulties when working with roller bit tools is the transportation of the chippings. If there is no transportation process the roller bits are cutting the same rock again and again until they turn into powder. In the event of water in the borehole the powder turns into a sticky mixture and jam the roller bits so they lose their cutting characteristic.

The solution is to use an air lift system in order to remove the chippings immediately from the cutting surface. That system usually requires special drill pipes. Due to the telescopic bars in the Kelly bar, this tool cannot be used for air flushing. There has been attempts manufacturing flushing telescopic Kelly bars but until now they were all in vain. Nevertheless, lots of energy has been put in developing a solution for how to get rid of the chippings from the bottom of the borehole not using special drill pipes.

Drill pipes would require a major modification and changes on the drill rig. The result of all these developments is the so called Down-the-hole Flushing head. This tool is located between a standard Kelly bar and the actual drilling tool. It allows flushing air down and avoids winding the air hose around the Kelly bar due to rotating and nonrotating parts. Three stabilizers extend when crowd is activated and they are braced against the borehole wall.

The air (<20 m^3/min at 10 bar) runs through the flushing head plus through the tool downwards and finally exits into one or two suction pipes. This inlet into the suction pipe is above the roller bits, so an air lift system is created which sucks up all the chippings and flushes them, together with the required fluid (either water or bentonite) into a calyx basket.

The actual drilling tool is either the core barrel or the latest development the Full Face Cutter (FFC). The FFC is similar to a tunnel boring machine. Big roller bits are arranged in different circular paths on a bottom plate and with a continuous rotation of the tool it is made sure that the complete surface is cut.

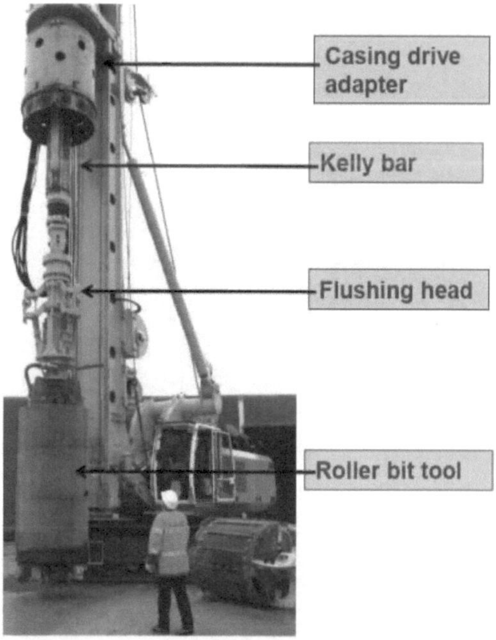

Figure 8. Complete set up of drill string.

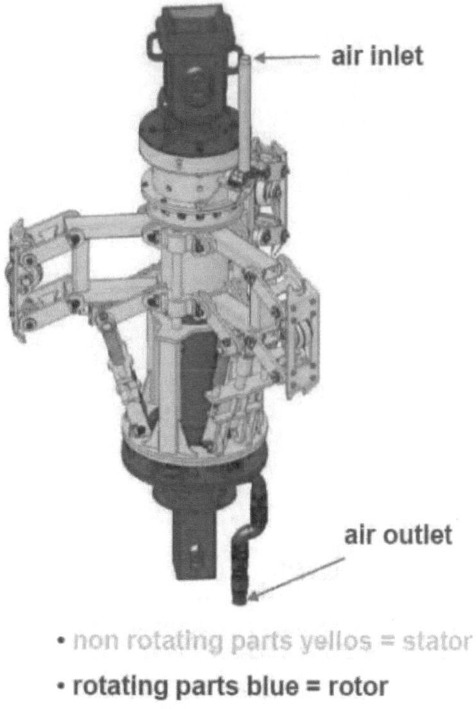

Figure 9. Down the hole flushing head.

Figure 10. Roller bit core barrel with 2 suction pipes.

Figure 11. Full Face Cutter with suction pipe.

The freshly cut chippings are sucked up by the air lift and get flushed through a suction pipe into the hollow space of the drilling tool. This procedure ensures not cutting the chippings twice and allows high drilling progress. The discharge is similar to conventional drilling buckets. A foldable bottom device ensures easy and quick discharge above ground level without additional cleaning activities.

One of the biggest strenghts of the Full Face Cutter is, there is no rock core left in the borehole therefore no exchange of drilling tool is required. Furthermore, due to the air lift the bottom of the borehole is always clean which makes the usage of special cleaing tools dispensable. Drilling diameters for roller bit tools start at 780 mm and go up to 2000 mm.

Figure 12. Site overview – Inclined foundation piles.

2.1 *Case Study: Full Face Cutter for bridge foundation in Finland*

In cooperation with SKANSKA Finland the Full Face Cutter was used for drilling rock sockets for foundation piles diameter 1500 mm in February 2010. The pile depth varied between 6 to 25 m whereas the rock socket was between 2 to 3 m. The rock strength was >150 MPa and the rock surface often occurred very slopy which made it difficult to screw the casing into the rock. In such cases the cross cutter was used in advance and levelled the sloping rock surface. The idea worked but the Cross Cutter pretty soon reached it's application limits due to the rock strength. As soon as the rock surface had been levelled the Full Face Cutter could be used. At the beginning the drilling performance was 30 cm/h which steadily increased after the crew gained experience with the new tool. Performance in rock increased to up to 1 m/h during the project. When the drilling was finished a underwater camera was lowered to inspect the bottom of the borepile. It could be clearly seen the bottom is free of chippings and the pile ready for concreting.

The most important factor when using roller bit tools is the crowd pressure. According to our experience it makes sense to cut rock up to strength of 200 MPa. Above that the drilling rig respectively the available crowd pressure is the limiting factor. For that reason it is recommended to go for hammer tools.

3 HAMMER TOOLS FOR ROCK AND BOULDERS

Single down the hole hammer are usually used up to a diameter of 600 mm in Europe. Single down the hole hammer are well known for anchor drilling or for steel core piles. In both cases the air runs through a drill pipe to the bit and flushes all the chippings upwards between drill pipe and borehole wall. Bigger diameters

Figure 13. Discharge of the Full Face Cutter.

Figure 14. Multi Hammer Drill (MHD).

haven't been used a lot due to the enormous amount of air which used to be required to run a bunch of hammers. This weakness has now been eliminated with a new type of low pressure hammer which require only 11.3 m^3/min at 10.3 bar per one hammer.

The basic principle for a hammer is a piston driven by air pressure which hits a bit at frequent intervals. The carbide bits inside the drilling bit crack the rock.

A further weakness used to be the occurrence of boulders embedded in soft material which are only half way in the bore hole. A hammer requires a rebound otherwise the bit cannot retract and all the air would exit at the "open" dropped out bit. The hammer which is now sitting on the boulder wouldn't get sufficient air because the air is lost somewhere else. Center Rock Inc. has developed and patented a valve which shuts off the air at the hammer with the dropped bit in order to supply sufficient air to the hammers which require air to drill the boulder.

Another issue was how to bring the air down to the hammers? The most economical way of drilling overburden is the Kelly drilling method. When it comes to drill boulders or rock we cannot expect the piling contractor to change the Kelly bar into drill pipe with all its related accessories.

All these issues as well as the worldwide increasing demand for large diameter rock sockets were the driving factor to develop a large diameter hammer tool.

Center Rock Inc. started using such tools in 2001 already and since then the technology has been improved steadily. First projects were carried out in the United States, followed by Canada, South America and Russia. In 2011 the tool was introduced for the first time in Germany and piling contractors from all over the world were pretty impressed not only by the drilling speed.

The MHD consists of multi single down the hole hammers (depending on the drilling diameter) which are united in a canister. The air is supplied with a flexible air hose and connected on top of the canister just under the Kelly box. The tool is simply oscillated back and forth from right hand to left hand rotation. Doing so and the fact that the weight on bit and also the rotation speed is very little, a flushing head is not required. This means the standard Kelly bar can still be used only the drilling tool is exchanged. The air runs the hammer, travels through the drilling bits and flush the chippings between canister and borehole wall upwards. The air stream velocity suddenly drops as soon as the annulus increases consequently the chippings fall back into a calyx basket which is open at the top. After reaching 0.60 m drilling progress the calyx basket is full and the tool is lifted above ground level for emptying the calyx.

The highest drilling performance is reached in case the borehole is dry and this is another big difference to the roller bit tools. In case heads of water greater than 12 m are encountered, high pressure air will be required. This typically will also require an increase in volume. In dry conditions dust can be reduced using a water pump which is plugged into the air line. Dust can be reduced but not eliminated.

Due to the fact that these hammer are working in a low pressure range much less air volume, compared to other down the hole hammer, is needed to run the tool. For instance a 1060 mm MHD loaded with 7 hammer requires an air volume between 60 to 80 m^3/min at 10 to 12 bar. This advantage makes it now possible to use a MHD even for large diameter. So called Full Face MHD range from 600 mm up to 1500 mm, above that hole openers are used after a pilot hole has been drilled. The biggest hole opener on site was a 3000 mm used for foundation piles in New York for the Trump Towers (project carried out by Cased Foundation).

When comparing mono hammers to the Multi Hammer Drill, the rotating direction of the MHD is not directional (can be rotated right hand and left hand), whereas the mono hammer can only be rotated right hand. Furthermore, with the MHD the rebound energy up the drill string is almost nothing which ensures no damages to the Kelly bar and rotary drive. Large mono hammers and bits can destroy rotary heads if the very cost intensive, high maintenance shock subs are not installed.

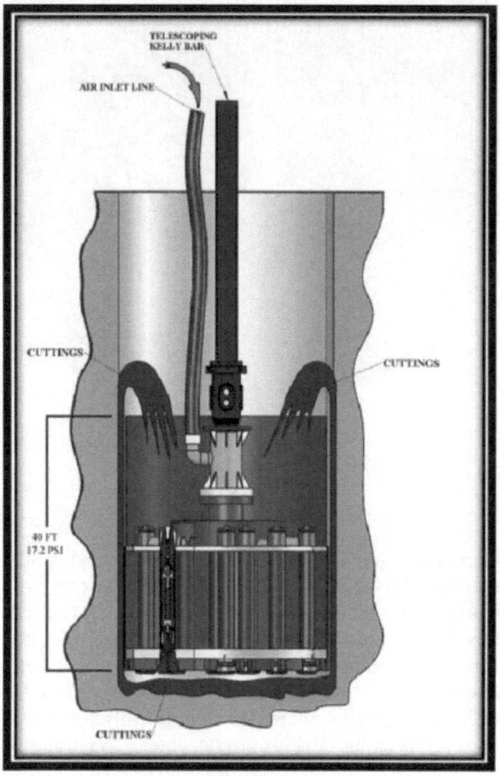

Figure 15. Working procedure of the MHD.

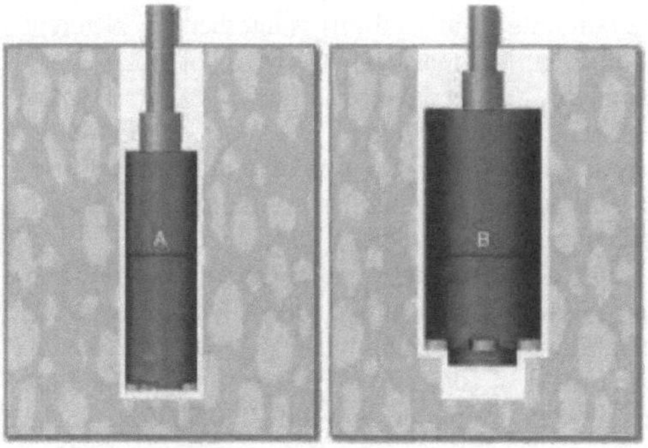

Figure 16. Working procedure of the MHD – Full Face MHD left, Hole Opener right.

3.1 *Case Study: Multi Hammer for drilling through boulder layers in Sochi*

In Sochi Russia, many piling contractors have been drilling since 2008 for the foundation works for the Olympic Winter Games 2014. Today there are still more than 60 drilling rigs working. Main part of the drilling project is to drill 1500 mm diameter piles in a former riverbed for an elevated railway line.

The subsoil consists of boulders different sizes and strength which are embedded in soft material. One challenge is to cope with boulders the other challenge is to bring the casing to final depth between 20 to 30 m. Conventional drilling tools (core barrel, rock augers etc.) in combination with powerful drilling rigs are mainly used but since 2009 Center Rock Inc. supplied one contractor with two Multi Hammer Drills. They found out the best drilling progress can be achieved with a combination of conventional tools for overburden and soft soil, the hammer tool is used for destroying boulders and doing the rock socket. Drilling buckets for rock achieve a performance of 3 to 4 m/h in the overburden. The penetration speed using the MHD in boulder and rock is between 1 to 2 m/h. In order to screw the casing in, powerful rigs with high torque are used and in addition casing oscillators.

Figure 17. BG 28 with Multi Hammer Drill in Sochi.

Figure 18. Lifting up the Calyx basket for dumping.

3.2 *Case Study: Multi Hammer for drilling under power lines in the US*

A project where the Multi Hammer Drill was the only chance to cope with the extremely hard rock is currently going on in Providence in the United States. The contractor has to drill piles under existing high

Figure 19. MHD for the Chilean mine rescue.

Figure 20. Tools for the Chilean mine rescue.

voltage power lines, so the usage of big machineries is not possible. The max. height is only 8 m and the piles are rather big in diameter, between 1.20 m and 2.70 m. Depths are shallow up to 6 or 8 m. They encounter rock right from ground level, which makes it very difficult to set a casing. The rock has a very high silica content, is very abrasive and the strength is extremely high between 200 to 220 MPa. First piles were drilled with conventional drilling tools (rock auger, core barrel) and one pile took between four to five weeks. The only chance to stay within the time schedule and to get these piles done economically was to change to the Multi Hammer tools different diameters. The MHD enables them to do one pile within 2 days.

3.3 *Case Study: Most famous hole for the Multi Hammer Drill in Chile*

The rescue drill for the trapped miners in Chile in 2010 is undoubtedly the most famous borehole for the MHD. 33 miners were trapped in 700 m depth and three teams tried with different tools and machines to drill down in order to lift each person up in a capsule through that bore hole. The geology was pure rock with high silica content all the way to the miners. Center Rock Inc. supplied their tools and onsite supervision with two drilling experts.

They started following a pilot hole with a single down the hole hammer diameter 300 mm. In the next step they opened up the hole to the final diameter using a MHD diameter 710 mm, loaded with four hammers. The team (Plan B) started off second, but were the first crew who entered the rescue room in 700 m depth.

The Chilean mine rescue points out there is equipment and knowledge available who can cope with challenges which some people might have thought this is impossible to do.

4 SUMMARY

The paper highlights, tools for hard rock have reached an highly sophisticated stage of development. Three different types of tools are introduced and case studies from all over the world back up their suitability. Basically it can be said, if production is the driving force, and rock hardness is above 100 MPa the Multi Hammer Drill will always drill faster. If silica content and or Quartz inclusions are prominet, the MHD bit life is reduced but it is then not only faster the bit cost per meter is generally less than cost of round shank chisels, pockets, building auger flight back to standard etc. In terms of energy the MHD requires the least amount of energy from the drilling rig compared to the methods listed, which results in longer lifetime of the rig's components.

One part of the overall success on a project is certainly to choose the right tools in advance, bearing in mind there is such a variety. The idea to combine the tools for hard rock with the standard Kelly bar makes it easy and appealing for the user to make a step into new "rocky" dimensions.

REFERENCES

Grundbau Taschenbuch "Bohrtechnik", Gordian and Georg Ulrich (2009) 7. Auflage Ernst & Sohn Verlag. Technical article about different drilling techniques and tools.

Various material and pictures from Center Rock Inc., www.centerrock.com

Baltic Piling – Mets & Raudsepp (Eds)
© 2013 Taylor & Francis Group, London, ISBN 978-0-415-64334-4

New technologies of pile apparatus and their test in ground conditions of Belarus

Mikhail Ivanovich Nikitenko, Igor Leonidovich Boiko & Babak Moradi Sani
Department of Geotechnical and Ecology in the construction of National Technical University, Belarus

Nikolai Vasilyevich Chernoshey
"Drilling Company" Delta, Belarus

ABSTRACT: Describes the essence of the new piles device technologies and the results of their tests in the ground conditions in Belarus. Obtained experimental values of pile load capacity are compared with the predicted settlement of the existing Belarus standards.

1 INTRODUCTION

Traditional solutions pile foundations based on the use of driven piles, which, because of the limited cross-sections provided for the transfer of design loads, without exceeding the strength of the material. However, recently it has tended to increase in the number of buildings stories and loads on the foundations that resulted in a process of intensive use of piles with an increase in their diameters and lengths, and therefore the strength of the material and the bearing capacity of soil. The pile load capacity predicted in the design calculations considering the tubular resistance of soils, or data of static and dynamic cone penetration tests, but the actual values are determined based on the results of static load tests. The essence of the new bored pile's technologies and devices used in this equipment, as well as the comparison of the experimental values of bearing capacity these foundations with predictable calculations are explicated low.

2 NEW PILE DEVICE'S TECHNOLOGY WITH HIGH LOAD CAPACITY

When the bored piles reached to high load capacity their bases generally tend to increase due to the lengths and diameters of their shafts.The shaft lengths designate for piling in the bearing stratum, although do not consider that the increase in load bearing of a homogeneous soil after the depth of 6–8 m, rapidly slowing down. For example, the average medium sand compression resistance under the tip of a pile at the depth 24 m by 1.5 times as compared with 5 m length. However, the soil is always non-uniform and in-depth the resistance of individual layers can be abruptly reduced. It's simple to increase the soil compression resistance by 1.5 times with increasing the diameter of the pile's base only by 1,25 times or by the conical shape of the shaft [4, 8].

May also increase the soil strength with ramming dry concrete mix under the tip [4, 5, 6, 7, 8] or the displacement of soil in other ways, for example, injection of mortar mix or concrete. Such methods are used in the new pile device's technologies, which appeared today in Geotechnical practice in different countries, including Belarus.

Injection technology [5, 8, 9] for many years used ByBelorussian Geotechnical Companies: UM "Minskmetrostroy", LLP *(osaühing)* "Anker", a joint-stock company "VIZBAS", ODO "Foundations", SCHUP "Spetsinzhstroy", LLP "OptioStroy", LLP "Ataviya", LLP "Spetsgeostroy" and others.

3 IMPLEMENTATION OF PILES BY USING VIBRATORS

The Effective were conical piles concreted in the wells when immersed in the clay soil (Fig. 1) Conforming to form [4, 7, 8, 9].

These piles are found at several objects. Their advantages are transferring the compression to the ground along the entire length of the shaft, inclusion the interaction of the upper compacted of the thrust embankment soil among the piles to the exclusion of negative friction, limiting the active strata below the piles, possible not to cut with short piles the deep weak organic layers.

Displace the soil with compaction allows for installation the end plugged pipe (tube). This prevents the seepage failure of ground under the base and the penetration of the mass of diluted into the pipes even under the water, providing concreting of the pipe as the lifting casing. In the sands by an installation pipe

a) b)

Figure 1. Dipping the steel plunger suspended a vibrator when creating wells for bored piles (a) And a general view of the excavated stamped heels (b).

a) b)

Figure 2. The device vibratory driving piles: a – driving in water-saturated soil casing with tip lost by using an ABI machine with a vibrator rail derrick, b – concreting of the pile shaft.

with an open-ended may be created compacted soil plug, which should keep its height, with eliminating its erosion and loss of strength at the tip.

Technology devices bored piles with vibratory driving casing pipes (Fig. 2) In Belarus successfully applied specialized companies by using domestic and foreign equipment. In addition, companies from Lithuania and Estonia are using this technology in our country.

For dignity of this technology should be referred to increase the speed of pile installation and its bearing capacity in any geological conditions. But its major defect is the adverse dynamic effects on surrounding buildings and underground communications, is too dangerous in flooded soils with long-range spreading due to resonance in the process of starting and stopping the mechanism.This was manifested as deformation at the store "Romashka" when constructing near it pile for plate-pile foundation under the height-rise building [9]. However, here with vibratory driving casing pipe with Ø630 mm with an open-ended and punching the top of earth fill and organic soil, due to displacement of the soil plug in the inner pipe with Ø 530 mm (Fig. 3) It has provided the first in Belarus with the lengths of shafts bored piles by 8 m and for water carrier sand are getting resistance over 2800 kN.

To mitigate the effects of dynamic water carrier ground by the first author of this article suggested to pump the air through the pipe from the outside of casing with dipping in the power saving mode of

Figure 3. The device bored piles in the left – driving bailer for excavation of casing of weak soils above the compacted soil plug, on the right replacement plug with a blind end.

Figure 4. The immersion tube suspended without resonant vibrator.

vibration. Also greater affect ensures the application of purchasing from Western manufacturers of so-called "nonresonance" vibrators (Fig. 4).

Using **the technology of CFA** (Fig. 5). In the performance of bored piles with Italian equipment (Fig. 6) in all grounds, even water carrier soils, wells are drilled with a continuous flight auger, through which, with the lifting into the well is pumped under pressure from the concrete, and in it immediately reinforcement cage is immersed to the required depth. An important advantage of this technology is high-speed pile devices, and compressing of soil under pressure along the shaft helps eliminate sliming and to achieve a higher bearing capacity of foundation.

The speed of recovery of the auger and injection the concrete or grouted-aggregate concrete into the bore hole, the average cross sectional area of pile and it's broadening along the length is determined by using the on-board computer. With the help of this equipment now possible to devise CFA piles with a diameter between 0.3 m and 1.0 m at length to 24 m.

3.1 *CSP technology*

(Cased Secant Piles – Secant piles with casing) provides for using the drilling rig with two rotators for piles, including pile walls.This technology has developed special drilling rigs with two dipping rotators the casing pipe and within the continuous Auger (Fig. 7), which allows to do without bentonite slurry in any soils, including inundated.CSP technology allows to create secant piles with deviation from the vertical to 1.0–1.5%. With the help of existing equipment, "Drilling Company" Delta can perform a pile diameters: 660 mm, 820 mm and 1020 mm in length to 21.5 m. A large load-bearing capacity of these pile foundations help reduce the necessary number of piles and working time.

3.2 *Bored piles, which are made without excavation*

The formation of bore holes in porous low moisture loess soils due to displacement of soil through swing was suggested in the Soviet Union by V.I. Feklin. The German company "Bauer" in the development of this technology has developed a set of equipment for rolling and concreting bore holes in weak saturated soils. Thus, the technology of the system, "Bauer" includes the performance of wells for piles without removing the soil due to its compaction of working executive device as a cone with a helical vane at the end of the drill pipe (Fig. 8).

In the dipping process of the working body of the soil is displaced to the side and around the well compacted zone is formed, the size which depends on soil properties, the construction of a working body and the amount of injected concrete.

Due to pressing the soil around the bore hole bearing capacity of piles is greatly increased because the friction along the shaft is 30% more than the boring, and compressive strength at the tip by 50 ... 70%.

The set of equipment of the Italian company Soilmec includes the raskatchiki(injector) diameter 450 mm, 650 mm, 800 mm (Fig. 9) on the drilling machine SR-65.

The piles with large bearing capacity can be performed using inject technology (jet-grouting) and without digging in its erosion and mixing with the injected under pressure of 40–60 MPa cement grout to the extent lifting of the monitor. Partial excavation occurs only at drifting the leader wells drill head on the end of the hollow monitor rods with a jet at the

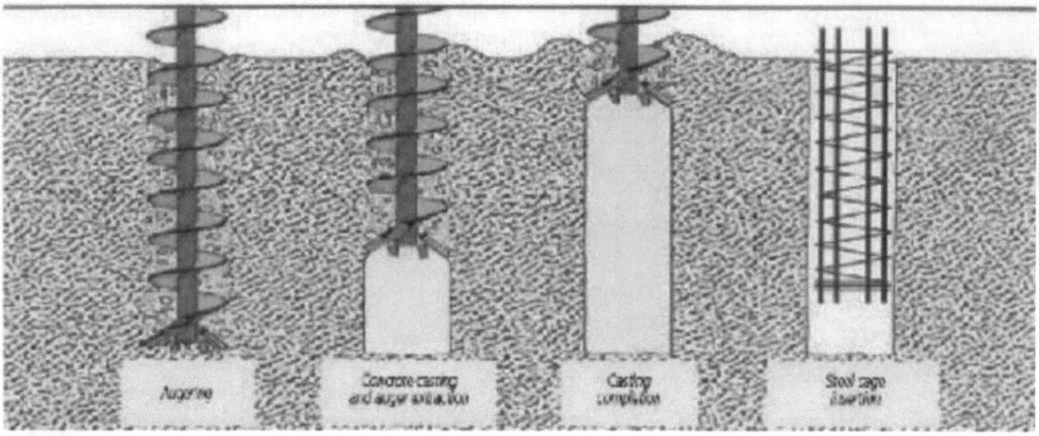

Figure 5. The device for bored piles with a complete continual auger.

Figure 6. Drilling vehicle to perform pile the system **CSP**.

bottom to upload the injection mixture. The diameter and shape of the pile depends on the injection pressure, injection volume of the mixture, the lifting and the rotational speed of the monitor, which is automatically controlled by the onboard computer. Work on the inject technology in Belarus performed with the use of foreign equipment sets (Fig. 9).

The piles with large bearing capacity can be performed using inject technology (jet-grouting) and without digging in its erosion and mixing with the injected under pressure of 40–60 MPa cement grout to the extent lifting of the monitor. Partial excavation occurs only at drifting the leader wells drill head on the end of the hollow monitor rods with a jet at the

Figure 7. Drilling vehicle to perform pile the system **CSP**.

bottom to upload the injection mixture. The diameter and shape of the pile depends on the injection pressure, injection volume of the mixture, the lifting and the rotational speed of the monitor, which is automatically controlled by the onboard computer. Work on the inject technology in Belarus performed with the use of foreign equipment sets (Fig. 10 and 11).

It is important to remember that all the new Geotechnical technology requires a thorough their observance and each of them has its own rationally area of use, but at the appropriate equipment needed to work trained cadres Otherwise it is possible ruin the expensive equipment and the technology will be ineffective.

3.3 *Normative basis about the structure or the design of bored piles*

In Belarus, there is a normative basis for the design and use of bored piles with molding surrounding soil due to its displacement to the side when rolling or when pumping concrete under pressure. For piles with the displacement of soil by valid orders [2, 4, 6]. The use of injecting technology is supported [4].

Figure 8. Injector for bore holes the "Bauer" system.

3.4 *Scheme of static loading tests of the piles*

The static loading tests of piles they are usually created with the help of jacks, disposable under anchor persistent structures (Fig. 12, a–c).

In the beam scheme of testing for fixing such structures often use adjacent drawing pile, boring pile, or injection anchors spaced at a distance from the test piles to avoid interference. Drawing piles held by friction along the shafts, so they need more. The preferred bored and bored injection anchors or club-footed piles.

At large spans resistant beams should be enlarged cross-sectional, making it difficult and expensive test system. The use of inclined boring anchors or bored injection anchors simplify the design of the support, but it also complicates the testing.

When using the loading platform (Fig. 13) requires a lot of loads, but unloading surface around piles compresses the shaft and increases the resistance of the soil shear strength, overstating the results.

In testing piles of high bearing capacity the traditional schemes create serious problems with the transfer to the piles the required pressing load. In our publication [9] called attention to the necessity of strict compliance with the requirements of the standard [8] in terms of bringing the testing loads to the pile settlement at least 40 mm. Noncompliance to do this requirement does not give objective information about the actual bearing capacity of pile foundations, contributing to over consumption of material and labor resources, but does not guarantee the required degree of reliability

Figure 9. Firm Soilmec injector.

Figure 10. Pile for end support chair lift slopes snowboarding in Silichi made by inject technology (in the background can be seen the machine company "Kasagrande" to perform pile).

Figure 11. The equipment of Italian company Soilmec for inject technology.

of pile foundations. The situation is aggravated by a small number of pile load test according to the prescriptions rules, in contrast to the full amount of test bored injection anchors.

To simplify pile load testing which with large bearing capacity, will allow offered and reflected in [4] scheme (see Fig. 12d). It is thanks to the use of two coaxially placed in the body of pipe piles can separately transfer a compressive force the inner of the pipe through the tip, and shear force along the shaft rendered monolithic or screwed outside. In this case, requires a smaller jack's load carrying capacity without using the insistent system or loading platform. At an insufficient shear strength along the shaft is required only partially of his unloading. To reduce the time of bored cast-in-place piles test have suggested steel pipes, and the outside casing acts as the shift and the inner plugged ends – the tip of the piles. Separate fixing shares of soil resistance under the tip and along the shafts increase the informativeness of the obtained results, especially in the presence of negative friction.

In the pile – slab foundations offered to test pile with emphasis in the base plate held by adjacent piles. To determine the proportion of soil shear strength resistance, including the negative along the shaft, you can test pulling out pile after the loading.

Estimate the proportion of soil resistance to compression under the slab foundation or bearing pile cap can through testing fragments, falling on individual piles and autonomously produced before performing respectively slabs or pile cap. In that case the fragments can be connected to slab or a pile cap after the test.

For testing piles of high bearing capacities for indentation force to 5000 kN OAO "Drilling Company" Delta has produced a special stand (Fig. 14) as retained helical anchors with crossbeam system to stop loading jacks.

Piles erected using the CFA technology were tested at site in Minsk while building the headquarter of National Olympic committee of Republic of Belarus.

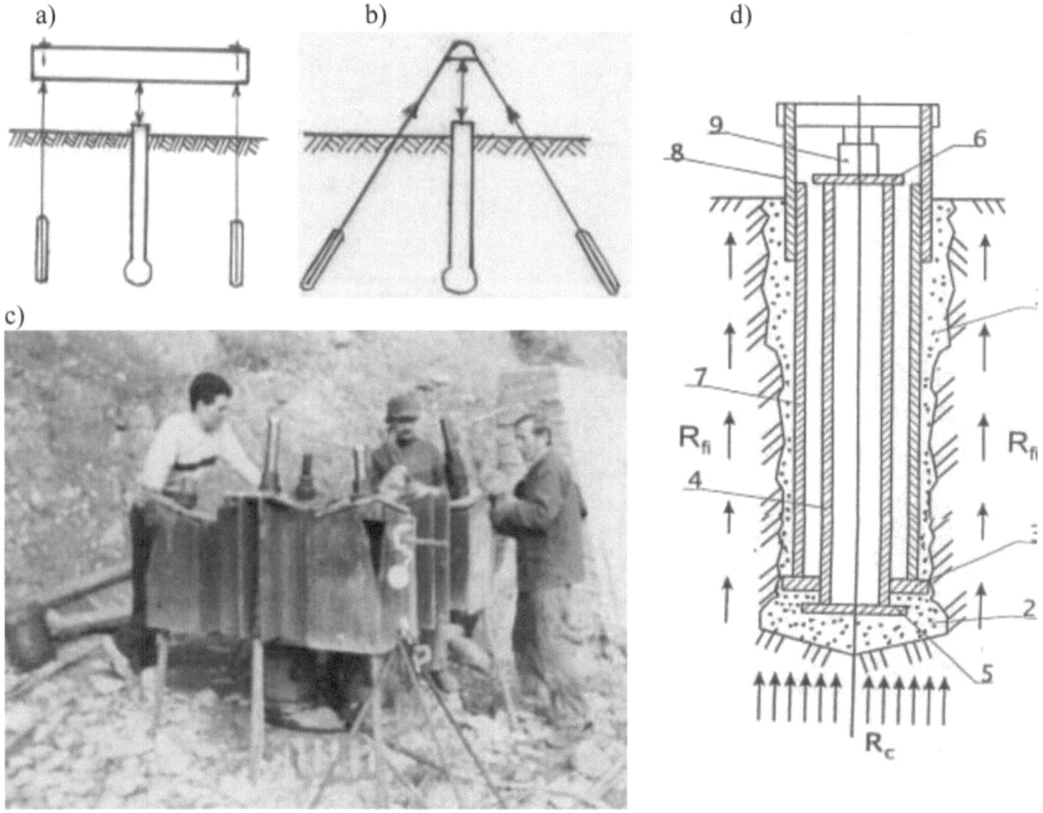

Figure 12. Schemes test pile for indentation: a – a persistent system of beam formed, b, c – the same Beam less; 1 – tampon cage at the top of the shaft, and 2 – foot; 3 – To be easily compressible elastic strip for separating feet and the shaft of the piles, 4 – the internal pipe to transmit pressed force to the heel, 5, 6 – persistent plate; 7 – Outer pipe with filling between it and the walls of the well to transfer shear compressing on the top barrel, 8 – draft stops, 9 – hydraulic jack.

Figure 13. Platform for testing piles with loads in the form of concrete slabs and blocks.

Figure 14. The loading test stands for pile by pressing loads to 5000 kN.

Pressing test loads of piles were led up from 1800 kN to 3000 kN (Fig. 15) due to the soil properties variability (Tab. 1), ground depth position (Fig. 15) and pile variable lengths character (length was about 7.5: 8.5 and 9.5 meters, diameter 0.7 and 0.9 meters). Thus values of pile bearing capacity according to SNB 5.01.01-99 were taken from 1606 kN to 2844 books at norm settlements of 24 mm.

Settlements schedules, given on Fig. 16 have a wide spacing, but at relative sizes in uniform system of coordinates (Fig. 17) the given schedules are grouped closely. Thus skilled ratios are accepted for the limited values of the settlements about 24 mm (the great number reached settlements were in the range from 28.32 mm to 52.14 mm). These values didn't affect nature of schedules curvature practically. This schedules property of in relative sizes allows to predict bearing capacity of the bases of piles using the

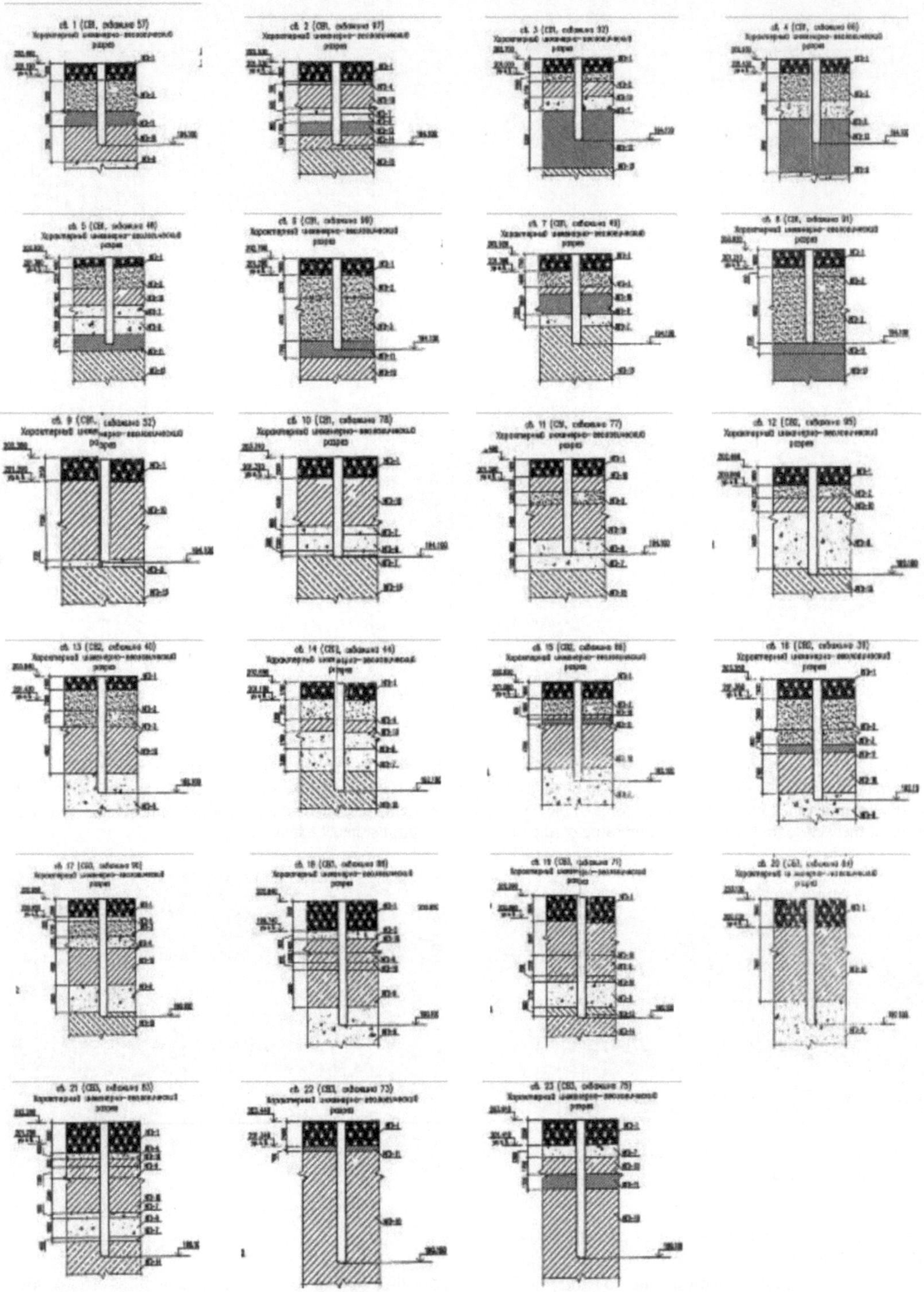

Figure 15. Site geological conditions at the place of pile erecting.

nonlinear extrapolation in case of achievement at site pile tests loading with the values less according to requirements of GOST 56-86-94 and even acceptable according to SNB 5.01.01-99.

The design pile bearing capacity of the bases is forecasted by the different ways [2, 3, 4, 5], and the reliability of such forecast is estimated by comparison the results with the dynamic or static tests sounding [8].

Table 1. Site soil characteristics.

No. gr.	Ground name	γ, kN/m^3	c_{II}, kPa	c_I, kPa	φ_{II}, degree	φ_I, degree	E, MPa
1	Bulk soil	18,7/9,3	–	–	–	–	–
2	Dusty average durability sand	17,6/10,3	4,0	3,0	30	27	18
3	Dusty strong sand	17,7/10,4	6,0	4,0	34	31	36
4	Small average durability sand	17,8/10,3	2,0	1,0	32	29	25
5	Small strong sand	–/10,3	4,0	3,0	36	33	38
7	Coarse, gravel average durability sand	–/10,6	1,0	0,7	39	35	25
8	Coarse, gravel strong sand	–/10,6	1,0	0,7	41	37	41
9	Dusty weak loam	20,1	19	14	18	17	4,6
10	Dusty average durability loam	20,2	30	27	21	20	15
11	Dusty strong loam	20,4	39	35	21	20	21
12	Dusty very strong loam	20,4	42	36	21	19	27
13	Sandy average durability loam	21,9	45	42	27	26	12
14	Супесь прочная Sandy loam strong	21,9	44	39	27	26	22
15	Супесь очень прочная Sandy loam very strong	21,5	48	45	31	30	26

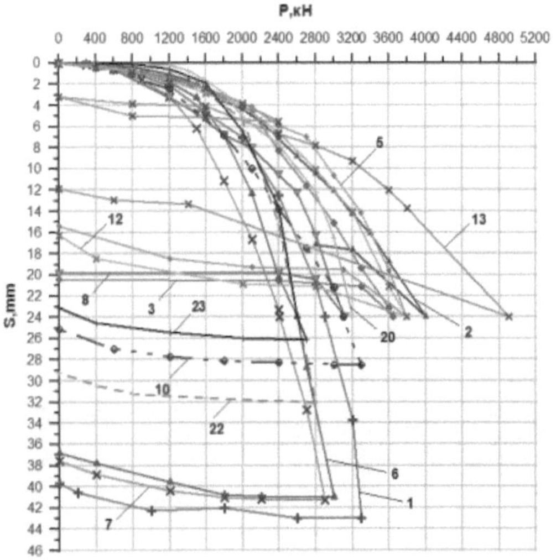

Figure 16. The settlement dependence schedules from test loadings for piles of No. No. 1, 2, 3, 5, 6, 7, 8, 10, 12, 13, 20, 22, 23.

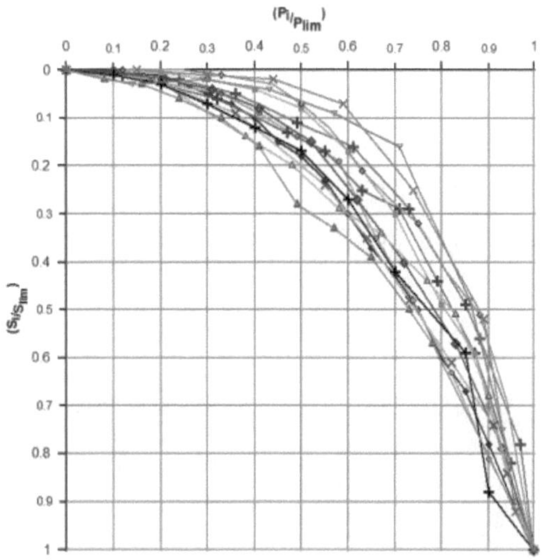

Figure 17. The dependences schedules of relative values S_i/S_{lim} depended from P_i/P_{lim} for the piles No. No. 1, 2, 3, 5, 6, 7, 8, 10, 12, 13, 20, 22, 23.

Soil resistance values along lateral surfaces and under the bottom ends of piles by the results of their static and dynamic sounding tests, the calculated pile bearing capacity using sounding [2], and also by settlement resistance agrees [4] are given in Table 2.

Ratios of the calculated resistance and the piles site test are given in Table 3.

4 CONCLUSION

1. New technologies allow to erect piles in any soil conditions and achieve great values of bearing capacity of their bases/This is especially important for the high-rise construction and loadings increase on the bases.
2. The proposed schemes allow to test piles of the great bearing capacity of the bases, including grillage or base plates bearing capacity in work. Without beam persistent systems with separate fixing of shares of resistance of soil to shift and base compression are preferable. That is especially important in the presence of a negative friction along trunks of piles.
3. Predicted design pile bearing capacity can be underestimated or overestimated in comparison with the actual values in real soil conditions. Therefore further specification of available correlation dependences for different types and properties of soil taking into account transformation at the expense of technology of the device of piles is required.

Table 2. Soil resistance values in kN on lateral surfaces, under the bottom ends of piles and total, defined by different methods

No. pile[1]	L, m	Designations of resistance	Resistance values in kN, certain			
			Static in site tests	by sounding[2] [2]		by settlement resistance [4]
				static	dynamic	
1	2	3	4	5	6	7
1	8,7	$F = F_b + F_c$	2900 = 1800 + 1100	F_c = 2405 (1360)	–	1623 = 462 + 1161
		$N = F/\gamma$	2417	1925 (1090)	–	1160
2	8,4	$F = F_b + F_c$	3600 = 2200 + 1400	F_c = 2400 (1930)	–	2433 = 811 + 1622
		$N = F/\gamma$	3000		–	1740
3	8,6	$F = F_b + F_c$	3600 = 2200 + 1400	2685 = 925 + 1760 (1990 = 900 + 1090)	–	1875 = 781 + 1094
		$N = F/\gamma$	3000 (3042)	2150 (1590)	–	1340
4	8,7	$F = F_b + F_c$	3767 = 2100 + 1667	F_c = 1685 (1000)	2815 = 860 + 1955 (2466 + 803 + 1663)	1799 = 692 + 1107
		$N = F/\gamma$	3056	1350 (790)	2165 (1895)	1285
6	8,7	$F = F_b + F_c$	2600 = 1600 + 1000	2635 = 1060 + 1580 (1905 = 1010 + 895)	-	2060 = 707 + 1353
		$N = F/\gamma$	2167	2195 (1525)	–	1470
7	8,7	$F = F_b + F_c$	2405 = 1400 + 1005	F_c = 4695 (2914)	3180 = 1255 + 1920 (2780 = 1160 + 1620)	2560 = 827 + 1733
		$N = F/\gamma$	2004	3755 (2330)	2445 (2140)	1610
8	8,9	$F = F_b + F_c$	3000 = 1500 + 1500	3615 = 1160 + 2455 (2660 = 1210 + 1450)	–	1891 = 809 + 1082
		$N = F/\gamma$	2500	2890 (2130)	–	1350
9	9,2	$F = F_b + F_c$	2200 = 1400 + 800	F_c = 2525 (1890)	–	4593 = 644 + 3949
		$N = F/\gamma$	1833	2020 (1510)	–	3280
10	9,2	$F = F_b + F_c$	3100 = 1600 + 1500	–	3000 = 900 + 2100 (2575 = 795 + 1780)	–
		$N = F/\gamma$	2583	–	2310 (1980)	–
11	9,2	$F = F_b + F_c$	3067 = 1600 + 1467	4855 = 890 + 3965 (2900 = 815 + 2085)	4490 = 1345 + 3145 (4030 = 1210 + 2820)	4739 = 768 + 3971
		$N = F/\gamma$	2556	3885 (2320)	3455 (3100)	3385
13	11,0	$F = F_b + F_c$	4900 = 2400 + 2500	F_c = 5255 (2860)	–	5611 = 1009 + 3971
		$N = F/\gamma$	4085	4205 (2290)	–	4008
14	10,6	$F = F_b + F_c$	2700 = 2000 + 700	F_c = 5255 (2860)	–	3316 = 1344 + 1972
		$N = F/\gamma$	2250	4205 (2290)	–	2370
15	10,6	$F = F_b + F_c$	3600 = 2100 + 1500	4445 = 945 + 3500 (3930 = 895 + 3035)	–	5409 = 944 + 4465
		$N = F/\gamma$	3000 (3500)	3555 (3143)	–	3860
16	10,6	$F = F_b + F_c$	3300 = 2100 + 1200	F_c = 1915 (1215)	–	2170 = 834 + 1336
		$N = F/\gamma$	3750 (3167)	1530 (970)	–	1550
17	12,9	$F = F_b + F_c$	3600 = 2100 + 1500	4070 = 1495 + 2575 (2860 = 1310 + 1550)	–	3147 = 1429 + 1718
		$N = F/\gamma$	3000 (3167)	3260 (2290)	–	2250
19	12,9	$F = F_b + F_c$	2500 =	3040 = 1390 + 1650 (2220 = 1260 + 960)	–	3250 = 1278 + 1972
		$N = F/\gamma$	2083	2430 (1780)	–	2320
20	13,0	$F = F_b + F_c$	3000 = 2000 + 1000	5320 = 1400 + 3920 (3180 = 1355 + 1825)	–	6058 = 1064 + 4994
		$N = F/\gamma$	2500 (2667)	4225 (2545)	–	4325
18, 21	13,0	$F = F_b + F_c$	2500 = 1200 + 1300	2960 = 1360 + 1600 (2150 = 1250 + 905)	4835 = 1360 + 3475 (4395 = 1235 + 3160)	3511 = 1348 + 2163
		$N = F/\gamma$	2083	2365 (1720)	3720 (3380)	2510
22	13,4	$F = F_b + F_c$	2600 = 2000 + 600	2635 = 1295 + 1340 (1975 = 1260 + 715)	–	2513 = 1012 + 1501
		$N = F/\gamma$	2167	2110 (1580)	–	1795
23	13,6	$F = F_b + F_c$	2600 = 1600 + 1000	2730 = 1430 + 1300 (2100 = 1430 + 670)	–	2437 = 1101 + 1336
		$N = F/\gamma$	2167	2180 (1680)	–	1740

Table 3. The ratios of the calculated resistance and the piles site test.

No. pile	Designations	Ratios of resistance, determined by: by sounding[2] [2] static	by sounding[2] [2] dynamic	conculations [4]	No. pile	Designations	Ratios of resistance, determined by: by sounding[2] [2] static	by sounding[2] [2] dynamic	conculations [4]
1	F	–	–	0,56	13	F	–	–	**1,15**
	F_b	–	–	0,26		F_b	–	–	0,42
	F_c	**2,19(1,24)**	–	1,06		F_c	**2,1 (1,14)**	–	**1,59**
	$N=F/\gamma$	0,8 (0,45)	–	0,48		$N=F/\gamma$	**1,03** (0,56)	–	0,98
2	F	–	–	0,68	14	F	–	–	**1,23**
	F_b	–	–	0,37		F_b	–	–	0,67
	F_c	**1,71(1,38)**	–	**1,16**		F_c	**7,51 (4,09)**	–	**2,82**
	$N=F/\gamma$	0,64(0,51)	–	0,58		$N=F/\gamma$	**1,87 (1,02)**	–	**1,05**
3	F	0,75(0,55)	–	0,52	15	F	**1,23 (1,04)**	–	**1,5**
	F_b	0,42(0,41)	–	0,36		F_b	0,45 (0,45)	–	0,45
	F_c	**1,26**(0,78)	–	0,78		F_c	**2,33 (1,91)**	–	**2,98**
	$N=F/\gamma$	0,72 (0,53)	–	0,45		$N=F/\gamma$	**1,19** (1,0)	–	**1,29**
4	F	–	0,75(0,65)	0,48	16	F	–	–	0,66
	F_b	–	0,41(0,38)	0,33		F_b	–	–	0,40
	F_c	1,01 (0,60)	**1,17**(1,0)	0,66		F_c	**1,6 (1,1)**	–	**1,11**
	$N=F/\gamma$	0,44 (0,26)	0,7 (0,62)	0,42		$N=F/\gamma$	0,41 (0,26)	–	0,41
6	F	1,01 (0,62)	–	0,79	17	F	**1,22** (0,89)	–	0,87
	F_b	0,66 (0,63)	–	0,44		F_b	0,87 (0,79)	–	0,68
	F_c	**1,5** (0,9)	–	**1,35**		F_c	**1,72 (1,03)**	–	**1,15**
	$N=F/\gamma$	1,01 (0,7)	–	0,68		$N=F/\gamma$	**1,17** (0,85)	–	0,75
7	F	–	**1,32 (1,16)**	**1,06**	19	F	**1,34** (1,01)	–	**2,50**
	F_b	–	0,9 (0,83)	0,59		F_b	–		-
	F_c	**4,67 (2,9)**	**1,91 (1,61)**	**1,72**		F_c	–		-
	$N=F/\gamma$	**1,87 (1,16)**	**1,22 (1,07)**	0,80		$N=F/\gamma$	**1,28** (0,97)	–	**1,11**
8	F	**1,3** (0,87)	–	0,63	20	F	**1,77 (1,06)**		**2,02**
	F_b	0,96 (0,81)	–	0,54		F_b	0,7 (0,68)		0,53
	F_c	**1,64** (0,97)	–	0,72		F_c	**3,92 (1,83)**		**5,0**
	$N=F/\gamma$	**1,5** (0,85)	–	0,54		$N=F/\gamma$	**1,69 (1,02)**		**1,73**
9	F	–	–	**2,09**	18, 21	F	**1,28** (0,97)	**1,93 (1,76)**	**1,40**
	F_b	–	–	0,46		F_b	**1,34 (1,27)**	**1,13 (1,03)**	**1,12**
	F_c	**3,16 (2,36)**	–	**4,94**		F_c	**1,23** (0,7)	**2,67 (2,43)**	**1,66**
	$N=F/\gamma$	**1,1** (0,82)	–	**1,79**		$N=F/\gamma$	**1,23** (0,93)	**1,79 (1,62)**	**1,20**
10	F	–	0,97 (0,83)	–	22	F	**1,08** (0,84)	–	0,97
	F_b	–	0,56 (0,5)	–		F_b	0,74 (0,74)	–	0,51
	F_c	–	**1,4 (1,19)**	–		F_c	**2,23 (1,19)**	–	**2,50**
	$N=F/\gamma$	–	0,89 (0,77)	–		$N=F/\gamma$	**1,04** (0,81)	–	0,83
11	F	**1,49** (0,95)	**1,46 (1,31)**	**1,55**	23	F	0,99 (0,74)	–	0,94
	F_b	0,56 (0,51)	0,84 (0,76)	0,48		F_b	0,8 (0,78)	–	0,69
	F_c	**2,7 (1,42)**	**2,14 (1,92)**	**2,71**		F_c	**1,3** (0,67)	–	**1,34**
	$N=F/\gamma$	**1,52** (0,91)	**1,35 (1,21)**	**1,32**		$N=F/\gamma$	0,95 (0,71)	–	0,80

Notes to tables No. 2 and No. 3:
[1]Piles No. 1, 5–7, 9–14, 18, 19, 21 were tested by UE "Stroikompleks", No. 2–4, 8, 15–17, 20 – JSC "Delta Drilling Company" (all the piles had diameters equal 0.9 m);
[2]Values given the brackets are calculated in the absence of a soil pressing, and without brackets – with pressing existence as for probe piles;
[3]The overestimated values of resistance are printed bold.

4. Static in site pile tests are the main and serve for the subsequent updating of the project with specification of their quantity, the sizes and erecting technology.

REFERENCES

[1] SNB 5.01.01-99. Bases and foundstions of buildings and constructions. – Minstroyarkhitektury RB. – Minsk, 1999. – 36 pages.

[2] Posobije P2-2000 to SNB 5.01.01-99. Design of probed and stuffed piles by results of sounding of soil. – Minstroyarkhitektura RB. – Minsk, 2001. – 23 pages.

[3] Posobije П 4-2000 to SNB 5.01.01-99. Design of probe piles. – Minstroyarkhitektura RB. – Minsk, 2001. – 68 pages.

[4] Posobije П 13-01 to SNB 5.01.01-99. Design and device of drilled piles. – Minstroyarkhitektura RB. – Minsk, 2002. – 43 pages.

[5] Posobije П 18-04 to SNB 5.01.01-99. Design and device of drilinjected anchors and piles. – Minstroyarkhitektura RB. – Minsk, 2005. – 79 pages.

[6] Posobije П 19-04 to SNB 5.01.01-99. Design and the device of the bases from piles with the condensed basis. – Minstroyarkhitektura RB. – Minsk, 2006. – 88 pages.
[7] TKP 45-5.01-45-2006 (02250). The bases and the underground constructions erected with use of jet technology. Design and device rules. – Minstroyarhitektura RB. – Minsk, 2006. – 33 pages.
[8] Nikitenko, M.I. Drillinjected anchors and piles at construction and reconstruction of buildings and constructions/M. I. Nikitenko. – Minsk: BNTU, 2007. – 580 pages.
[9] Nikitenko, M. I. Some problems of the pile bases in geotechnical practice of Belarus/M. I. Nikitenko, V.Yu. Zhuravsky//Construction science and equipment. – Minsk, 2008. – No. 4 (19). – Page 44–51.

Baltic Piling – Mets & Raudsepp (Eds)
© 2013 Taylor & Francis Group, London, ISBN 978-0-415-64334-4

Defining boundaries of efficiency for pile-raft foundations

V.M. Ulitsky
State Transport University, Saint Petersburg, Russia

V.A. Shashkin
Georeconstruction Engineering Co, Saint Petersburg, Russia

ABSTRACT: The paper presents results of numerical modeling of high and low raft piled foundations, featuring an elasto-plastic subsoil model with independent strain hardening. The analysis is based on modeling foundation loading up to a limit state, realized by means of FEM-models software. Likewise, the paper evaluates the extent to which low rafts contribute to bearing capacity of piled foundations. Additionally, the paper studies safety factors prescribed for piled foundation design by both Russian and international codes.

1 INTRODUCTION

Over the past decade researchers in and outside Russia have shown a heightened degree of interest to piled foundations with a low raft, traditionally referred to as "the pile-raft foundation". There have been a number of works focusing on bearing capacity of such foundations, e.g. those by R. Katzenbach, H. Puolos, V.G. Fedorovsky and a good number of other leading geotechnical specialists [1, 3, 4].

There exist several fundamental approaches to design of pile-raft foundations, differing in safety factors applied to a single pile, a pile group or a foundation as a whole. All such approaches presuppose the entire load to be assumed by the piles. Their difference, however, lies in the degree of the piles' bearing capacity to be reflected as the safety factor. There is a veritable spectrum of approaches to calculations and analyses – those based on elementary analytical methods featuring a large number of conditional limitations and simplifications, and those based on numerical codes. The fundamental problem compromising correct use of the latter is the well-known difficulty of having to unite both pile and raft foundation types within a single calculation profile, each type having a significantly different limitation of compressible stratum thickness.

Consequently, the present paper has the objective to define the conditions in which a piled foundation with a low raft turns into a pile-raft foundation, or the conditions in which the raft of a pile foundation directly engages with the soil. Differently put, we intend to define the boundaries beyond which this structure, depending on what kind of interaction with subsoil it has, degrades either into a raft foundation on natural subsoil, or into a classical piled foundation. Thereat our numerical experiment is based on a soil model unrelated to artificially imposed limitations of the compressible stratum.

2 DEFINING THE TASKS FOR NUMERICAL MODELLING

For the needs of the present research pile-raft foundations were analyzed by means of *FEM models* software [2]. The elasto-plastic model with independent strain hardening, featured in the software, enables one to localize in a natural way the active strain development zone in subsoil, being devoid of artificial limitations [2]. Collating the results with *in situ* observations shows the model as capable of higher accuracy settlement predictions compared with the known engineering methods. The model was also verified, in particular, for being true to pile load tests and plate tests results [2]. The soil model in question allows localization of the active zone in subsoil to happen naturally and automatically in light of intensive decay of volumetric and shear strains with depth reflecting, respectively, lowering of gradient and deviatoric stress; this removes the necessity to impose artificial limitations on the active zone. Employing this model to represent a pile-raft foundation seems reasonable, as it allows a soft transition from a raft foundation settlement, with its companion strain zone, to that of a piled foundation, with its strain zone being located underneath pile toes.

Numerical modelling of various foundation types was performed for the same ground conditions in which, on the one hand, a shallow raft foundation was acceptable, and on the other hand, a piled or a pile-raft foundation could be seen as equally effective. Based on the most recent edition of the Russian national code SNiP "Piled foundations" (ref. СП 24.13330.2011), a pile-raft foundation can be considered as eligible in such cases when there is a subsoil comprising medium to high density sands or clays with yield number $I_L < 0.5$. That is to say, clays should have stiff, medium-stiff or low-stiff consistency. Another obvious condition for selecting a soil type for the modelling was a double-layer subsoil with the upper layer of a higher compressibility and the bottom layer of a lower compressibility; this choice was made to demonstrate efficiency of a pile-raft foundation in a more lucid manner.

For the modelling needs, the upper layer was taken as being 19 m thick and featuring the following properties: strain modulus E = 10000 kPa, weight of soil $\gamma = 20\,\text{kN/m}^2$, unit cohesion c = 100 kPa, angle of internal friction $\varphi = 15°$; the properties of the bottom layer were as follows: strain modulus E = 20000 kPa, weight of soil $\gamma = 20\,\text{kN/m}^2$, unit cohesion c = 200 kPa, angle of internal friction $\varphi = 15°$. In nature, double layer soils with similar properties are prevalent in such cities as Frankfurt, London or southern districts of Saint Petersburg. A certain degree of generalization notwithstanding, this example can still be seen as typical for pile-raft foundations being prescribed in the currently applicable codes.

In light of the instructions contained in SNiP СП 24.13330.2011 "Piled foundations", when calculating large piled foundations, strain properties of pile, raft and superstructure materials may be taken as elastic, provided the loads are restricted by limits of linear proportionality. The raft and the piles were modelled with first order elastic finite elements, their non-linear action disregarded. Pile-raft coupling was taken to be of hinged type. Plan dimensions of the raft were 10 by 20 m, pile cross-section 400 × 400 mm, pile length 20 m. Mechanical properties of finite elements representing the piles were equivalent to those of class B25 concrete, whereas elelements representing the raft featured infinite rigidity in order to provide uniform redestribution of loads onto the pies. Infinite rigidity in this case is not a mathematical abstarction but is quite in line with contemporary research in the area of soil-structure interaction [2], according to which tower-type buildings impose a characteric contact pressure epure on the edge of raft, resembling the epure underneath a stiff plate. Such approach is fully compliant with the requirements contained in СП 24.13330.2011 "Piled foundations", allowing subsoil rigidity definition to be performed substituting spatial calculations of the piled foundation as a whole for calculations of its individual elements, as it is permissible to assume a flexible raft as absolutely rigid.

In all problems the loads were assumed kinematically, i.e. we modelled movements of tops of piles, plates or rafts. When building the elements mesh for the subsoils, as per stipulations of СП 24.13330.2011, the required dimensions of the area to be calculated were assigned in such a way so as to ensure that the boundary conditions should not interfere with the results.

In design of any piled foundation the defining factor is bearing capacity of a single pile, therefore, first and foremost, we modelled static load pile tests in the ground conditions under considerations. In modeling a pile load test by means of finite elelemts it is necessary to consider specific strength parameters of the contact area between the shaft and the surrounding soil. Both concentration of shear strains and plastic flow of soil along "pile-soil" interface, as per СП 24.13330.2011 "Piled Foundations" should be represented by means of special "interface" (or contact) elements or, alternatively, by means of requisite densification of finite elements (or finite differences) mesh.

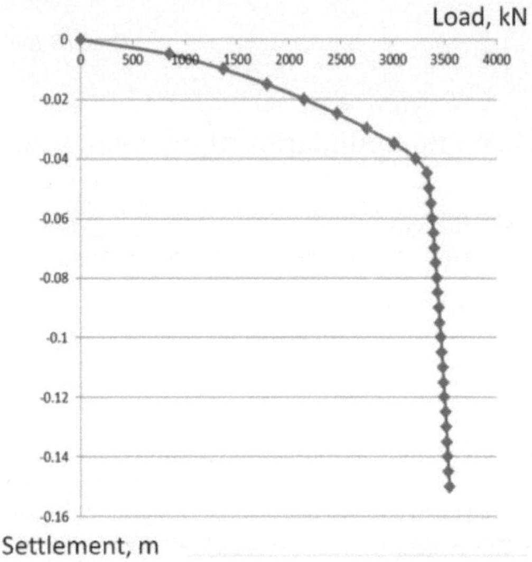

Figure 1. "Load-settlement" plot for a single pile.

As is well known, bearing capacity of a skin friction pile, and it is precisely this type of pile that was being considered for the assumed ground conditions, depends on toe resistance, pressure and shear resistance of soil developing along the shaft skin. As a pile is being constructed significant disturbance is rendered to structural bonds in texture of soil adjacent to the shaft, which if disregarded in calculations can lead to a fictional overstatement of skin friction related bearing capacity. In this paper cohesion value was assumed as being lower to that of the surrounding soil. Based on our experience in modeling static load tests [2], the interface elements in the "pile-soil" area had 80% cohesion value of undrained soil strength. Based on the calculation results a "load-settlement" plot was drawn, as contained in Figure 1. There, a vertical "drop" is clearly visible, as the pile's bearing capacity plummeted abruptly, yielding the numerically calculated bearing value of 3200 kN.

We followed with calculating "boundary" problems, which provided an insight into some trends in changing foundation behaviour as it gradually progressed from an embedded plate to a surface plate by means of reducing pile number (viz. increasing spacing between the piles). The surface plate arrangement can be theoretically considered as a pile-raft foundation with pile distances being infinitely long, whereas in the embedded plate variation the piles would be represented as positioned at infinitely small distances (the embedment level of plate would be the same the piles toe level). Following that, we sought to solve two

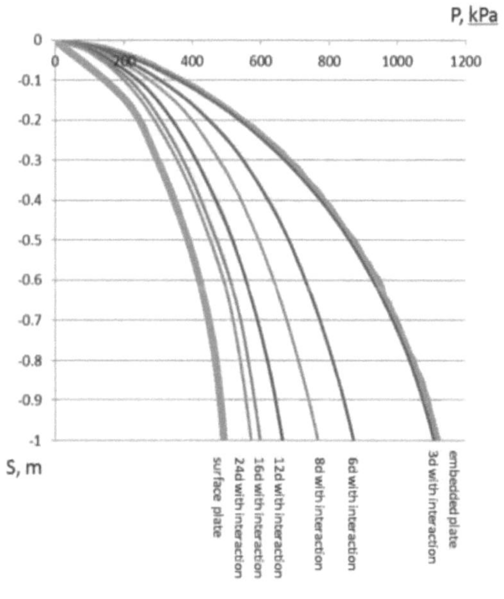

Figure 2. "Load – settlement" curves for the low raft foundation options.

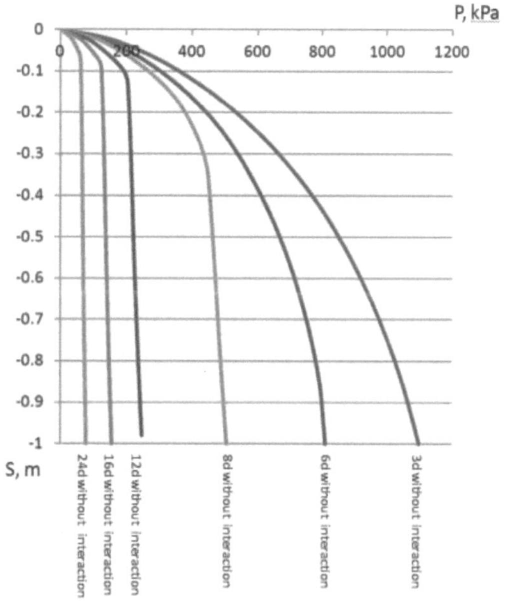

Figure 3. "Load – settlement" curves for the high raft foundation options.

sets of problems: that for a low raft (there being a contact between the raft and the soil) – provided the raft is engaged, such foundation can be considered as being of the pile-raft type and that for a high raft (there being an air cushion between the raft and the soil) – a pile group (a cluster of piles). Each calculation set contained a series of problems with distances between piles varying from 3d to 24d (where d is the cross-sectional dimension of a square pile). Based on the obtained results "load-settlement" curves were plotted for all types of problems (see Figures 2, 3).

As had been initially assumed, the load-settlement curves for the pile-raft foundations with varyingly paced piles were located in the interval limited by the "boundary" problems – the curves for the surface plate and those for the embedded plate. Below we shall compare bearing capacities of foundations with settlements being equal to 1 m and 20 cm. Such arbitrary choice of the settlement values to be used in definition of the corresponding bearig capacity of soil is conditioned by the fact that the curve for the embedded plate, piled and pile-raft foundations does not have a clearly defined "drop", visible on the analogous curves for a single pile and the surface plate. In such cases one may follow recommendations on how to define ultimate bearing capacity of a pile, whose load-settlement plot has a similar curve but no steep dive in bearing value: in such cases it is assumed that the ultimate static load resistance of the tested pile is the load, whereat this pile's settlement is equal to a certain degree of the mean ultimate settlement of the newly constructed building. Usually, in Saint Petersburg ground conditions the limit value of 4 cm is assigned for a single pile situation. In our case, the settlement value being *that* insignificant, collation of various options would be quite difficult, because their difference will not yet have manifested in full. It is precisely for that reason that we chose the two above settlement values, at which bearing capacities of foundations with varyingly paced piles are considered.

Bearing capacity value for the embedded plate, a pile-raft foundation and a pile group with 3d spacing is practically the same, viz. 1100 kPa at settlement of 1 m and 540 kPa at 20 cm. Bearing capacity values of the surface plate and a pile-raft foundation with 24d spacing are likewise similar, being 500 kPa at settlement of 1 m and 280 kPa at 20 cm.

It is due to the identical foundation behaviour patterns that the "load-settlement" curves for those solutions coincide. At 3d spacing soil and piles work like one unified "theoretical" foundation (Figure 4).

In all cases underneath the foundation footing (or at the piles toe-level) a sort of a compacted nucleus with characteristic slip surfaces is formed.

At spacing of 24 d the nature of strains of a pile-raft foundation is similar to the surface plate: formation of the compacted nucleus occurs directly underneath the raft, whereat similar slip surfaces also appear, leading to appearance of multiple heaves (see Figure 5).

When bearing capacity of a pile-raft foundation and of a pile group is lost the piles behave similarly, in both cases slipping down the soil. The difference in bearing capacity of a pile-raft foundation relative to a pile group at the given spacing is defined by the function of the raft, which limits free slipping of the piles and transfers a significant portion of the load onto the soil. In a pile group the maximum movement is localized and concentrated underneath each individual pile toe.

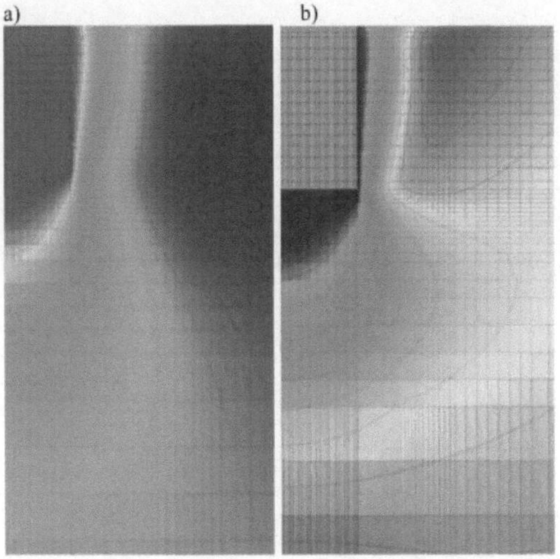

Figure 4. Contours of vertical movements of a plate at depth (a), contours of maximum movements of a plate at depth (b).

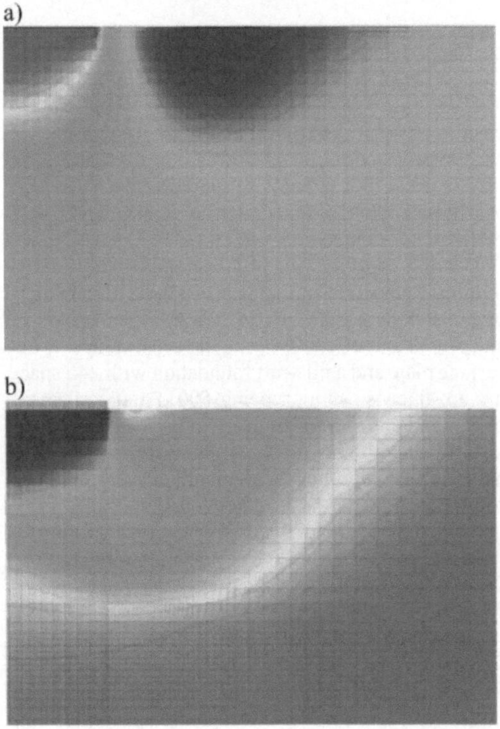

Figure 5. Contours of vertical movements of a plate at depth (a), contours of maximum movements of a plate at depth (b).

3 ANALYZING EFFICIENCY OF PILE-RAFT FOUNDATIONS

The most lucid efficiency representation of a pile-raft foundation can be rendered comparing its bearing capacity with that of a raft or a piled foundation (Figure 6a,b).

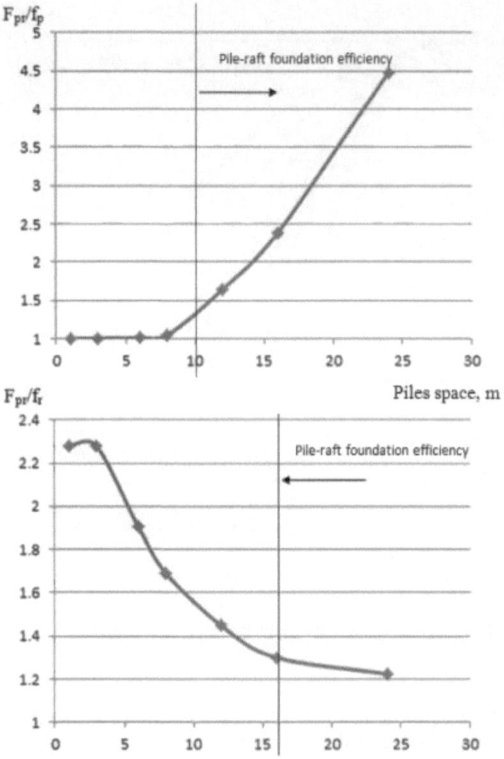

Figure 6. Evaluating efficiency of a pile-raft foundation in comparison with a piled foundation (a) and a raft foundation (b) (F_{pr}/f_p – relation of bearing capacity of a pile-raft foundation to that of a pile group, F_{pr}/f_r – relation of bearing capacity of a pile-raft foundation to that of a raft foundation).

Increasing pile spacing by over 16d, bearing capacity of a pile-raft foundation turns out to be only 30% higher than that of a raft foundation (Figure 6b). Piles spaced at 8d increase bearing capacity of a pile-raft foundation merely by 5% in excess of the corresponding piled foundation (Figure 6a). If one considers efficiency of a pile-raft foundation as its superior bearing capacity (over 30% in relation to a raft and a piled foundation options), then one will be confined to a relatively narrow spacing, roughly from 10d to 16d. In all other cases the pile-raft foundation will degrade either into a piled or a raft foundation.

This assumption agrees well enough with the diagram of active boundaries of a pile-raft foundation, proposed by a professor of the Technical University of Darmstadt R. Katzenbach [1] (see Figure 7), which was based on his high-rise buildings data analyses, gleaning material from raft, pile-raft and piled foundation types.

In this diagram factor $\alpha_{ПСФ}$ establishes dependency between pile load $\Sigma R_{СВаИ}$ and total load from superstructure $R_{общ}$:

$$\alpha_{ПСФ} = \frac{\Sigma R_{сваи}}{R_{общ}}$$

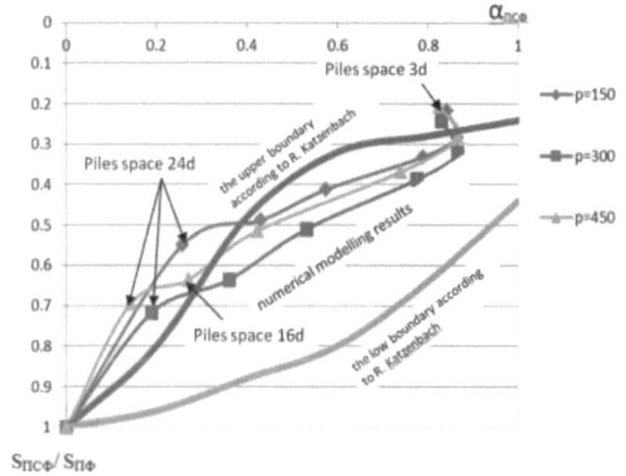

Figure 7. Plot showing loads distribution between raft and piles – comparison of analyzed and measured results according to R. Katzenbach ($\alpha_{\Pi C \Phi}$ – factor determining relationship between load on piles and toal load on building, $S_{\Pi C \Phi}/S_{\Pi \Phi}$ – relation of bearing capacity of a pile-raft foundation to that of a raft foundation).

where $\alpha_{\Pi c \Phi} = 0$ corresponds to a pile-free raft foundation, and factor $\alpha_{\Pi c \Phi} = 1$ reflects a pile group.

4 SAFETY FACTORS IN DESIGN OF PILE-RAFT FOUNDATIONS

In Germany, at three technical universities (those of Darmstadt, Leipzig and Berlin) groups led by R. Katzenbach and G. König developed the guidelines on construction of combined pile-raft foundations [5]. In those guidelines the ultimate bearing capacity of a pile-raft foundation and the permissible load thereon is defined by dividing the ultimate bearing capacity by safety factor of 2. It is to be noted at this point that the Russian codes do not stipulate a possibility of applying a safety factor in relation to the ultimate resistance of a pile-raft foundation in general. As per the code ref. СП 24.13330.2011 "Piled foundations", when defining a rated load, the safety factor of 1.2 should be introduced, provided that bearing capacity of piles had been established by means of a static load test.

As was sated above (see Figure 6a), the actual bearing capacity of a pile-raft foundation with piles distanced at (10...12) d is 1.25...1.65 times higher than bearing capacity of a piled foundation. Thus, to the safety factor of 1.2, used in defining bearing capacity of piles, a multiplier of 1.25...1.65 is added, representing the so called "hidden" safety factor. This multiplier is *de facto* present in Russian codes because, by default, the positive contribution of a low raft is assumed as a reserve to the general bearing capacity of the foundation. As a result, combination of those two factors produces the generalized safety factor of 1.5...2.0. Correspondingly, the actual safety factor applied in the Russian codes to the "raft – piles – soil" system is not in excess of the value used in Europe.

Moreover, it is obvious that historically the appearance of the whole pile-raft foundation idea was conditioned by those higher safety factors in European codes. Indeed, with the pile safety factor of 2.0 and the "hidden" reserve value of 1.25...1.65 the overall actual safety factor according to the European codes would have climbed to as high as 2.5...3.3, which would have been clearly overcautious – hence the idea of a pile-raft foundation to correct the issue.

5 CONCLUSIONS

1. As shown by a numerical experiment, performed for ground conditions characteristic e.g. for southern areas of Saint Petersburg, a pile-raft foundation has a pronounced effect within the range of its pile spacing being from 10d up to 16d, where d – cross-sectional dimension of a square pile. At spacing between the piles equal to, or inferior of, 8d, bearing capacity of a pile-raft foundation does not differ from that of an analogous piled foundation, whereas at spacing equal to, or in excess of, 24 d it approaches bearing capacity of a raft.
2. The actual safety factor for bearing capacity of a pile-raft foundation applied in Russian and European design practice is the value of 2.0. Please note, that in Russian codes this value is derived from the safety factor for piles in soil, as well as from the "hidden" reserve value representing a raft.
3. As stipulated by the Russian codes, there is no need to introduce a specific design procedure for pile-raft foundations. They should be designed as piled foundations with a raft, envisaging that the piles assume the entire rated superstructure load, using the prescribed safety factors and reserve values. In this case the actual bearing capacity reserve, contributed by the raft, is used to harmonize both Russian and European code stipulations.

REFERENCES

Aleksandrovich V.F. Kurillo S.V. Fedorovkiy V.G. K voprosu o vsaimnom vliyanii svay i pliti v osnovanii svayno-plitnogo fundamenta.// Reconstrukcia istoricheskih gorodov i geotechnicheskoe stroitelstvo. Trudi konferencii. 2003. C.125–143 (in Russian).

Hanisch J., Katzenbach R., König G. KombiniertePfahl-Plattengründungen. Berlin: Ernst & Sohn. 2002.

Katzenbach R., Shmidt A. Ramm X. Osnovnie principi proectirovania I monitoringa visotnih zdaniy Frankfurta-na-Mayne. Sluchai iz praktiki.// Rekonstrukcia gorodov I geotehnicheskoe stroitelstvo. 2005, No. 9. c. 80–99 (in Russian).

Poulos H.G. Piled Raft Foundations – Design and Applications.Geotechnique, Vol. 50 (2). 2001. P. 95–113.

Ulitskiy V.M., Shashkin A.G., Shashkin K.G. Geotechnicheskoe soprovozdenie razvitia gorodov. – Stroyizdat Severo-Zapad, "Georekonstrukcia", SPb, 2010. 551 c (in Russian)

Baltic Piling – Mets & Raudsepp (Eds)
© 2013 Taylor & Francis Group, London, ISBN 978-0-415-64334-4

Investigation of piled-raft foundations in soil conditions of Minsk

V. Sernov
Belorussian National Technical University, Minsk, Belarus

ABSTRACT: A raft being in direct contact with the soil transmits a part of the applied load on the base. Taking into account the raft-soil interaction increases the bearing capacity of pile foundations and reduces the cost of the foundations due to decrease of the pile number and their length. The results of model and field tests of piled-raft foundations in the sites in Minsk are shown. Methods of analysis of piled-raft foundations are described.

1 INTRODUCTION

Of late years pile foundations are used extensively in connection with load increment on the soil. Often pile foundations are applied in comparative favorable ground conditions at the surface of the base. In such cases increase the bearing capacity of foundations and reduce construction cost are questions of first-rate importance. One of the ways to increase the bearing capacity of a piled foundation is taking into account soil resistance in the base of the raft. Piled raft is similar to shallow foundation and can transfer significant part of the load into the base. It allows reducing number of pile in the foundation or shortening their length. As a result building terms and foundation costs reduce considerably. However nowadays there is no reliable and suitable in design practice methods of calculation pile foundations with bearing rafts. In order to devise such methods we have analyzed the results of piled foundation tests fulfilled by various authors up to date and have carried out series of laboratory and field investigations.

Prof. A. A. Bartolomey had tested more than 70 pile foundation of natural size in different ground conditions. The length of piles was 3...12 m (Бартоломей 1995). The observed results show that the raft carries about 10...12% of the total load applied when length of piles was over 9 m and space between piles a = 3d. Bearing capacity of the raft is about 35...40% when lenth of piles is 5...8 m and a = 6d. Prof. Golubkov has tested 2 prismatic piles (cross section: 350 × 350 mm, L = 3800 m) and 2 pyramidal piles (cross section: head – 600 × 600 mm, foot – 100 × 100 mm, L = 3000 m). One of the piles was tested with a raft and the other without a raft. Significant influence of the raft on total bearing capacity was determined, the more settlement the more influence of the raft. Bearing capacity of the raft was 27,5% in foundation with prismatic pile and 42% with pyramidal pile (Голубков et al. 1971).

Dr. J. Sh Zijazov had tested pile foundation (cross sections of the piles were 0.2 × 0.2 and 0.3 × 0.3 m, length: L = 6 m) with raking piles (slope angle was 0, 10, 15, 20 and 22°). It was determined that when the pile space was increased from 3d to 4d the raft resistance increased about 25% and 20 and 11% with raking piles with slope angle was 10 and 15° correspondingly (Зиязов 1983).

The part of the load that the raft carries grows up when the pile length is reduced and the space between piles is increased. Some test results carried out by different authors confirm it.

2 MODEL TESTS

The aim of the model tests was to determine the main factors of pile-raft-soil interaction. Model investigations were divided into three groups.

The object of the first test series is to research the soil deformations under the raft in pile space. Models of driving piles (cross section 20 × 20 mm and length 150 and 300 mm) were tested in the sand-filled test rig with transparent front wall (Fig. 1, 2). The soil was laid in the tray with layerwise compaction. Indicative strips were made of chalk through 50 mm. The load was applied with leverage and weights.

Deformations of the soil in the base were fixed persistent by means of photofixation in the time of pile driving and loading of the groups. Significant surface warp of indicative strips was detected in process of pile driving (Fig. 3).

After driving the pile tops were combined with the raft. Pile groups were subjected to static load. In the first place the high-raft pile foundation (without contact between the raft bed and the soil) were tested. Observational results show slipping of the lateral surface of the piles relative to surrounding ground. Additional minor displacements of the soil were found out only in the immediate vicinity of piles and away from the lateral surfaces were not occurred (Сеорнов 2008).

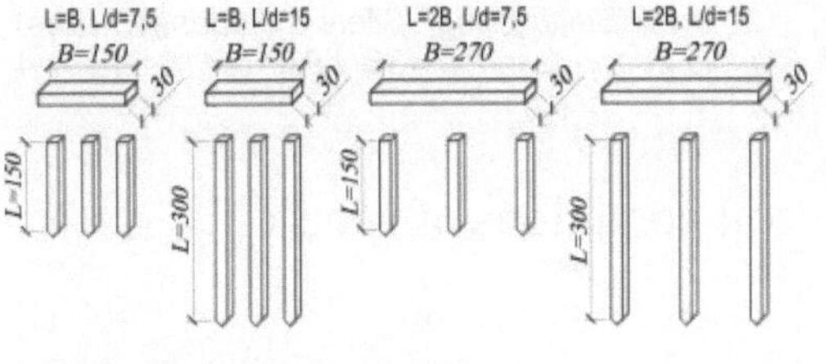

Figure 1. Experimental models.

Figure 2. Test rig.

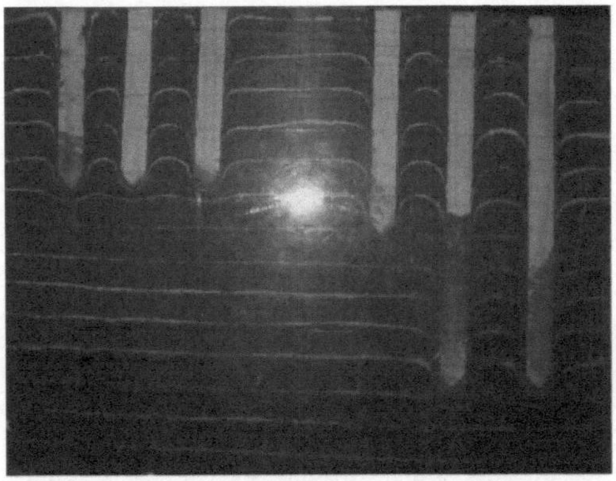

Figure 3. Indicative strips warp in process of pile driving.

In the case of testing of the pile foundations with load-bearing raft the significant additional displacements of the soil in the pile space were observed. The settlement of the ground layers are diminished with depth similarly to shallow foundation (Fig. 4).

The results of research show that the soil under the raft in the pile space does not settle side by side with a high-raft pile group and can resist a considerable part of the external load.

The object of the second test series is to research interaction between elements of piled-raft foundation.

As a rule, bearing capacity and settlement of single pile are different from comparable magnitudes of a pile in a group. This effect can be explained as interaction between the piles in the group. The interaction

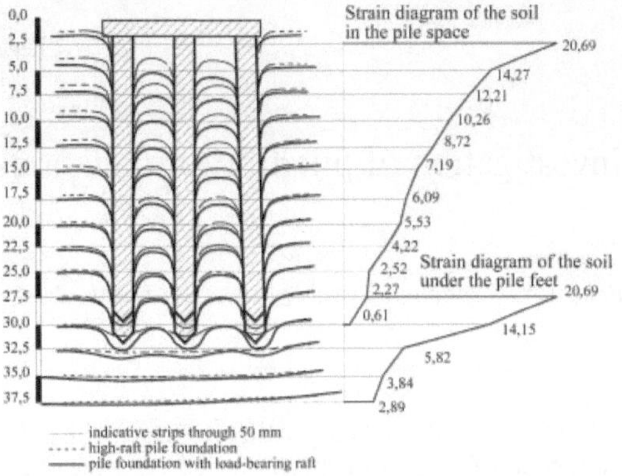

Figure 4. Ground deformation in the base of pile foundation.

depends on pile length and space, number and type of the piles and ground conditions. Influence of number of piles in group on bearing capacity of pile foundation was investigated in the Laboratory of Geotechnics of Belorussian National Technical University. The results of model tests of pile foundations consist of driven piles in sand show that bearing capacity of the pile in the group greater than the same single pile in the same ground conditions. Increase of bearing capacity is the result of pile shaft squeezing and ground conditions improvement along the pile shafts and under the pile bases.

The two groups of model piles were investigated to find out interaction between the piles and the raft. Pile length was 500 mm and pile space was 2 and 3d (Fig. 5).

The order of pile driving is shown on Fig. 5. The corner piles were rammed in the first place. Every pile and pile groups with and without rafts were tested by static load. The results of investigation are sown on Table 1.

One can see in Table 1 that the average bearing capacity of the pile in group is increased by 32.5% (model 1) and 18.4% (model 2). Bearing capacity of pile group is increased by 49.5% (model 1) and 21.4% (model 2) when the raft is interacting with the ground. In terms of our investigation we can conclude that pile-pile and pile-raft interaction in sand significantly increases bearing capacity of the pile in a group (Nikitenko et al. 2006).

Our investigations of piled raft foundations include: the character of work of a pile along and piles with the raft in all range of loading; the basic factors affecting the bearing capacity of the raft. A model of piled-raft foundation was tested in our research laboratory. The model consists of collapsible square raft (160 × 160 mm) and driving piles (cross section 20 × 20 mm and length 100, 200, 300, 400 mm). The lower part of the raft had 9 square holes (20.5 × 20.5 mm) (Fig. 6a) which were guiding lines while driving piles and were used for fixing pile heads while loading the foundation. The top part without

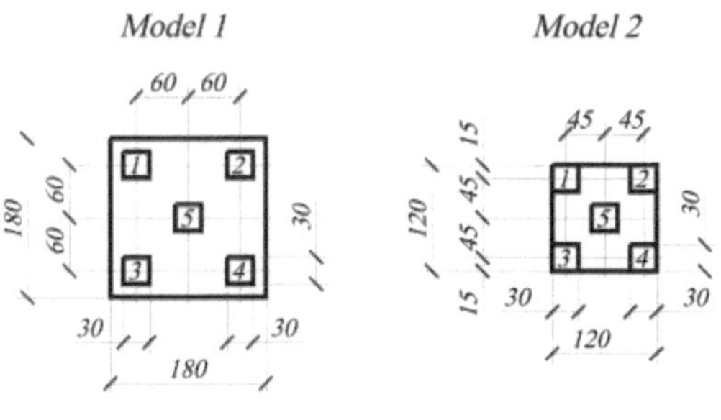

Figure 5. The layout of piles in the group.

Table 1. Bearing capacity of piled raft foundation elements

	Model 1	Model 2
	Bearing capacity, kN	
Pile 1	4,00	5,10
Pile 2	4,60	5,27
Pile 3	4,77	6,00
Pile 4	6,25	6,75
Pile 5	6,9	7,07
S	26,52	30,19
Group consist of 5 piles	22,75	28,00
5 piles + raft	34,00	34,00
Group consist of 4 piles		19,5
4 piles + raft		26,9
Raft	14,00	6,84

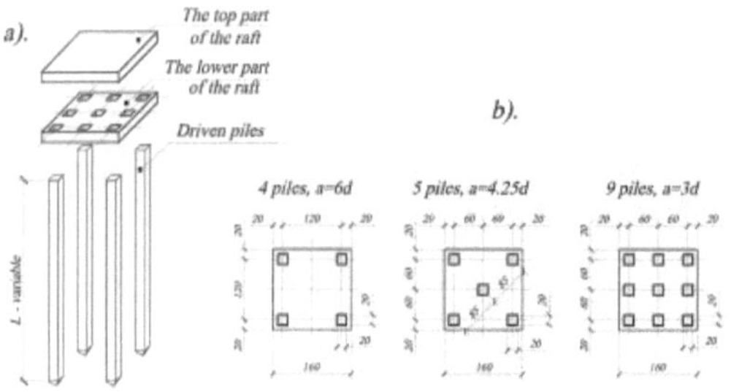

Figure 6. The elements of the pile foundation (a) and positioning of piles in the raft (b).

holes distributed the load on the piles (Nikitenko et al. 2005).

This construction of the raft allows to use it for testing foundations with different number of piles and pile spaces. Investigations were carried out in sand with average particle size. The results of 24 tests of pile foundations with raft and without it (Fig. 6b) are shown on Figure 7. The number of piles in groups are 4, 5 and 9; pile space accordingly — 6d, 4.25d, 3d; pile length — 100, 200, 300 and 400 mm.

The part of the load taken by the raft can be determined from the diagram S = f(N) (Fig. 7). One can see that bearing capacity of piled-raft foundation much more than pile foundation. Bearing capacity of

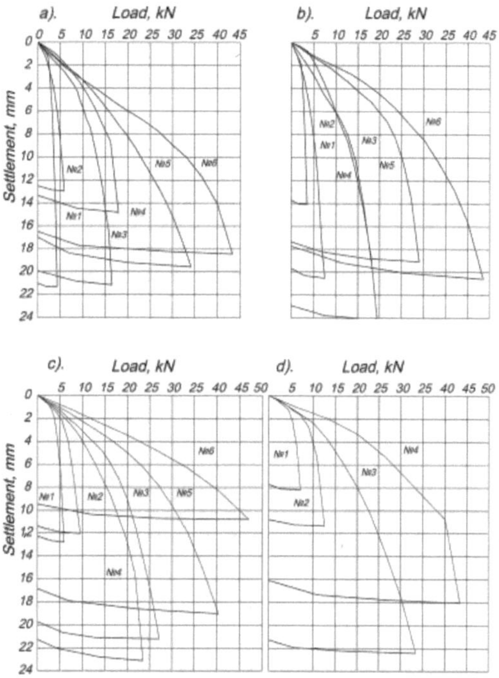

Figure 7. The results of tests of piled-raft and pile foundations models: a) — L = 100 mm, b) — L = 200 mm, c) — L = 300 mm, d) — L = 400 mm; No. 1 – 4 piles, No. 2 – 5 piles, No. 3 – 9 piles, No. 4 – 4 piles with the raft, No. 5 – 5 piles with the raft, No. 6 – 9 piles with the raft, (L — pile length).

the piled-raft foundation with 4 or 5 piles (L = 100–300 mm) more than bearing capacity of the pile foundation with the same 9 piles. On the other hand we can see that the settlement of the piled-raft foundation with 9 piles (L = 100–200 mm) less than settlement of the foundation with 9 piles (L = 400 mm). We have to note that the raft can take the load immediately after start of loading and it work is noticeable at the settlement of 0.5–1.5 mm. One of the reasons of this case is that the base under the raft was densify carefully.

The graphic charts: dependence of the load carried by the raft from the number of pile in the group *Kcyr* = f(n), and from the pile length Kr = f(L) can be seen in figures 8, 9.

One can see in fig. 8, that increasing of pile number in the group decrease the load carried by the raft and tend to 0, i. e. effectiveness of the raft is being reduced. Graphical char Kp = f(n) is similar to hyperbola. The part of the load carried by the raft is increased when load is rose and length of piles is reduced and graphical char Kp = f(L) became more flat. When the length of the pile is rose work of the raft is reduced and tend to 0 (Сернов 2003, Сернов 2004). Approximately Kr = f(L) can be represented as a line. To investigate these patterns of relationship more precisely the large-scale models have to be tested.

The aim of the third test series is to estimate efficiency of the raft in large-scale models of foundations

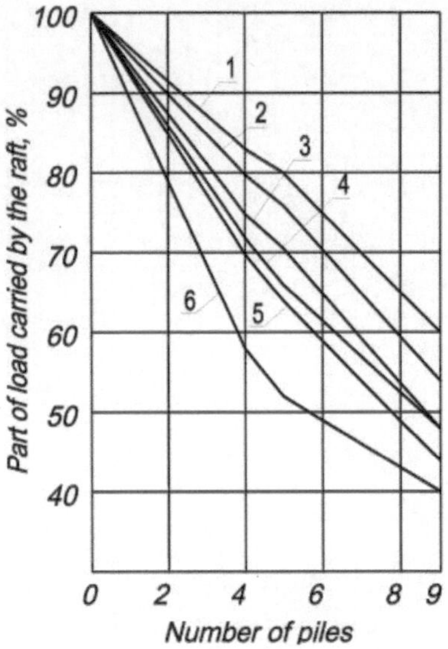

Figure 8. The graphic chart $K_r = f(n)$: 1 – N = 30, L = 100; 2 – N = 20, L = 100; 3 – N = 30, L = 300; 4 – N = 10, L = 100; 5 – N = 20, L = 300; 1 – N = 10, L = 300; (N — load, kN; L — pile length, mm).

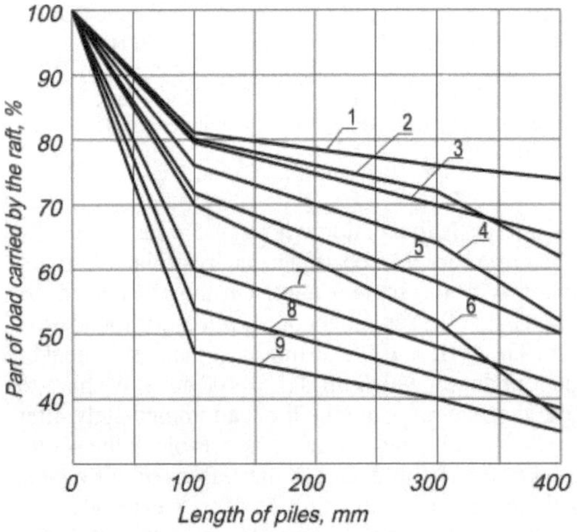

Figure 9. The graphic chart $K_r = f(L)$: 1 – N = 30, n = 4; 2 – N = 30, n = 5; 3 – N = 20, n = 4; 4 – N = 20, n = 5; 5 – N = 10, n = 4; 6 – N = 10, n = 5; 7 – N = 30, n = 9; 8 – N = 20, n = 9; 9 – N = 10, n=9; (N — load, kN; n — number of piles).

consist of bored piles in sandy soil. Investigations have been performed on the Test Site of Geotechnics of Belorussian National Technical University. The test rig is shown in Fig. 10. Large-scale model tests of deep-raft and high-raft pile foundations are shown in figures 11 and 12. The single piles, raft, groups of piles and piled-raft foundations were tested with static load.

The raft was made as rigid square metal stamp (b = 500 mm). Pile foundations consist of 1, 4, 6 and 9 bored piles with length of 1000 mm and 2000 mm and diameter of 60 mm were tested both with deep and high

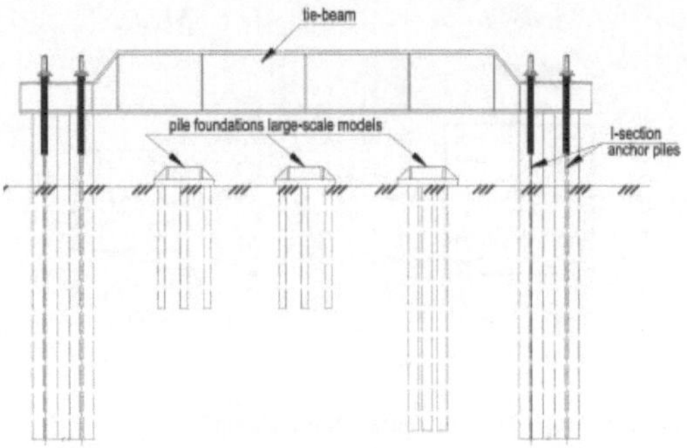

Figure 10. Test rig for large-scale model investigations.

Figure 11. Large-scale model test of deep-raft pile foundation.

Figure 12. Large-scale model test of high-raft pile foundation.

rafts. The layout of piles in the groups is shown in Figure 13. The results of pile foundation tests represented in Figures 14 and 15.

The large-scale model test results confirm high efficiency of raft-soil interaction in full band of size-shape factors of pile foundation. In case of deep-raft pile foundation (or piled raft foundation) bearing capacity is significantly increased. The major bearing factors

with length of 2000 mm (by 18%). In other cases bearing capacity increases with pile length shortening, pile space extension and settlement rising (Сернов 2005).

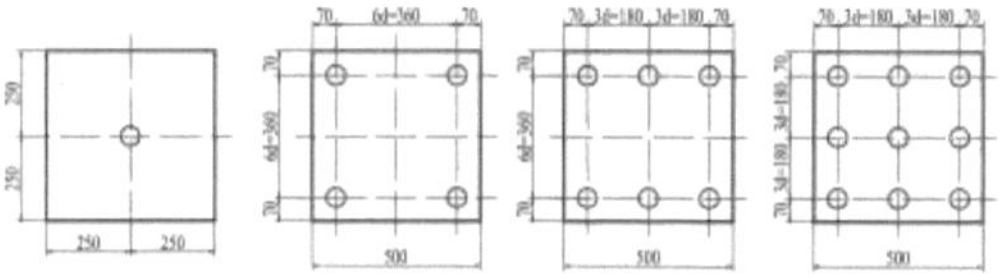

Figure 13. The layout of piles in the groups of large-scale models of pile foundations.

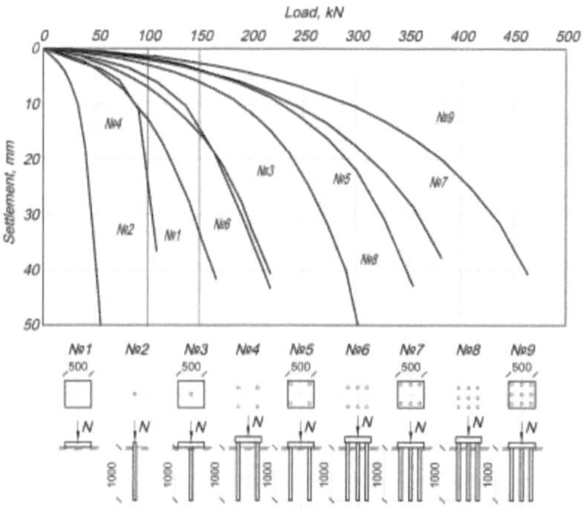

Figure 14. The results of large-scale model tests of high-raft and deep-raft pile foundations consist of bored piles with length of 1000 mm.

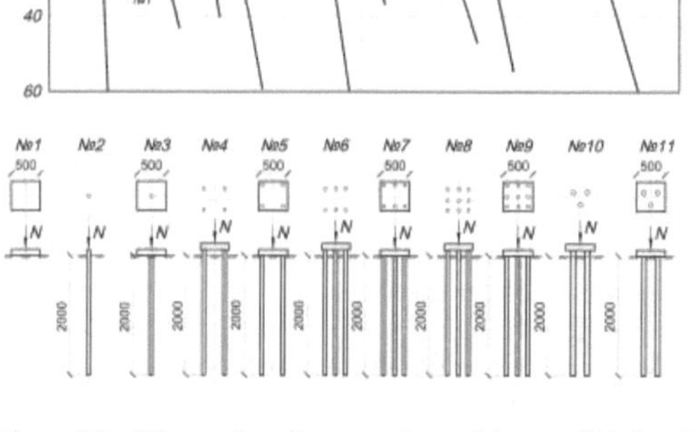

Figure 15. The results of large-scale model tests of high-raft and deep-raft pile foundations consist of bored piles with length of 2000 mm.

influencing the raft interaction are pile length, pile space and settlement on the foundation.

The load-settlement diagrams presented in Figures 14 and 15 show that minimal increment of bearing capacity of the foundation take place in case of 9 piles

3 FIELD TESTS

Building sites of Minsk are characterized by complicated geological conditions. Quite often the firm soils are located at the surface but at a depth of 5–10 m a soft ground layers occur. Frequently filled-up ground is located at the surface. In these cases conventional long driven piles are used to transmit the load at deep-laid firm soil. The length of such piles could be 20 m and more. The alternative foundation in such ground conditions consist of tapered piles with bearing raft. The sloping surfaces of tapered piles compact ground between piles and around the foundation. Bearing capacity of raft and lateral surface of piles considerably increase. Bearing capacity of raft with tapered piles much more than bearing capacity of raft with uniform cross-section piles. Stresses in the base of the foundation with tapered piles are dissipated in the top slices and do not reach soft ground layer at depth.

The results of tests prove efficiency of vibrostamped cast-in-situ tapered piles with bearing rafts when thick layer of filled-up ground is located at the surface. Strength characteristics of filled-up ground considerably improved due to consolidation. Sloping lateral pile faces eliminate negative friction. Compacted filled-up soil becomes the bearing layer both for piles and for rafts.

To estimate efficiency of piled foundations with bearing rafts static tests of single piles and couples of pile with the raft were carried out in different sites of Minsk in various ground conditions (Nikitenko et al. 2007).

At the site in Belsky street 148 in Minsk firm loam soils were located at the surface but at a depth of 10 m and more a soft loam soils with peat layers occurs. There were two alternate versions of foundation. The first was long driven prismatic piles with length of more than 12m to transmit the load at deep-laid firm soil. The second alternate was short vibrostamped cast-in-situ tapered piles (L = 3–4 m) with rammed dry concrete mix at the pile foot and bearing rafts to transmit the load at the surface layer of firm soil. Technical and economic assessment shows that the cost of the second alternate is far less. Bearing capacity of vibrostamped tapered piles was increased by 80% by means of ramming dry concrete mix at the pile foots and then by 30% by means of bearing raft (Fig. 16). Raft ratio ($K_r = F_{pr}/F_p$ = bearing capacity of piled-raft foundation/ bearing capacity of piled foundation) in such case $K_r = 1.30$.

The subsoil in the site in Pritytzky Street consists of loessial loam, fine sand and peat layers. The first alternate versions of foundation was driven prismatic piles with length up to 20 m. The second was vibrostamped cast-in-place tapered piles with widening at the pile

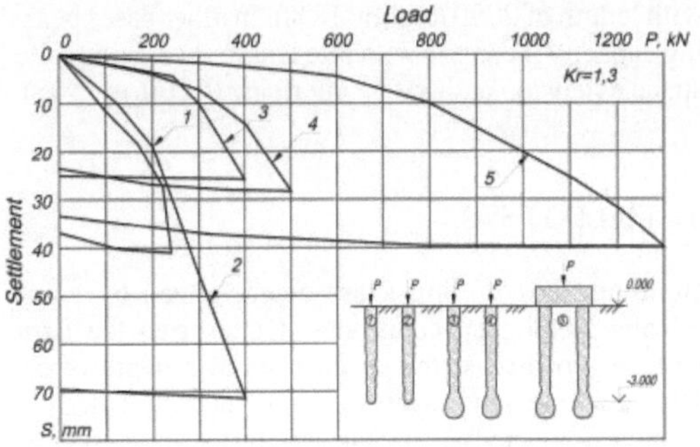

Figure 16. The results of tests of single piles and 2 piles with raft in Belsky Street 148: 1, 2 — single piles; 3, 4 — single piles with widening at the pile foot; 5 — 2 piles with a raft (1.8 × 0.9 m).

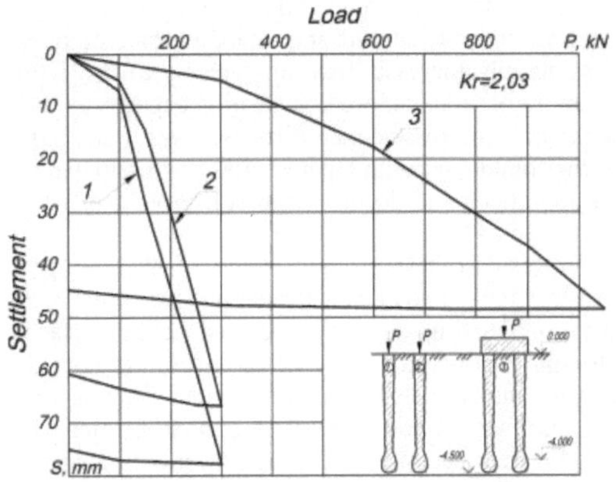

Figure 17. The results of tests of single piles and 2 piles with raft in Pritytzky Street: 1,2 — single piles; 3 — 2 piles with a raft (1.8 × 1.2 m).

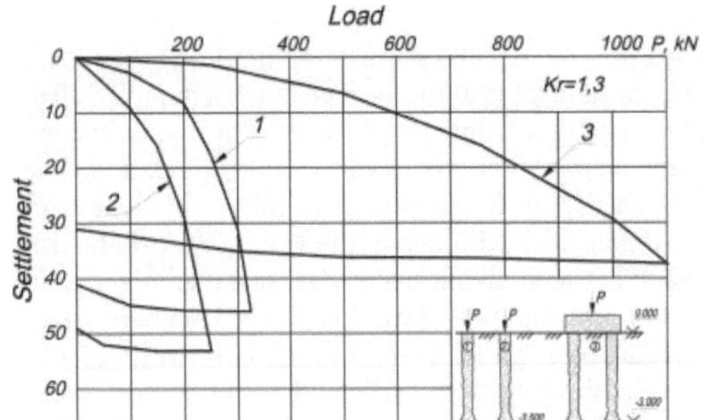

Figure 18. The results of tests of single piles and 2 piles with raft in Pritytzky Street: 1, 2 — single piles; 3 — 2 piles with a raft (1.9 × 0.4 m).

feet (the length of the piles is 3 and 4 m, diameter at the pile top 0.5 m, diameter at the pile foot 0.3 m).

Single tapered piles and two fragments of foundation were tested with static load (Fig. 17, 18). Load-settlement diagram S = f(N) shows that the raft interacting with subsoil increase bearing capacity of the foundation by 30–103% ($K_r = 1.30 \div 2.03$). The magnitude of raft ratio K_r depends on raft area and ground conditions in the base of the piles and the raft (Сернов 2005).

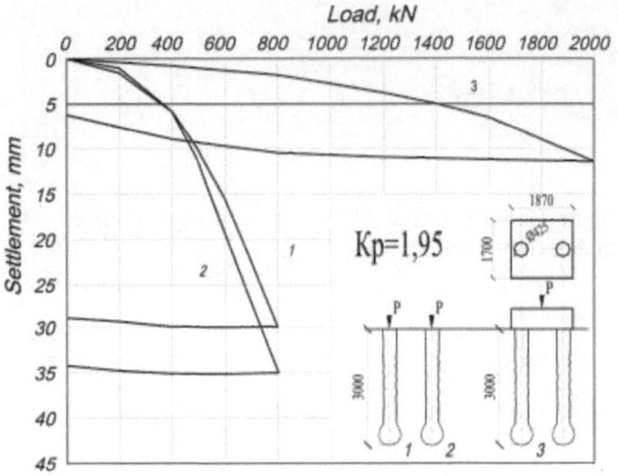

Figure 19. The results of tests of single piles and 2 piles with raft in residential quarter of Burdeinogo Street and Jakubovskogo Street: 1, 2 — single piles; 3 — 2 piles with a raft (1.7 × 1.87 m).

The similar results were obtained in the site in residential quarter of Burdeinogo Street and Jakubovskogo Street. The two single bored piles and the same piles with a raft fragment were tested with static load (Fig. 19).

Load-settlement diagram S = f(N) shows that the bearing capacity of the foundation was increased by 95% ($K_r = 1.95$).

4 METHODS OF ANALYSIS OF PILED-RAFT FOUNDATIONS

Analysis of model and field tests carried out in the Laboratory and Test Site of Geotechnics of Belorussian National Technical University as well as various building areas in Minsk and other cities allows to detect the basic factors which affecting the raft-soil interaction and to determine range of application of pile foundations with load-bearing rafts. Design methods of ultimate vertical load and settlement of piled-raft foundations were developed (Сернов 2008). The results of theoretical and experimental researches performed by various authors were taken into consideration in development of the methods.

It has been found experimentally that the soil under the raft in the pile space does not settle with a pile group and in spite of traditional design approaches can resist a considerable part of the external load. Experimental total vertical stress diagrams in the base of various deep-raft pile foundations were carried out by Kazachok in the 1970s (Казачок 1979). Contrastive analysis of experimental data and estimated stress diagrams received with linearly elastic semispace theory show that the experimental curves are similar to theoretical diagrams (Fig. 20). Calculating error is not more than 5%. On basis of results of our investigations we can draw a conclusion that stress and depth

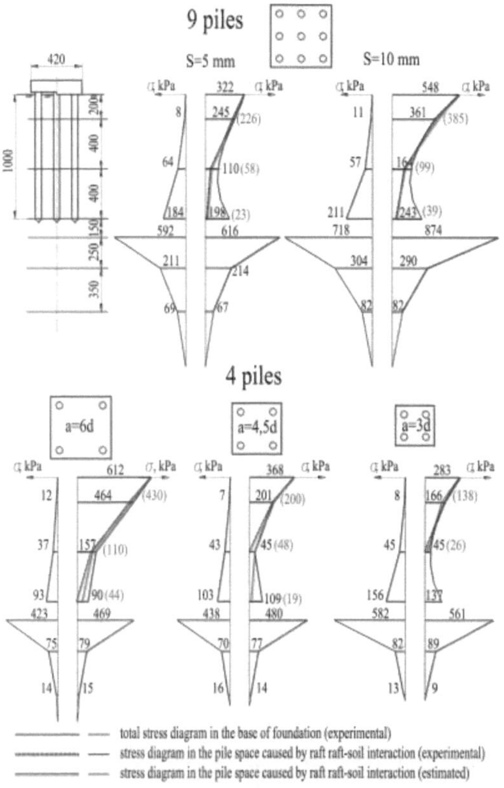

Figure 20. Experimental and estimated stress diagrams in the base of the rafts.

of compressed layer of the soil in the base of a raft and shallow foundation are similar and can be described with similar methods.

The ultimate vertical load carried by pile foundations with bearing raft is calculated:

$$N = N_r + N_{pf}, \tag{1}$$

$$N_r = \gamma_{cr} p_r A_r, \tag{2}$$

N_r — ultimate vertical load carried by pile group;
N_{pf} — ultimate vertical load carried by the raft;
γ_{cr} — factor depends on ground surface preparation under the raft bed:
$\gamma_{cr} = 1.0$ — compacted stone bed;
$\gamma_{cr} = 0.9$ — compacted sandy soil;
$\gamma_{cr} = 0.8$ — compacted clay soil;
p_r — average contact pressure under the raft bed depends on ground conditions and the assumed settlement of pile group;
A_r — raft area without total cross-sectional area of the underlying piles.

The settlement of pile foundations with bearing raft is calculated:

$$S_{pf} = S_p + S_{add} \tag{3}$$

S_p — settlement of pile group under the load N_r;

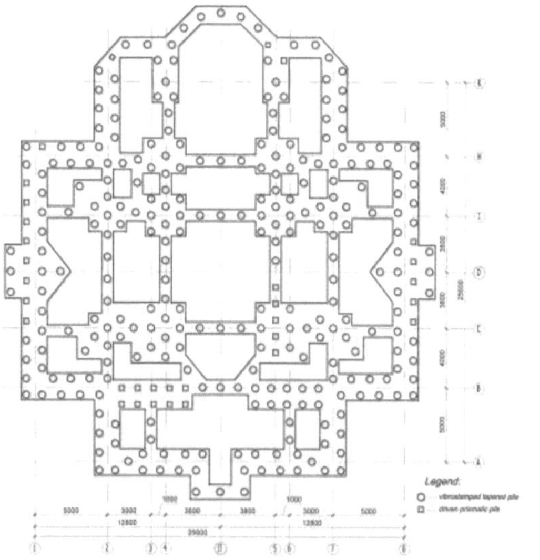

Figure 21. Orthodox Church: piled-raft foundation.

S_{add} — additional settlement of pile group arise from additional vertical stress under the pile feet in case of $H_c > L$ (H_c – compressed layer of the soil, L – pile length).

Detailed description of computing methods of pile foundations with bearing rafts, range of application, test and control methods are given in "Design recommendations of pile foundations with bearing rafts" (Никитенко and Сернов 2005).

5 REALIZATION OF THE INVESTIGATION RESULTS

The experience of use vibrostamped tapered piles was adopted in the site of Orthodox Church in Sukharevo district in Minsk (Fig. 21).

Ground conditions under the raft:

1 – dust clay sand (E = 18 MPa, h = 4 m);
2 – soft sandy clay (E = 6 MPa, h = 2 m);
3 – sandy clay with organic residues (E = 10 MPa, h = 1.5 m);
4 – peat (E = 3 MPa, h = 2.5 m);
5 – firm sand (E = 25 MPa).

The first alternate was foundation consist of 480 driven prismatic piles with length of 14 m and cross sections 0.3 × 0.3 and 0.35 × 0.35 m (Nikitenko et al. 2007).

The most reasonable decision in such ground conditions is load transfer onto relatively firm upper layers of base. Calculation of the foundation according to [20] has shown that the raft transmits 40% of the load onto the base and the rest part of the load is transmitted by vibrostamped tapered piles with length of 3 m, top diameter 0.5 m and foot diameter 0.3 m. The layout of the piles shown on Figure 16. The foundation consists of 285 tapered piles and 32 driven prismatic piles made before.

The cost foundation was reduced almost in two times by means of use short vibrostamped tapered piles with bearing raft.

6 CONCLUSIONS

1. The pile-pile interaction in sand significantly increases bearing capacity of the pile in a group.
2. The raft has significant influence on mode of deformation of the base. The large volume of the soil take part in the work and increase bearing capacity of the base about 10–100% depending on size and shape of foundation and ground conditions of the site. The large reserve of bearing capacity of the base has not investigated sufficiently.
3. When load is increased the settlement of piled-raft foundation take place more uniformly than piled foundation. On the graphical char Кр = f(N) there is no pronounced critical point and sudden loss of bearing capacity.
4. The basic factors which affect the work of the raft in foundation are ground conditions, quality and consolidation of soil under the raft, size and shape of the foundation (type of piles, pile length and space, width of the raft); load on the foundation.
5. Application of short vibrostamped tapered piles with widening at the pile foots and bearing rafts is more effective than conventional foundations consist of driven prismatic piles. Compressing stress transmits into ground along the whole length of lateral pile faces without negative friction and by the raft. Upper layers of filled-up soil are compacted by means of sloping lateral pile faces and bearing capacity of this soil increasing considerably.
6. Application of our investigation results in construction practice shows that taking into account soil-raft interaction can reduce the cost of pile foundation in certain cases in 30–50%.

REFERENCES

Nikitenko M.I., Sernov V.A., Protaschik O.N., Tisetzkaja M.S. 2005. Interaction between rafts and piles in calculation of bearing capacity and settlements of residential constructions and bridge footings. Zakládání staveb — foundations — grundbau: 33rd konferenci se zahraniční účastí. Brno. 2005, 311–316.

Nikitenko M.I., Sernov V.A., Kurilenko Y.A., Kunitzkaya K.S. 2006. Interaction between elements of piled-raft foundation. 34. Konf. Zakládání staveb – foundations – grundbau. Brno. 2006, 38–41.

Nikitenko M. I., Sernov V.A., Boguslavchik O.U., Shumilina A.V. 2007. Soil compaction in the bases of piles and rafts. Zbornik Medzinár. geotechnickej konferencie: Zlepšévanie vlastností základových pôd. Bratislava. 2007, 209–211.

Бартоломей А. А. 1995. Основы прогноза осадок свайных фундаментов [Basis of pile foundation settlement prediction]. Основания, фундаменты и механика грунтов. 3: 8–10.

Голубков В. Н., Тугаенко Ю. Ф., Хуторянский Б. О. 1971. Исследование процесса формирования зоны деформации в основаниях одиночных свай [Investigation of deformation zone forming in the base of single piles]. Основания и фундаменты. Киев: Будівельник. 4: 9–13.

Зиязов Я. Ш. 1983. Исследование вертикально и горизонтально нагруженных фундаментов из забивных наклонных свай [Investigation of vertical and horizontal loaded foundations consist of raking driven piles]. Свайные фундаменты Уфа. 43–60.

Козачок Л.Д. 1979. Исследование распределения вертикальных напряжений в основании кустов висячих свай с низким ростверком [Investigations of vertical stress distribution in the base of friction pile groups with deep rafts]. Ленинград. 174 с.

Никитенко, М.И., Сернов В.А. 2005. Рекомендации по расчету свайных фундаментов с несущими ростверками [Design recommendations of pile foundations with bearing rafts] Р5.01.015.05. Минск, 2005. 24р.

Сернов В. А., Лаврещук Е.В., Абдуллаев Р.В., Иванов А.С. 2003. Исследование моделей свайных фундаментов с низкими и высокими ростверками [Researches of the models of deep-grid and high-grid pile foundations]. Будаўніцтва – Строительство – Construction. Минск. 3-4: 132 – 139.

Сернов В.А. Лаврещук Е.В., Архангельская Е.В., Щербицкий Д.Л. 2004. Исследование взаимодействия элементов свайных фундаментов [Investigations of pile foundation elements interaction] Международная научно-практическая конференциия: Актуальные проблемы строительного и дорожного комплексов. 18–21 мая 2004, Йошкар-Ола, 245–250.

Сернов В.А. 2005. Расчет свайных фундаментов с несущими ростверками [Calculation of pile foundations with bearing rafts] Перспективы развития новых технологий в строительстве и подготовка инженерных кадров Республики Беларусь: сб. научн. трудов междунар. научно-метод. межвуз. семинара. Могилев. 2005, 432–435.

Сернов В.А. 2005. Взаимодействие групп свай и несущих ростверков с грунтом основания [Interaction between piled-raft foundation and the soil] Строительная наука и техника. 3: 56–62.

Сернов В.А. Расчет несущей способности и осадки свайного фундамента с несущим ростверком [Calculation of bearing capacity and settlement of deep-raft pile foundations] Геотехника: актуальные теоретические и практические проблемы: межвуз. тематич. сб. трудов. Санкт-Петербург. 2006, 110–114.

Baltic Piling – Mets & Raudsepp (Eds)
© 2013 Taylor & Francis Group, London, ISBN 978-0-415-64334-4

The underpinning of a timber pile foundation

S. Aksenov & A. Nevzorov
Northern (Arctic) Federal University named after M.V. Lomonosov,
Laboratory of engineering geology and foundation engineering, Arkhangelsk, Russia

ABSTRACT: In article the method of the underpinning of a timber pile foundation is considered. The rotten timber piles have been replaced by frost-stable ground pillow. Above a pillow concrete shallow foundation has been executed. The result of 3D-numerical simulation and in-situ observations have a good convergence.

In 2004 the staff of the department of engineering geology and foundation engineering made an investigation of a wooden residential building in Severodvinsk city which had suffered considerable deformation. The following works were done: exploring shafts excavation, cone penetration tests (see Figure 1), drilling of borehole laboratory tests of soils.

The following soils underlie the site:

- filled sand, thickness from 0.8 to 2.2 m;
- peat, thickness from 0.2 to 0.8 m;
- water-saturated fine sand, thickness more than 15.5 m;

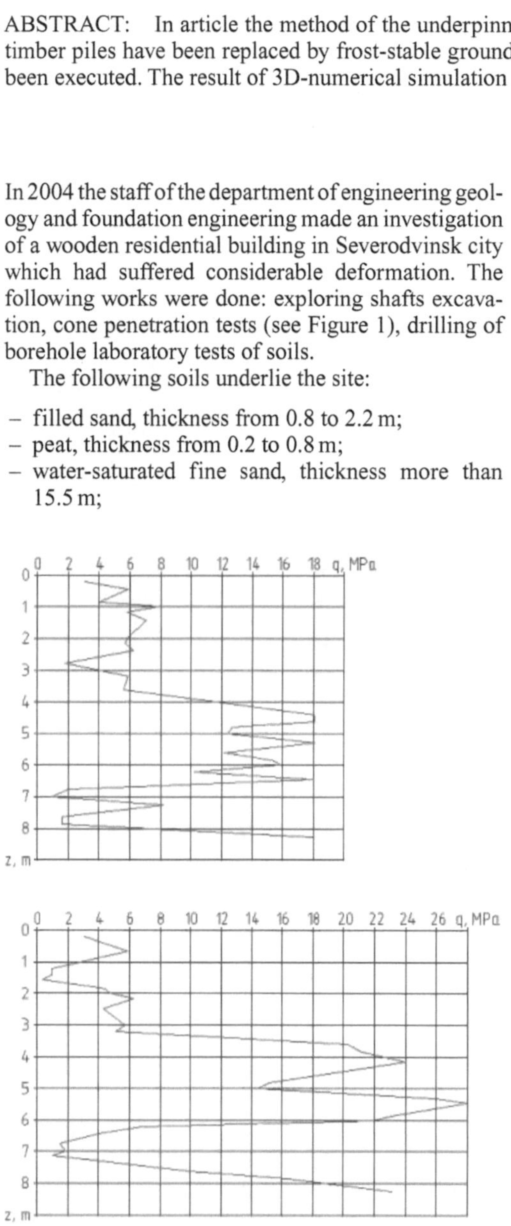

Figure 1. Results of cone penetration tests.

At the depth from 6.2 to 7.8 m there is a mud layer.

Physical and mechanical properties of soils are given in the table.

When the survey was being carried out on April, ground water level was at the depth of 0.8 m from the surface.

Table 1. Soil properties.

	Density, g/cm^3					
Soil	soil	particles	Void ratio	φ, degree	c, kPa	E, MPa
Filled sand	1,86	2,64	0,60	21,0	0	11,0
Peat	1,02	1,52	6,84	7	4,7	0,2
Fine sand	1,86	2,66	0,64	35,4	3,5	27,7
Mud	1,03	2,66	1,55	3,7	10,0	1,9

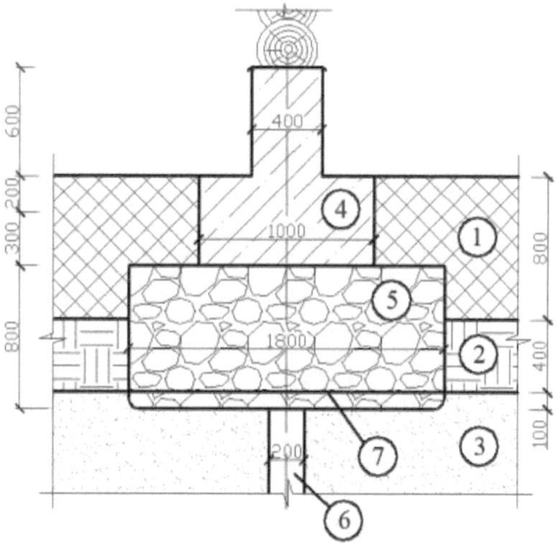

Figure 2. Foundation after underpinning. 1 – filled soil, 2 – peat, 3 – fine sand, 4 – shallow foundation, 5 – gravel, 6 – timber pile, 7 - geotextile.

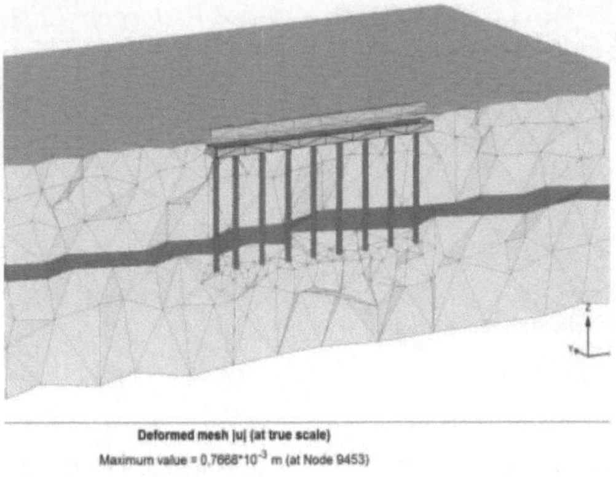

Figure 3. FE Model.

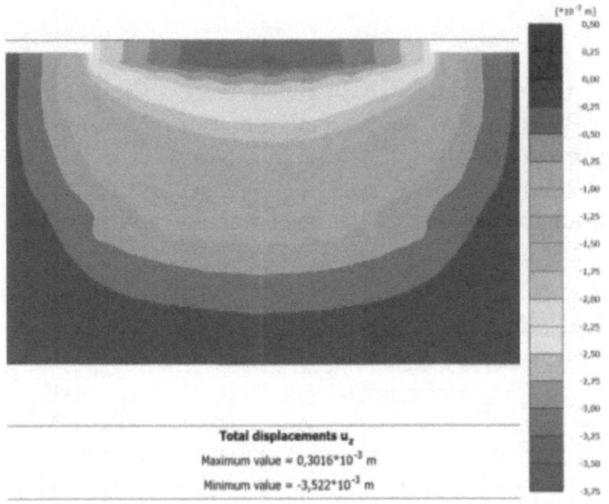

Figure 4. Results of vertical displacements of shallow foundation with timber piles.

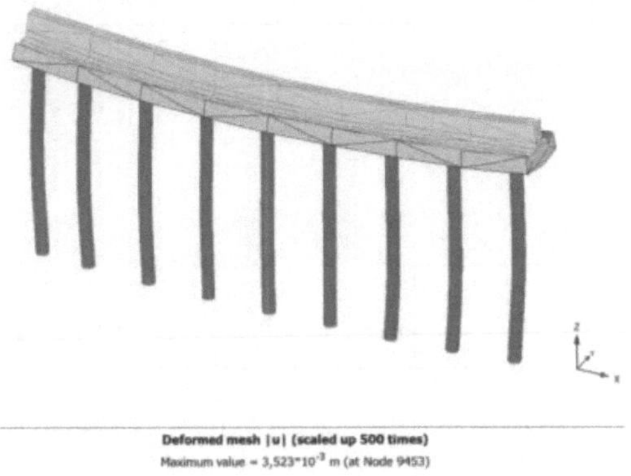

Figure 5. Deformed mesh.

Investigation showed that the building have a wooden pile foundation. Diameter of piles was from 180 to 220 mm, distance between piles ranges from 1.8 to 2.2 m. Building displacements were caused by pile heads rotten from the surface to the depth of 0.6...0.8 m. Up to 80% of pile cross-section was rotten.

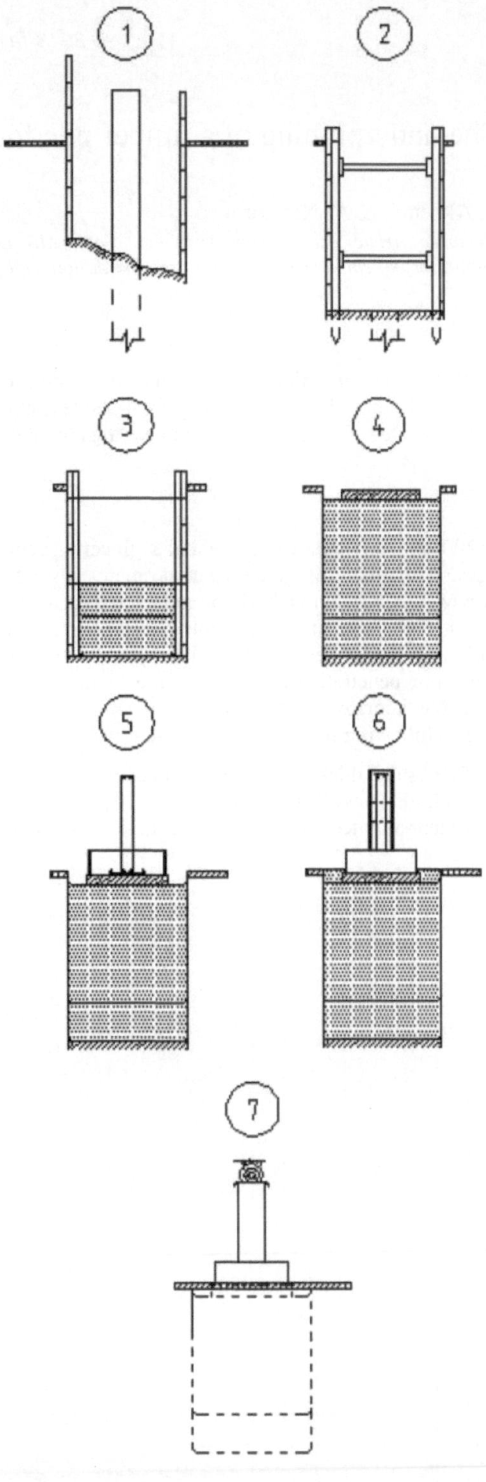

Figure 6. Working phases (cross-section). 1 – soil excavation and cutting of timber pile heads, 2 – strengthening of trench surface, 3 – implementation of compacted gravel bed, 4 – implementation of foundation mat, 5, 6 – concreting of reinforced-concrete continuous spread footing, 7 – implementation of waterproofing.

Figure 7. Phases of foundation reconstruction.

In case like this reinforced concrete caps are usually applied instead of rotten pile heads or remaining piles are banded by low reinforced-concrete beam where reinforced-concrete columns are installed between beam and the wall. When the first method is applied there can appear problems to provide stability of 'pile-cap' joint. When the second method is applied wooden piles are under additional load from reinforced-concrete constructions.

It was decided to use a concrete strip foundation. Footing depth was accepted as 0.5 m. Gravel bed

Figure 8. Reconstructed building.

(fraction 20...40 mm) was set up under the foundation. Wooden piles were cut at the depth of 1.3 m, at the level of top fine sand. Geotextile was laid near the bottom of gravel bed to smooth the displacements.

Thus, as a result of the reconstruction there appeared shallow strip foundation on the gravel bed and on the footing reinforced by wooden piles (see Figure 2).

Before the project was developed the calculation of foundation settling had been made in PLAXIS. The FE model is shown in Figure 3, the calculated results are shown in Figures 4 and 5.

According to the calculations made in PLAXIS 3D the settlement of foundation was about 4 mm.

Foundation reconstruction was implemented in the following stages:

1st stage – preparatory works disassembling of: the floor and the beams;
2nd stage – the hanging of the walls;
3rd stage – excavation of ground;
4th stage – filling the trench of gravel, and a laying the geotextile;
5th stage – erection of strip foundation;
6th stage – transferring the load to the new foundation.

Technology of foundation construction is shown in Figures 6 and 7.

Baltic Piling – Mets & Raudsepp (Eds)
© 2013 Taylor & Francis Group, London, ISBN 978-0-415-64334-4

Designed model of the basis of deep foundations with account of spatial nonlinear long deformation of soils

I.T. Mirsayapov & I.V. Koroleva
Kazan State University of Architecture and Engineering, Kazan, Russia

ABSTRACT: The spatial model of dilating soil is proposed. Hypothesis which says that strength dry friction of the Coulomb deflect from platforms of a limiting condition and function on surface slipping was taken as a principle. The strain creep of soil is described according to hereditary creep theory. Volumetric deformation is summed up change of shape and volume.

The experience of construction of high-rise buildings, which, as a rule, have a well-developed underground part in the form of deep foundation, shows that in many cases the negative processes occurring in the individual elements, as well as in the building as a whole, connected with the lack of study of engineering-geological conditions of the construction site and incorrect prediction of geomechanical processes in the ground arrays ground foundations.

For the qualitative quantitative forecasting the behaviour of soil ranges grounds deep foundations, first of all, it is necessary to determine the strength and deformability of parameters of soil grounds, taking into account their initial stress-strain state, the greater the depth of the active the zone of deformation; regularities of formation and transformation of the stress-strain state in the process of excavation of soil from excavation and subsequent settlement of underground and above-ground parts of the building.

In the basis of deep foundations, which are an integral part of a high-rise building, the formation of the stress-strain state differs significantly from the nature of the formation of the stress-strain state in the basis of the foundations of small medium only.

In these cases, depending on the depth of the basement and of the ratio d/b (d is the depth of the soles of the foundation, b – width of the foundation) change the conditions of deformation of cargo handling of the foundation. Loading of deep foundations (d/b > 2) protrude ground from under the soles do not occur and the limit of the stress state are localized inside the soils in the lower part of the lateral surfaces of the foundation (Figure 1).

Therefore, the loading of the foundation depth is accompanied by the movement of soil down and to the side. In the basis of deep foundation implemented the limit condition, due to which the soil array squeezed in the hand and is implemented by the stress-strain state, similar to the conditions of the true triaxial compression.

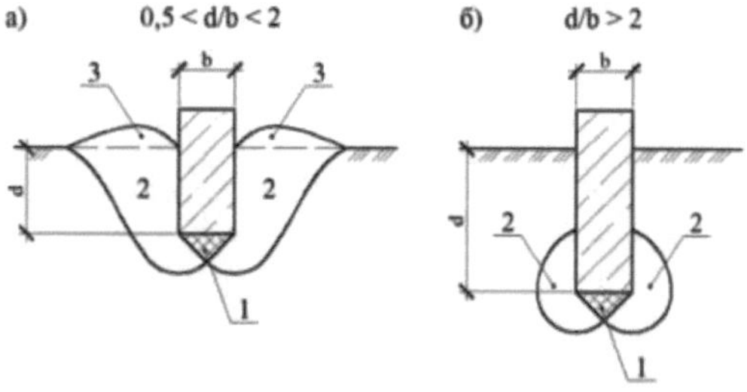

Figure 1. The formation of fields of limiting balance in the basis of the different relative depth of the basement: 1 – dense core; 2 – area of limiting balance; 3 – shafts shift.

The same results were obtained in carrying out numerical simulation of soil on the basis of deep foundation (Figure 2).

In connection with the above, there is a need to evaluation of the bearing capacity and deformations grounds deep foundations with consideration of space of the stress condition and rheological properties of soil under long-term load.

In standard environment, when ground is situated under the influence of it's own weight and external force, complex is formed where each soil element is under the influence of stress tensor. Soil's deflected mode is defined, if in each of it's volume elements with coordinates X, Y, Z components of normal (σ) and shearing (τ) stresses; angular (γ) and space coordinates (u, v, w) are defined. In a state when ultimate strength is reached shearing surface in every volume element's formed and oriented in particular way to direction of main stresses $\sigma_1 > \sigma_2 > \sigma_3$ [10].

Experimental studies of clay soil in long-term triaxial compression [8, 9] have defined the mechanism of tested specimen's failure. The process can be described in a following way: While pressure is applied, consolidated areas in a form of pyramids are formed, pyramids occur in specimen's upper and lower surface and at

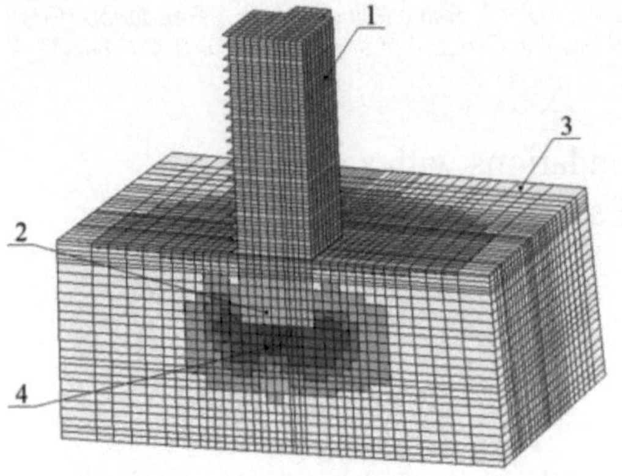

Figure 2. The results of numerical simulation 1 – the above-ground part of the building; 2 – the underground part of the building; 3 – active zone of deformation of soils in the grounds; 4 – zone of the critical state of the soil-base deep Foundation.

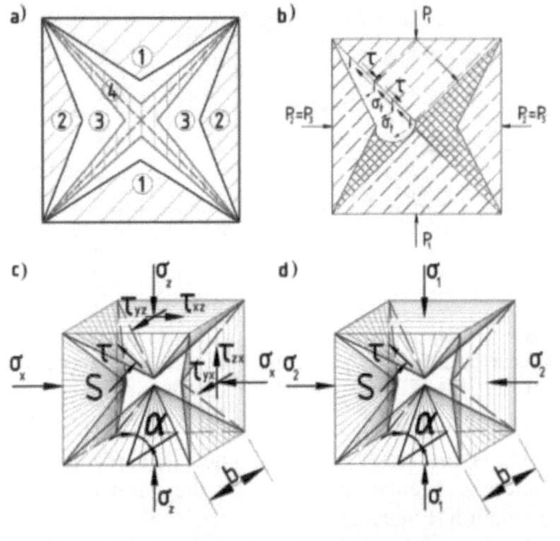

Figure 3. a) – consolidated areas of different density of a specimen in triaxial tests: 1 – vertical consolidated pyramids; 2 – consolidated pyramids at specimen's sides; 3 – uniformed deflected state area; 4 – area of dilatancy; b) – deformed state of clay between pyramids formed; c) – deflected state of volume element in space in random moment of time at preultimatecondition (stresses and tensions are not shown); d) – deflected state of volume element in space at ultimate condition (stresses and tensions are not shown).

specimen's sides (Figure 3a). Geometrical sizes of pyramids mentioned, depends on loading conditions.

To define physical-mechanical properties, soil samples were taken from relevant areas. [9].

As a result of analysis of failure process and areas of different densities, a suggestion was made. Authors state that triaxial compression leads to formation of areas of different deflected state. As a result of gradual increment of load, consolidated pyramids of different shapes and sizes are formed in specimen's upper and lower surfaces and at specimen's sides. Size and shape of pyramids depends on loading conditions (Figure 3).

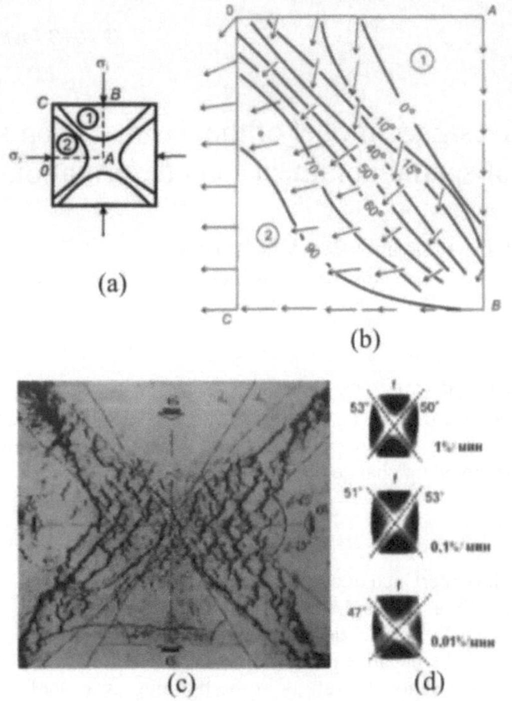

Figure 4. (a) – Loading and unloading soil specimen (data by Boldyrev); (b) – isolines of displacement vectors' angles of inclination: solid areas in soil specimen; (c) – soil specimen's failure picture in triaxial compression device with flexible walls (data by Kryzhanovskii); (d) – Dense clay test results (Higo).

Specimen's deformation occurs as a result of this pyramid's movement, where pyramids are considered as solid bodies. Authors observe improvement of physical-mechanical properties in consolidated areas mentioned above (increment of density up to 11%; φ – 88%; c – 138%). Negative processes related to reduction of physical-mechanical properties are located in areas between pyramids (area 4, Figure 3a) (density reduction up to 43%; ϕ – 45%; c – 67%). At the same time Mohr's circles drawn based on results of series of tests conducted, show reduction of internal tension angle ϕ up to 16% and cohesive force c – 6% in an integral volume. Visual investigation of shearing surface after specimen's failure, shows that soil at this area is subjected to both detachment and shearing (Figures 3a, 3b).

Similar schemes of failure were observed in surveys done by Boldyrev [1], Kryzhanovskii [5], Higo [12] that correspond to surfaces of normal and shearing stresses described in [11] (Figure 4).

Results of experimental surveys can be represented as shown below:

$$\varepsilon_V = \varepsilon_V^0 \pm \varepsilon_V^D \quad (1)$$

where ε_V^0 – volume strain during uniform pressure; ε_V^D – volume strain caused by stress deviator.

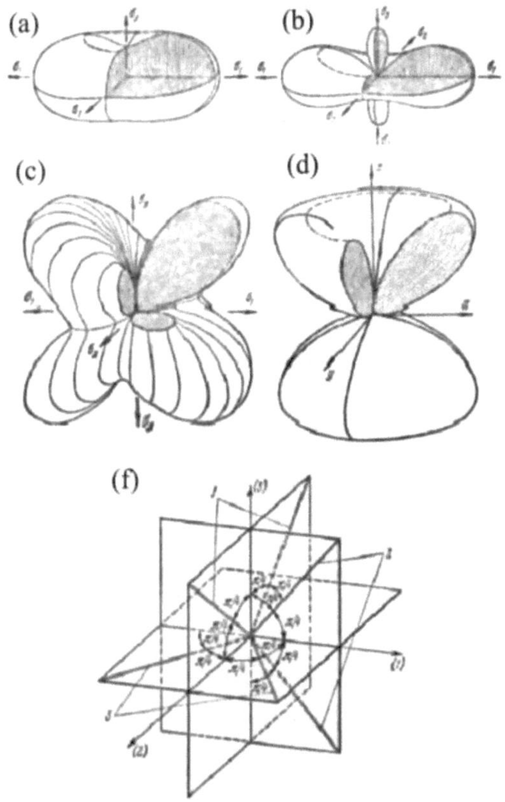

Figure 5. (a) – Normal stresses surface during general state of stress in case of same signs of all three main stresses; (b) – Normal stresses surface during general state of stress in case of different signs of main stresses ($\sigma_2 = 0.5\sigma_1; \sigma_3 = -0.5\sigma_1$); (c) – Normal stresses surfacewhen $\sigma_1 \backslash = \sigma_2 \backslash = \sigma_3$; (d) – Normal stresses surfacewhen $\sigma_1 \backslash = \sigma_2 = \sigma_3$; (f) – directions with maximal shear displacements: 1 – directions with shear displacement $\gamma_1 = \varepsilon_2 - \varepsilon_3$; 2 – directions with shear displacement $\gamma_2 = \varepsilon_3 - \varepsilon_1$; 3 – directions with shear displacement $\gamma_3 = \varepsilon_1 - \varepsilon_2$.

ε_V^0 value does not depend only on σ_m but also on shearing stresses τ, occurring as a result of difference between soil's resistance to compression and tension:

$$\varepsilon_V^0 = f_1^*(\sigma_m, \tau, t) \tag{2}$$

With the assumption that shearing displacements γ_i depend on intensity of shearing stresses τ_i, normal stresses σ_m and type of deflected state, displacements of forming can be represented as following:

$$\varepsilon_V^D = f_2^*(\sigma_m, \tau_i, t) \tag{3}$$

Taking in consideration the law of volume deformation change, found by authors [8, 9, 6], according to which Poisson's ratio in soil remains constant ($\nu = const$) during tests on creep age, we suggest that volumetric deformations change and deformations of forming develop in time similarly.

Volumetric deformation of soil in time is described in compliance with Bolzman-Valter's heredity theory, modified by Maslov-Arutunjan in respect to soil. According to this theory, total shear or volume deformation during random way of loading can be represented as following:

$$\varepsilon_i = \frac{\sigma_i(t)}{G(t)} + \frac{1}{G(t)} \int_{\tau}^{t} K_\gamma(t,\tau) \cdot \sigma_i(\tau) \cdot d\tau; \tag{4}$$

$$\varepsilon_v = \frac{\sigma(t)}{K(t)} + \frac{1}{K(t)} \int_{\tau}^{t} K_V(t,\tau) \cdot \sigma(\tau) \cdot d\tau; \tag{5}$$

where $G(t)$ and $K(t)$ – shear modulus and bulk modulus;

$K_\gamma(t, \tau)$ – creep kernel, represented as velocity of shear deformation with unit value of loading intensity;

$K_V(t, \tau)$ – creep kernel, represented as velocity of volumetric deformation with unit value of mean stress.

Analysis of test data found helped to define creep kernels $K_\gamma(t, \tau)$, $K_V(t, \tau)$ for soil in general state of stress.

While patterning equations of clay soil deformation in general state of stress processes of clay hardening (aging) were taken into account, that is the reason why aging function $\varphi(t)$ is used. Taking hardening process into account, creep kernels can be represented as shown below:

$$K_\gamma(\sigma, t-\tau) = \begin{cases} \varphi_\gamma(\tau) \cdot \dfrac{\delta_{\gamma 1}}{(t-\tau)} \cdot \delta_{\gamma 2} & npu \quad 0 \le t \le t_{нm} \\ \varphi_\gamma(\tau) \cdot C(\sigma) & npu \quad t > t_{нm} \end{cases} \tag{6}$$

$$K_V(\sigma, t-\tau) = \begin{cases} \varphi_V(\tau) \cdot \dfrac{\delta_{V1}}{(t-\tau)} \cdot \delta_{V2} & npu \quad 0 \le t \le t_{нm} \\ \varphi_V(\tau) \cdot C(\sigma) & npu \quad t > t_{нm} \end{cases} \tag{7}$$

where t_{Hm} – deformation speed hold start.

Based on suggestions stated above, we can assume that in the beginning stage of long-lasting triaxial stress deflected state of soil specimen is inhomogeneous. Deviator loading and long-term isolation during loading accompanied by initiation and development of set of shear surfaces and specimen uniformity break. Surface positions change in correlation with time and process of deviator increment. Therefore negative dilatancy (aeration) of clay soil during long-term triaxial stress is localized in the range of potentially available areas of limit equilibrium [9]. Since aeration of soil occurs in straitened localized areas, significant values of dilatancy stresses are reached in zones surrounding aeration area, which is considered to be buffer area where stresses are transferred from vertical pyramids to area of undisturbed soil (Figure 3a, 3b). Described above is the mechanism of shear and compacted areas correlation. When in limiting state clay soil's failure is localized in areas between pyramids. On this stage deflected state of soil specimen can be assumed as homogeneous.

Under the assumption of determined scheme of failure and deflected state areas general scheme of inelastic long-term clay soil deformation is suggested. Suggested scheme is based on modified model of non associated plastic flow proposed by Nicolaevskii [7], according to which Coulomb's dry friction force is deviated from the platform of limit equilibrium and acting in a direction of tangential glide of physical particles. These platforms' directions can only be defined if soils' deformed state is taken in account.

Taking into consideration, that unrelatedly to the degree of starting uniformity of soil specimen's deflected state, failure always takes place in the space of main stresses, overlapping the space of main stresses σ and space of main deformations ε remaining the principle of stresses and deformation speed's tensors coaxiality [7]. We will assume that Coulomb's dry friction law relates projections of forces acting on platforms of limited equilibrium on its normal and on the platform itself. Thus, condition of long-term loading flow can be represented as above:

$$|t| = S \cdot tg\varphi(t,\tau) + c_0(t,\tau), \tag{8}$$

where $S = \sigma_1 \cdot l \cdot l' + \sigma_2 \cdot m \cdot m' + \sigma_3 \cdot n \cdot n'$

$$t = \sqrt{(\sigma_1 \cdot l \cdot m' - \sigma_2 \cdot l' \cdot m)^2 + (\sigma_2 \cdot m \cdot n' - \sigma_3 \cdot m' \cdot n)^2 + + (\sigma_3 \cdot n \cdot l' - \sigma_1 \cdot n' \cdot l)^2}$$

$\varphi(t,\tau)$ – varying internal tension angle;
$c_0(t,\tau)$ – varying cohesive force;
$l,\ m,\ n$ – cosines to platforms of limit equilibrium;
$l',\ m',\ n'$ – cosines to platforms of gliding.

Special orientation of limit equilibrium platforms is defined by equations below:

$$l^2 = \frac{\overline{I_3}}{\overline{I_2} \cdot \overline{\sigma_1}}; \quad m^2 = \frac{\overline{I_3}}{\overline{I_2} \cdot \overline{\sigma_2}}; \quad n^2 = \frac{\overline{I_3}}{\overline{I_2} \cdot \overline{\sigma_3}}; \tag{9}$$

where $\overline{I_2} = \overline{\sigma_1} \cdot \overline{\sigma_2} + \overline{\sigma_2} \cdot \overline{\sigma_3} + \overline{\sigma_3} \cdot \overline{\sigma_1}$ and $\overline{I_3} = \overline{\sigma_1} \cdot \overline{\sigma_2} \cdot \overline{\sigma_3}$ – second and third invariants of tensors of modified main stresses $\overline{\sigma_i} = \sigma_i + H (i = 1,\ 2,\ 3)$.

$H = c/ctg\varphi$ – uniform compression defined by Mohr-Coulomb's hypothesis

φ – internal tension angle

Expressions for cosines to gliding platform's normal are as in [7]:

$$\left.\begin{aligned} (l')^2 &= \frac{1}{3} \cdot \frac{3 \cdot d\varepsilon_2 \cdot d\varepsilon_3 - I_2 + \sqrt{I_2^2 - 3 \cdot I_1 \cdot I_3}}{(d\varepsilon_1 - d\varepsilon_2) \cdot (d\varepsilon_1 - d\varepsilon_3)}; \\ (m')^2 &= \frac{1}{3} \cdot \frac{3 \cdot d\varepsilon_1 \cdot d\varepsilon_3 - I_2 + \sqrt{I_2^2 - 3 \cdot I_1 \cdot I_3}}{(d\varepsilon_2 - d\varepsilon_1) \cdot (d\varepsilon_2 - d\varepsilon_3)}; \\ (n')^2 &= \frac{1}{3} \cdot \frac{3 \cdot d\varepsilon_1 \cdot d\varepsilon_2 - I_2 + \sqrt{I_2^2 - 3 \cdot I_1 \cdot I_3}}{(d\varepsilon_3 - d\varepsilon_1) \cdot (d\varepsilon_3 - d\varepsilon_2)}, \end{aligned}\right\} \tag{10}$$

where $d\varepsilon_1,\ d\varepsilon_2,\ d\varepsilon_3$ – increments of main deformations.

$I_1 = d\varepsilon_1 + d\varepsilon_2 + d\varepsilon_3, I_2 = d\varepsilon_1 \cdot d\varepsilon_2 + d\varepsilon_2 \cdot d\varepsilon_3 + d\varepsilon_3 \cdot d\varepsilon_1, I_3 = d\varepsilon_1 \cdot d\varepsilon_2 \cdot d\varepsilon_3$ – first, second and third invariants of deformation increments

Based on the fact that the long-term worksites compression increment of inelastic deformation takes place mainly at the stages of девиаторного loading and long shutter, exposure of potential sites of limiting state (shift) in the process of inelastic deformation is defined in accordance with the formula:

$$\alpha(t,\tau) = \arccos \sqrt{\frac{1}{3} \cdot \frac{3d\varepsilon_2 \cdot d\varepsilon_3 - I_2 + \sqrt{I_2^2 - 3I_1I_3}}{(d\varepsilon_1 - d\varepsilon_2)(d\varepsilon_1 - d\varepsilon_3)}}, \tag{11}$$

where $d\varepsilon_1 = d\varepsilon_{1D} + d\varepsilon_1(t)$, $d\varepsilon_2 = d\varepsilon_{2D} + d\varepsilon_2(t)$, $d\varepsilon_3 = d\varepsilon_{3D} + d\varepsilon_3(t)$ – increment the main deformations;

$d\varepsilon_{1D}, d\varepsilon_{2D}, d\varepsilon_{3D}$ – increment the main inelastic deformations on the stage of the девиаторного loading; $d\varepsilon_1(t), d\varepsilon_2(t), d\varepsilon_3(t)$ – increment the main inelastic deformations at the stage of a long exposure due to the strain creep;

$$d\varepsilon_1(t) = \frac{1}{E(t)} \int_{\tau}^{t} K_0(t,\tau) \cdot \sigma_1(\tau) \cdot d\tau;$$

$$d\varepsilon_2(t) = \frac{1}{E(t)} \int_{\tau}^{t} K_0(t,\tau) \cdot \sigma_2(\tau) \cdot d\tau;$$

$$d\varepsilon_3(t) = \frac{1}{E(t)} \int_{\tau}^{t} K_0(t,\tau) \cdot \sigma_3(\tau) \cdot d\tau;$$

where $K_0(t,\tau)$ – the core of the volume creep; $E(t)$ – the module of deformation; $I_1 = d\varepsilon_1 + d\varepsilon_2 + d\varepsilon_3, I_2 = d\varepsilon_1 \cdot d\varepsilon_2 + d\varepsilon_2 \cdot d\varepsilon_3 + d\varepsilon_3 \cdot d\varepsilon_1, I_3 = d\varepsilon_1 \cdot d\varepsilon_2 \cdot d\varepsilon_3$ – the first, the second and the third invariants of incremental deformations.

Taking in account surveys conducted and model described above, long-term strength condition can be represented as follows:

$$4 \cdot [\sigma_V(t) \cdot A_{sh} \cdot \cos\alpha_1(t) + \tau_V(t) \cdot A_{sh} \cdot \sin\alpha_1(t)] \geq \sigma_1 \cdot A_1; \tag{12}$$

where $A_{sh} = \dfrac{b^2}{4\cos\alpha_2(t)}$;
$A_1 = b^2$ – cube's face area;
α_1 – corner incline of balance's limit plan;
α_2 – corner incline of shear's plan;

$\sigma_V(t) = \sigma_1 \cdot l(t) \cdot l'(t) + \sigma_2 \cdot m(t) \cdot m'(t) + \sigma_3 \cdot n(t) \cdot n'(t) + \sigma_d(t)$ – normal stresses; $\sigma_d(t) = E/(1+\nu) \cdot r \cdot \Delta\delta_d$ – dilatancy stresses; $\tau_V(t) = S \cdot tg\varphi_0(t,\tau) + c_0(t,\tau)$. – shear stresses.

Thus, soil's strength, during long-term triaxial compression, depends on variation of internal tension angle, cohesive force and angle of limit equilibrium surface incline.

According to kinetic theory of soil deformation, proposed by Vyalov, Zaretskii [2, 4], failure takes place, when damage level by microfractures in limit equilibrium area reaches its critical value.

Vyalov [2] states that soil strength decrease in time generally occurs as a result of cohesive force lowering, while internal tension angle change is insignificant.

Based on test results [1, 2, 4, 8, 9] following scheme of creeping deformation development and long-term strength change can be suggested. Depending on magnitude and duration of load appliance, two mutually compensated processes occur in multistate clay soil – strengthening conditioned by defect treatment and more compact integrity of particles and weakening conditioned by re-orientation of particles and forming and development of micro and macrofractures (Figure 3b). When weakening prevails over strengthening, phase of failure and progressive creeping occurs. At this stage intensive disintegration of microstructure takes place and particles start to re-direct, though these processes take place onle in limit equilibrium areas where values of strength are lower and fractures are developed.

Taking in account statements above, expression for varying cohesive force can be represented as follows:

$$c_0(t,\tau) = K_{Ict}^M \cdot q(S), \tag{13}$$

where $q\ (S)$ – summary fracture length function; K_{Ict}^M – soil fracture tip stresses intensity coefficient.

Finding S from (11), taking in account that $K_{Ict}^M = \sqrt{2E\gamma}$ according to [3] S can be represented in relation to external force T as follows:

$$S_* = T^{-1} \cdot \left(T \Big/ \sqrt{2E\gamma} \right), \tag{14}$$

where T^{-1} is rising function inverse to T.

Using procedure, suggested in [3] function for soil weakening can be represented as follows:

$$\eta(t,\tau_1) = m(t,\tau_1) \cdot \lambda(t,\tau_1) \cdot \sqrt{\frac{K(\tau_1)}{K(t)} \cdot \frac{1}{1 + K(\tau_1) \cdot C(t,\tau_1)}}. \tag{15}$$

Cohesive force varying in time can be represented as follows:

$$C_0(t,\tau_1) = C_0(\tau_1) \cdot m(t,\tau_1) \cdot \lambda(t,\tau_1) \cdot \sqrt{\frac{K(\tau_1)}{K(t)} \cdot \frac{1}{1 + K(\tau_1) \cdot C(t,\tau_1)}}, \tag{16}$$

where $C(t, \tau_1)$ – volumetric creeping value; $C_0(\tau_1)$ – initial value of cohesive force during short-term loading; $m(t, \tau_1)$ – function of soil strengthening in expense of water-colloid bonds; $\lambda(t, \tau_1)$ – function of soil strengthening in expense of soil bonds during long-term deformation.

Internal tension angle change is defined according to orientation of limit equilibrium platforms' change during long-term inelastic deformation.

REFERENCES

Boldyrev G.G. 2008. Methods definition mechanical properties of soils. Sostoyanie voprosa. Penza.

Kryzhanovskii A. L., Vil'gel'm Yu. S., Rakhmanov T. 1983. Determination of the angle of friction of granular soils in triaxial apparatus and shear devices // Soil Mechanics and Foundation Engineering. – No. 6. – P. 24–27.

Meschyan S.R. 1985. Experimenal rheology of Clayey Soils. – Moscow: Publishers Nedra.

Vyalov S.S. 1978. Rheological principles of soil mechanics. – Moscow: HSPH.

Y.V. Zaytsev. 1991. Mechanic destructions for builder. – Moscow: HSPH.

Zaretskii Yu.K. & Vyalov S.S. 1971. Structural mechanics of clay soils. // Soil Mechanics and Foundation Engineering. – No. 3.

Baltic Piling – Mets & Raudsepp (Eds)
© 2013 Taylor & Francis Group, London, ISBN 978-0-415-64334-4

Instrumentation of timber piles with removable extensometers – Description of the experimental site

J. Christin & A. Le Kouby
University Paris Est, IFSTTAR, France

P. Reiffsteck
Terrasol, Paris, France

M. Grenier
University of New Hampshire, Darham, USA

ABSTRACT: In France, construction on timber piles foundation was given up at the end of the 19th century. This abandonment is partly due to the invention of immersed concrete in the middle of this century. Nowadays, deep foundations in France are mainly designed based on the results obtained from pressuremeter tests. Shaft friction capacity factors are determined based on the measured limit pressures and empirical relations are used to obtain the unit shaft friction. These empirical relations were developed from the results of load tests carried out on concrete and steel piles instrumented with removable extensometers. Currently, there is still no available modern method to estimate the shaft friction on wooden piles from in-situ tests. The "Pieux Bois" research project (Timber Piles in English), funded by French ministry of sustainable development, aims to reintroduce this technique in current French piling practice.

This paper explains the adaptation of removable extensometers for this application. The experimental site on which the timber piles were driven is described. Timber pile driving curves are also presented and discussed.

1 INTRODUCTION

The construction of the first deep foundations on timber piles in France dates back to the Neolithic period, 6000 years ago. During this period, timber pile foundations were driven in the soil using human strength. Timber piles dated from 600 B.C. were notably discovered in France and Switzerland, near the lakes of Neuchatel and Chalain (Maillard 2006). The first records of this technique were written by the Romans, who were the first to develop pile driving. From the Renaissance period to the industrial era, this particular type of foundations evolved: timber piles were driven before the construction of the mattress and the grillage on the top of piles. During the 19th century, in France, mattress and grillage were given up and replaced by submerged concrete, before the definitive abandonment of this technique at the end of this century (Levillain 1980).

In spite of the abandonment of timber pile construction, many buildings and bridges which were built on timber pile foundations, still stand in France. The "Belle Croix" or "Rousseau" bridges on the river "Loire", the castle of Chambord, the "Grand Palais" in Paris, or the port docks in Rouen are a few examples of French monuments built on timber pile.

This abandonment happened before the invention and development of the prebored pressuremeter by L. Menard in the 1950's. The results of pressuremeter tests are the basis of the French design method of deep foundations. The French design method of deep foundations is written in the French standard "Fascicule 62 Titre V". This direct design method is based on the calculation of the unit shaft friction of the pile versus the limit pressure (Fig. 1).

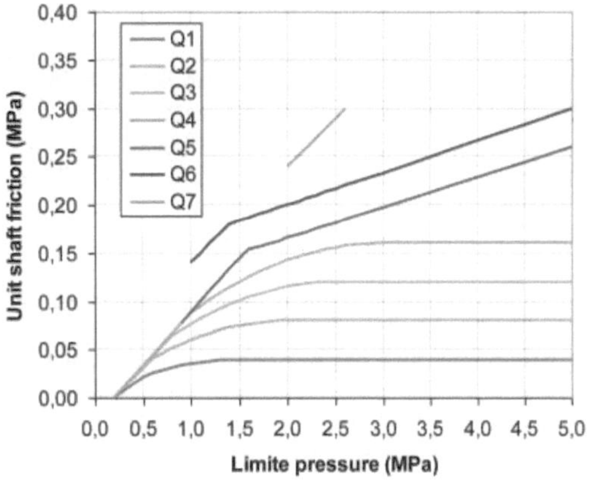

Figure 1. Unit shaft friction versus limit pressure.

Each curve (Q1 to Q7) depends on the pile type and the nature of the soil fixing the soil shaft interaction curves. These graphs were fitted using the results of static load tests made on concrete and steel piles. These piles were instrumented with removable extensometers and the shaft friction was measured at different depths. The case of timber piles is not taken into account.

So the lack of information in the French design method relative to timber pile foundations does not allow the reintroduction of this technique in the current practice.

Nevertheless, the US and Holland set up a national design method (respectively the Timber Pile Design and Construction Manual (American Wood Preservers Institute 2002) and the NEN 6743 (Dutch Standardization Institute, 1991)) which allow the design of timber piles.

Constituted of five partners (3 companies and 2 laboratories), the "Pieux Bois" research project (Timber Piles in English, 2010–2013) aims at setting up a national design method based on the Menard pressuremeter. This new design method will add a curve on (Fig. 1) for the timber piles based on the results on a set of pile static load tests.

The present work aims, in a first part, to present the timber piles geometry and mechanical characteristics; in a second part, explain the instrumentation of timber piles with removable extensometers; and in the third part, describe the driving of the piles on an experimental site.

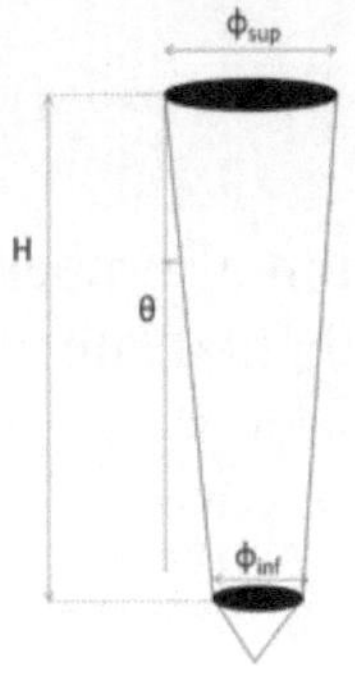

Figure 2. Timber pile taper (American Wood Preservers Institute 2002).

Table 1. Timber piles geometry [(i) represents the order of instrumentation in laboratory].

Specie (number)	Length m	Top diam cm	Tip diam cm	Taper mm/m
Beech (1)	5	27.2	24.4	2.8
Beech (2)	5	32.3	29.4	2.9
Pine (3)	5	26.7	24.2	2.5
Pine (4)	5	24.8	23.1	1.8
Oak (5)	5	27.1	23.9	3.2
Oak (10)	5	26.9	23.6	3.3
Acacia (8)	5	22.1	19.6	2.5
Acacia (9)	5	23.1	21.2	1.9

2 CHARACTERISTICS OF TIMBER PILES

2.1 *Piles geometry*

Timber piles tested in the research project are of four different species: oak and acacia because they are considered the species least susceptible to rot; pine because it is considered the specie most susceptible to rot; beech because its mechanical properties are between those of acacia and pine. Eight instrumented timber piles will be tested.

The natural taper of timber piles was conserved, so the piles were not squared. The pile shape was circular, their length was 5 m, and the piles were stripped of bark. Pile taper was calculated from the formula (American Wood Preservers Institute 2002):

$$\tan(\vartheta) = \frac{\Phi_{sup} - \Phi_{inf}}{2 * H} \quad (1)$$

The taper θ or $\tan(\theta)$ can be expressed in degrees or in mm/m.

Table 1 gives the timber piles' tip and top geometry and its taper.

Each specie is represented by two piles on the experimental site. This will allow evaluation of the influence of piles taper and diameter on the unit shaft resistance.

2.2 *Piles mechanical characteristics*

The density of the timber piles was derived using a volume measurement method in the laboratory by sampling a slice of wood. The volume of the slice was calculated by measuring the displacement of the volume of water in a beaker. The wood was considered "green", or that its moisture was approximately the same as its natural moisture (about 30%).

The wood is natural material, and its internal structure and shape are heterogeneous. Thus, the volume and the modulus of elasticity of the pile could not be calculated accurately with theoretical formula. So the latter was determined from the natural frequency of free vibration of piles. Three accelerometers (one at the quarter, one in the middle and one at three quarters of the pile length) were placed on each timber pile (Fig. 3). The hammer strike created a vibration within the pile and the free vibration acceleration time records collected were stored in a computer. Frequency domain analysis was performed on the free vibration measurements through Fast Fourier Transforms (FFT) (Fig. 4).

Then timber piles and their boundary limits were modelled on the FEM software CASTEM and the value of the modulus of elasticity selected was the one for which the resonant frequency given by the model was equivalent to that measured by accelerometers.

Table 2 presents the mechanical characteristics of the eight timber piles.

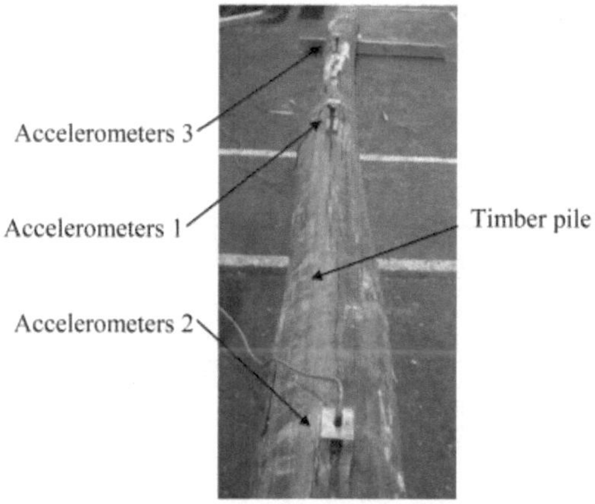

Figure 3. Accelerometers on timber pile.

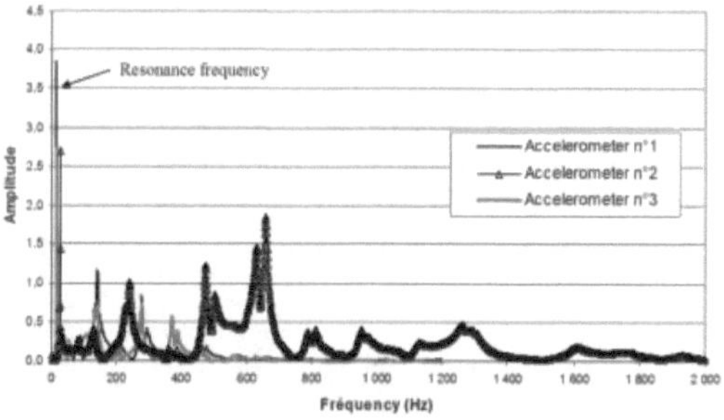

Figure 4. Frequency domain.

Table 2. Timber piles mechanical characteristics.

Specie (number)	Weight kg	Density kg	Resonance frequency Hz	Modulus of elasticity MPa
Beech (1)	136	0.655	18.2	6090
Beech (2)	275	0.709	21.1	8650
Pine (3)	153	0.828	19.9	9050
Pine (4)	120	0.587	20.5	9440
Oak (5)	201	0.723	16.4	7850
Oak (10)	192	0.723	14.6	6700
Acacia (8)	150	0.806	15.2	10700
Acacia (9)	153	0.757	18.8	12900

The density of timber pile 5 (oak) was not measured and was taken as equal to the density of timber pile 6 (same specie).

3 INSTRUMENTATION OF TIMBER PILES

3.1 *Objectives of instrumentation*

The aim of the instrumentation of timber piles with removable extensometers is to measure at different depths, the elastic shortening of the piles versus the load applied on the top of the pile

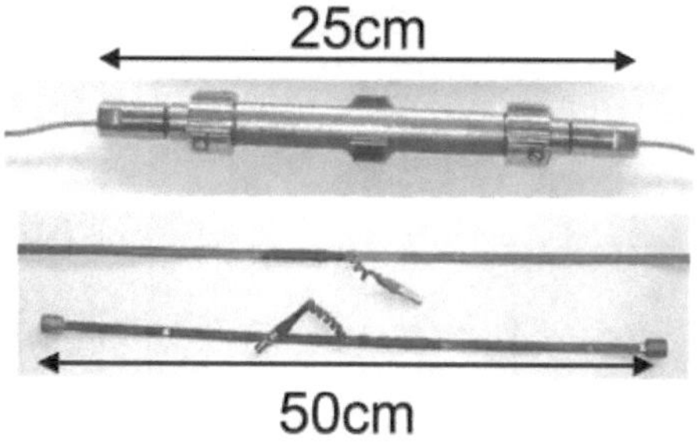

Figure 5. Packers and ribbons with strain gauges (Le Kouby & Rocher-Lacoste, 2012).

(Bustamante & Jézéquel 1975). The values of pile shortening allow, through the modulus of elasticity, calculation of the distribution of the load along the pile and then determination of the unit shaft friction between the pile and the soil at different depths and the load transfer at the tip. Knowing the limit pressure of the soil at the same depths, the curve "timber pile unit shaft friction versus limit pressure" can be drawn.

3.2 *Removable extensometers*

The elastic shortening is measured with a chain of removable extensometers. This chain consists of metallic ribbons on which strain gauges are placed. The ribbons are connected together by inflatable packers (Fig. 5).

In practice, the set of ribbons and packers is placed in a steel tube attached to the pile. Once the lowest packer is inflated, the system is placed under mechanical tension, and then the other packers are inflated. They follow the displacement of the steel tube (and also the pile) and the strain gauges measure the vertical strain at each level limited by two packers.

The non-linearity, the natural taper of the timber pile, and the difficulties in attaching the steel tube to a pile timber do not allow use of this practice. So a new guiding system for the removable extensometers was designed and fixed to the pile.

3.3 *Instrumentation of timber piles*

A trough was cut into the entire length of the most linear side of each timber pile. The width of the trough was 80 mm, the depth was 60 mm, and the trough was cut with a six-axis robot. Metal pieces, called "fixed pieces", 60 mm wide and 80 mm long were attached to the trough, at right angles with packers. The spacing between two "fixed pieces" was 750 mm (Fig. 6).

Once the "fixed pieces" were attached to the trough, cylindrical metal tubes, 25 mm in diameter and 25 cm long, with chamfered edges and diametrically drilled 3 times, were affixed to the "fixed pieces". The metal tubes fit into the V-groove in the "fixed pieces" (Fig. 7).

Rectangular metal tubing was placed along the length of the trough to protect the cylindrical metal tubes and fixed pieces during pile driving. Holes 26 cm

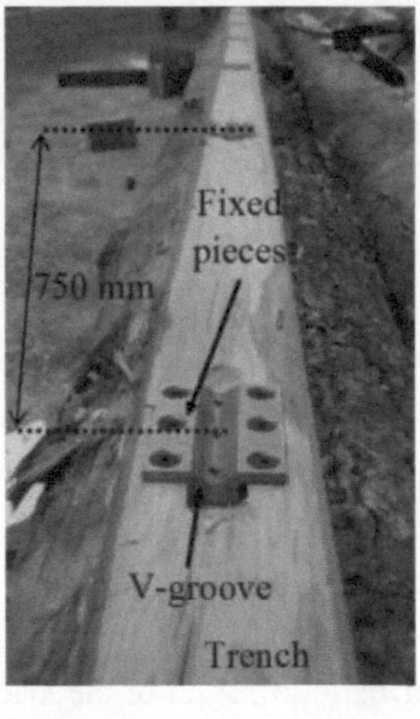

Figure 6. Fixed pieces.

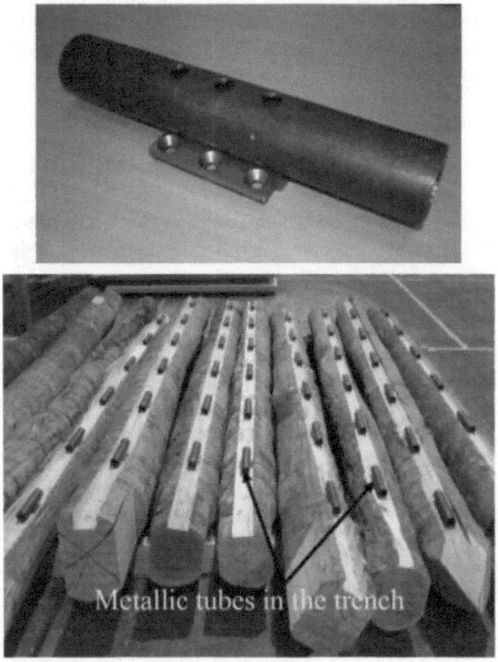

Figure 7. Metallic tube on the "fixed piece" and in the trench of the pile.

Figure 8. Rectangular tubes protecting "fixed pieces" and cylindrical metal tubes.

Figure 9. Reinforcement of the piles tip with metallic strips.

Figure 10. Pile head machined.

long and 80 mm wide were machined in a rectangular metal tubing which fit over the cylindrical metal tubes. The rectangular metal tubing was attached to the timber pile with wood screws (Fig. 8).

The pile tip was machined with the six-axis robot in a pyramidal shape. One of the extremities of the rectangular metal tube at the pile tip was bevelled to match the shape of the pile tip. The four faces of the pile tip were protected with metallic strips (Fig. 8 and Fig. 9).

The pile head was also machined with the robot. Two diametrically opposed cuts, 20 cm apart, were cut on the pile head. These cuts made the attachment of the pile driving helmet easier (Fig. 10).

4 EXPERIMENTAL SITE

4.1 *Ground conditions*

The site is located about 20 km from Rouen, near the Seine river (France). It belongs to Rouen Harbour Authorities.

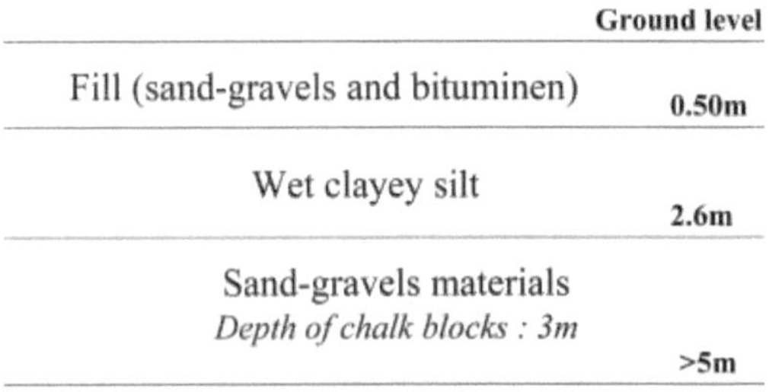

Figure 11. Ground conditions.

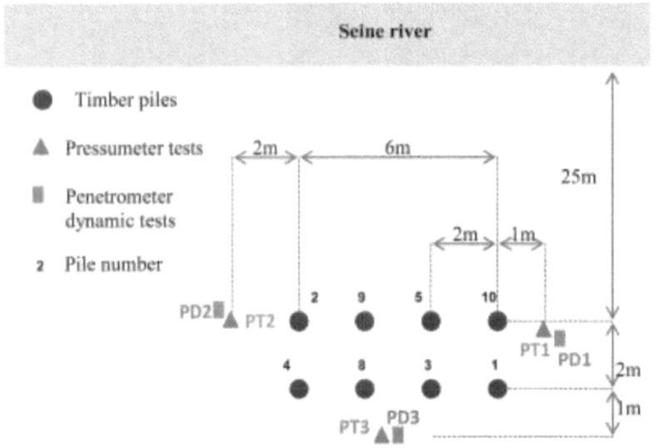

Figure 12. Location of pressuremeter and dynamic penetrometer tests.

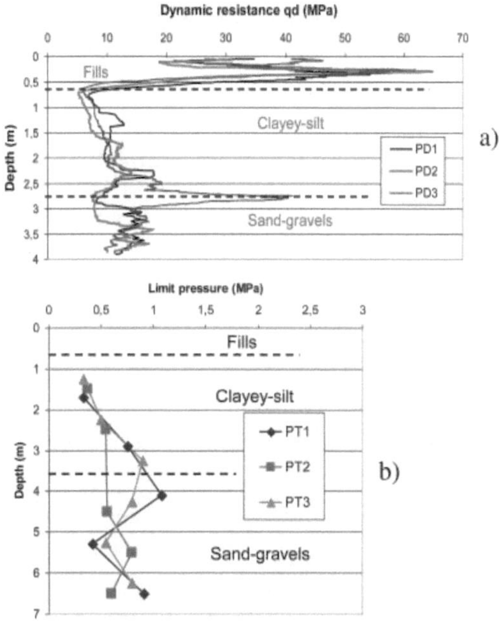

Figure 13. Dynamic penetrometer (a) and pressuremeter (b) tests results.

The geology of the site is detailed on Fig. 11. Ground conditions consist of about 0.50 m of fill (sand-gravels with bitumen) over 2.60 m of wet clayey silt, underlain by sand and gravel materials. Chalk blocks were encountered at a depth of 3 m in the sand and gravel materials. Soil investigations conducted with a shovel did not allow encounter groundwater. Nevertheless, as the Seine river is close to the site, the depth of the water level can be estimated at about 5 m.

4.2 *Soil investigations*

Soil investigations consist of pressuremeter and dynamic penetrometer tests. Three pressuremeter tests were performed to depths of 6.5 m. Three dynamic penetrometer tests were performed next to the pressuremeter tests, in order to determine correlations between the limit pressure and dynamic resistance. The location and results of the tests are presented on Fig. 12 and Fig. 13. This first campaign will be enriched with SPT and CPT sounding.

The three soil layers can be identified from the dynamic penetrometer tests results. The dynamic resistance of the first layer (fill) is about 60 MPa. The mean resistance of the second layer (clayey-silt) is about 10 MPa. The increase of the resistance measured from a depth of 3–3.5 m identifies the third layer (sand-gravels). The high resistance measured on the dynamic penetrometer no. 3 is probably due to a chalk block.

Table 3. Mechanical characteristics of soil layers.

	Mechanical characteristics MPa
Fill	$20 < q_d < 65$
Clayey silt	$5 < q_d < 15$
	$0.3 < p_l < 0.8$
	$3.6 < E_M < 4.6$
	$7.5 < E_M/p_l < 10.5$
	$q_{d,mean} = 10.1$
	$16.7 < q_d/p_l < 18.7$
Sand-gravels	$10 < q_d < 17$
	$0.4 < p_l < 1.1$
	$6 < E_M < 11$
	$7.5 < E_M/p_l < 9.5$
	$q_{d,mean} = 14$
	$18 < q_d/p_l < 25$

Table 3 presents minimum and maximum values of the dynamic resistance q_d, the limit pressure p_l, the pressuremeter modulus E_M and the E_M/p_l ratio measured for each layer.

The high values of the resistance in the sand and gravel materials due to the chalk blocks were not considered in Table 3.

4.3 *Timber piles driving*

4.3.1 *Pile driving resistance*

After the timber piles were instrumented in the laboratory, they were transported to the experimental site.

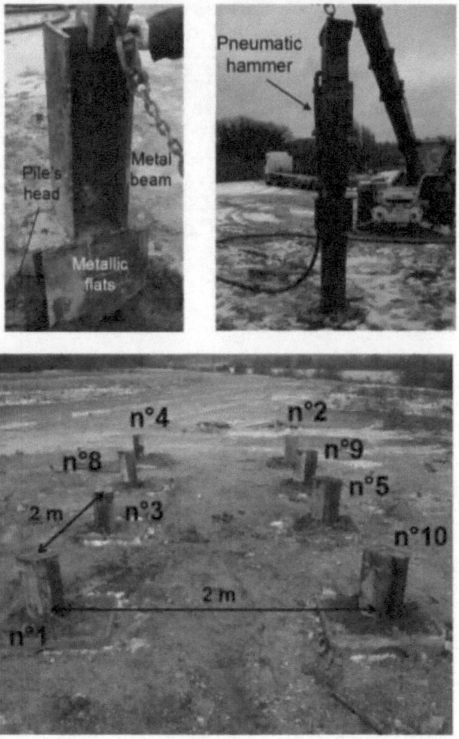

Figure 14. Timber piles driving.

Before driving the timber piles, the fill layer (0.5 m thick) around the piles was removed and replaced by clayey silt materials in order to increase the thickness of the clayey silt layer, and to reduce the chance of damaging the timber piles during driving.

Timber piles were driven with a pneumatic hammer without leads. The pile head was protected from the impact of the hammer with a helmet made with three metallic plates. These plates were attached to the pile head with wood screws. The helmet was extended with a metal beam, acting as a connection between the pile helmet and the hydraulic hammer (Fig. 13). Longitudinal and transverse distance between adjacent timber piles was approximately 2m.

The frequency of the pneumatic hammer was measured during the driving in order to record the number of strokes during timber pile penetration in the soil. (Fig. 14) presents the pile driving resistance of four timber piles and dynamic penetrometer test result.

4.3.2 *Effect of pile size*

The driving resistance increases with the pile size and with depth. Top and tip diameters of timber pile 9 are smaller than those of timber piles 1 and 3. The driving resistance of the timber pile 9 is lower than those of timber piles 1 and 3.

4.3.3 *Effect of pile taper*

Many authors have studied the effect of pile taper on the pile driving resistance. Tapered piles show an

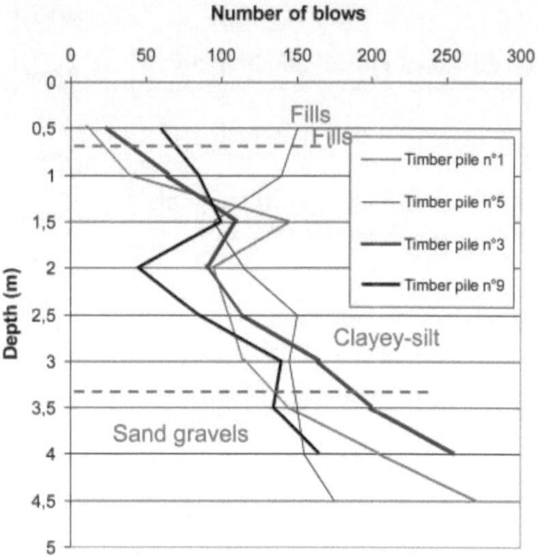

Figure 15. Timber piles driving resistance.

increase in adhesion due to the continual closing up of gaps in sand during the driving. This closing is the result of the transverse vibrations and the compaction of the soil around the pile (Diekmann 1979).

In clay, the high excess pore pressure created during timber pile driving was attributed to the strains generated in the expansion of a cylindrical cavity, this being due to the natural taper of timber pile (Blanchet et al. 1980). The taper of timber piles used in this study was about 9.9 mm/m. The increase of the volume in the cavity changes in the soil structure and increases the total stresses, and results in a higher piles driving resistance than for a straight-sided pile.

In this present study, timber pile 9 has the lowest driving resistance (number of blows per depth), independent of the soil layers. Timber piles 5 and 3 have the highest driving resistance in clayey silt layer and in the underlying layer (sand-gravels), respectively. One possible explanation of this difference is presented in the following paragraph.

In clayey silt layer, timber pile 5, which has the highest taper (3.2 mm/m, see Table 1), has the highest driving resistance. Then, timber piles 1 and 3, the taper of which is 2.8 and 2.5 mm/m respectively, have lower driving resistance than timber pile 5. In this layer, pile taper is a prominent part in the pile driving resistance.

In sandy-gravels (2.5–3 m depth), timber pile driving resistance is different than in clayey-silt layer. Timber piles 3, 1 and 5 respectively have the highest driving resistance. The effect of pile taper on driving resistance is less pronounced in sand and gravel than in clayey-silt because pile driving in sand and gravel occurs higher vibrations than in clayey silt. These vibrations are caused by the coarse particles presents in sand and gravel, and chalk blocks in this layer. The vibrations result in a gap between the pile and the surrounding soil. These gaps should close up in sand during pile driving, but not in clayey-silt. This phenomenon could be the origin of a decrease of pile taper influence in clayey-silt. (Seed & Reese 1955) observed the formation of the gap during pile driving in clay and showed that the upper two-thirds of the pile carried practically no load in adhesion. The biggest influence of pile driving resistance in sand and gravel is likely the pile size. The timber pile no. 3, no. 1 and no. 5 have, in ascending order, the highest pile tip diameter.

5 CONCLUSIONS

Timber pile taper and shape did not allow the use of the usual system of pile instrumentation with removable extensometers. The new system developed aims at measuring the load transfer at shaft soil interface during the static compressive load test without influencing the mechanical behaviour of the pile.

Once the timber pile are driven on the experimental site, a chain of removable extensometers was introduced in the tubes. At first glance, the system of instrumentation does not appear to have been damaged by pile driving. Nevertheless, this new system will be validated only when the packers are inflated and timber piles loaded.

This study highlights the influence of the taper and the size of timber pile in clayey-silt and sand and gravel. Pile taper and pile size, respectively, are the main factors in the increase of pile driving resistance in clayey-silt and sandy-gravel.

ACKNOWLEDGMENT

The authors are grateful for the funding of PhD and research project by French ministry of sustainable development.

REFERENCES

American Wood Preservers Institute. 2002. Timber Pile Design and Construction Manual: 145

Blanchet R. Garneau R. Tavenas F.1980. Behaviour of friction piles in soft sensitive clays. *Canadian Geotechnical Journal*. (Vol. 17): 203–224

Bustamante M. Jézéquel J.-F. 1989. Essai statique de pieu isolé sous charge axiale. *Méthode d'essai LPC* (31): 12p.

Diekmann E.F. 1979. Timber piles in standards, codes, and practice. *ASTM Special Technical Publication* (670): 264–281

Dutch Standardization Institute. 1991. NEN 6743: Geotechnics Calculation method for bearing capacity of pile foundation Compression piles: 30

Le Kouby A. Rocher-Lacoste F. 2012. Essai statique de pieu isolé instrumenté avec l'extensomètre amovible LPC sous charge axiale. Méthode, Recommandations et Interprétation – Méthode d'essai LPC

Levillain J-P. 1980. Fondations courantes des ouvrages d'art anciens. *Fondaconcept*: 57

Maillard J-C. 2006. Les fondations sur pilots de bois des ponts d'Emiland Gauthey en Bresse Bourgogne: 119

Reese L.C. Seed H.B. 1955. The action of soft clay along friction piles. Procedings American Society Civil Engineer (81): 842

Tavenas F.A. 1971. Load test results on friction piles in sand. *Canadian Geotechnical Journal*. (Vol. 8): 221–226

Baltic Piling – Mets & Raudsepp (Eds)
© 2013 Taylor & Francis Group, London, ISBN 978-0-415-64334-4

Behavior of piles in clay during tunneling processes

Mustafa H. Abdullah & Mohd Raihan Taha
Department of Civil and Structural Engineering, Universiti Kebangsaan Malaysia, Malaysia

ABSTRACT: The tunneling processes in urban area generate stresses and deformations in the soil around the tunnel. These stresses and deformations are transferred to the nearby piles that may cause danger to the structures on which their loads are carried by those piles. This study deals with the behavior of piles under the effects of nearby tunnel construction in soft clays focusing on the vertical and lateral displacements of the piles in various conditions. Two effects were studied in this paper. First effect is the horizontal and vertical distances from pile tip to the tunnel axis. The second is the sensitivity of changing the parameters of the Soft Soil Creep Model (c, ϕ, λ^*, and κ^*). The results show a great influence of the pile distance from tunnel axis. Tunneling also cause the piles to experience uneven settlements. The pile suffers high displacements and additional axial forces when the distance was less than tunnel's diameter. The pile experiences high lateral and vertical displacements with increasing the values of modified compression index λ^* and modified swelling index κ^* and decreasing the values of cohesion intercept c and internal friction ϕ.

1 INTRODUCTION

The need to construct tunnels in the big cities becomes a priority these days due to built-up area crowd. These tunnels have large utilities for transportation, sewerage, telecommunication purposes, etc., however they have negative impacts on the surrounding buildings and their piled foundations. Hence, it is necessary to study the effects of tunnel excavation on the adjacent piles and to predict the behavior of these piles subjected to additional stresses and forces from tunneling processes. Many studies have dealt with the behavior of piled foundations due to tunneling. Some of these studies focused on the numerical analyses (e.g. Mroueh and Shahrour, 2003, Surjadinata et al., 2006, Huang et al., 2009, etc.), while others (e.g. Standing and Leung, 2005, Lee and Yoo, 2006, Lee and Bassett, 2007, Meguid and Mattar, 2009) used experimental methods. This paper models the tunneling effects on adjacent pile group using three-dimensional finite element analysis. The tunnel, piles and pile cap were modeled using PLAXIS 3D TUNNEL software version 2. Soft Soil Creep (SSC) model is used as the constitutive model of the soil which describes the behavior of normally consolidated clay.

2 SOFT SOIL CREEP MODEL

The Soft Soil Creep (SSC) model was developed to model the normally consolidated soil especially for settlement problems of foundations. This model considers the creep effect of the soil due to primary and secondary compressions. This model can replace the Mohr-Coulomb for unloading problems e.g. tunneling and open excavation problems. The one dimensional creep model for the SSC model is formulated based on the studies of Buisman (1936), Šuklje (1957), Bjerrum (1967) and Garlanger (1972), of which the total strain consists of two parts; elastic and creep strains. Vermeer & Neher (1999) adopted this model and proposed its one and three-dimensional constitutive equations. The creep strain occurs under constant effective stresses and during the consolidation phase. In addition, the pre-consolidation pressure is related with creep strain accumulated during the time. Eq. 1 shows the one-dimensional strain (Vermeer & Neher, 1999).

$$\dot{\varepsilon} = \dot{\varepsilon}^e + \dot{\varepsilon}^c = -A\frac{\dot{\sigma}'}{\sigma'} - \frac{C}{\tau}\left(\frac{\sigma'}{\sigma_p}\right)^{\frac{B}{C}} \quad (1)$$

where: A, B and C are oedometer constants, σ' is the effective stress, σ_p is the primary stress, and τ is the consolidation time.

The one-dimensional model is then extended to derive the three-dimensional model. The principal stresses were replaced by the stress invariants for isotropic stresses, p′, and deviatoric stresses, q. The volumetric creep strain is obtained from Eq. 2.

$$-\dot{\varepsilon}_v^c = \frac{\mu^*}{\tau}\left(\frac{p^{eq}}{p_p^{eq}}\right)^{\frac{\lambda^* - \kappa^*}{\mu^*}} \quad (2)$$

where: p^{eq} is an equivalent pressure and p_p^{eq} is a generalized pre-consolidation pressure.

Table 1. Tunnel lining and pile cap properties.

Properties	Unit	Tunnel lining	Pile Cap
Model		Elastic	Elastic
EA	[kN/m]	8200000	1.00E+10
EI	[kNm^2/m]	83800	1.00E+10
w	[kN/m^2]	38.15	25
M_p	[kNm/m]	1.00E+15	1.00E+15

The elastic 3D strain may calculated by Eq. 3.

$$-\dot{\varepsilon}_v^e = \kappa^* \frac{\dot{P}'}{P'} = \frac{1+\upsilon_{ur}}{1-\upsilon_{ur}} \frac{\kappa^*}{1+2K_0} \frac{\dot{\sigma}'}{\sigma'} \quad (3)$$

The parameters of the model are divided into three groups (PLAXIS 3D TUNNEL V2):

- Failure parameters
  c: Cohesion intercept (kN/m^2)
  ϕ: Friction angle (°)
  ψ: Dilatancy angle (°)
- Basic stiffness parameters
  $\kappa*$: Modified swelling index
  $\lambda*$: Modified compression index
  $\mu*$: Modified creep index
- Advanced parameters
  ν_{ur}: Poisson's ratio for unloading-reloading (default 0.15)
  K_o^{NC} : $\sigma'_{xx}/\sigma'_{yy}$ stress ratio in a state of normal consolidation
  M : K_o^{NC}-related parameter which obtained by Eq.4.

$$M = \sqrt[3]{\frac{(1-K_0^{NC})^2}{(1+2K_0^{NC})^2} + \frac{(1-K_0^{NC})(1-2\upsilon_{ur})(\lambda^*/\kappa^*-1)}{(1+2K_0^{NC})(1-2\upsilon_{ur})\lambda^*/\kappa^* - (1-K_0^{NC})(1+\upsilon_{ur})}} \quad (4)$$

3 PROBLEM DEFINITION

The diameter of the tunnel is assumed to be 7 m and has a volume loss 1%. The properties of the tunnel's lining are described in Table 1. The pile group is composed of four square piles (1 m × 1 m) connected to the pile cap. The center to center distance between the piles is 7 m. The length of the pile is 25 m. the piles have poperties of: $\gamma = 24$ (kN/m^2), $\upsilon = 0.2$ and $E_{ref.} = 2.6 \times 10^7$ (kN/m^2). Table 1 shows the properties of the tunnel lining and the pile cap. The soil is soft clay and modeled using the Soft Soil Creep model, see Table 2, while the piles and the pile cap is assumed to be elastic. The pile cap is subjected to a continuous distributed load of 50 (kN/m^2). See Figures 1 and 2.

4 FINITE ELEMENT ANALYSIS OF TUNNEL-SOIL-PILE INTERACTION

In this study the pile tip is varied at different horizontal and vertical positions. The horizontal distance of the pile tip to the center of the tunnel (hd) is assumed to be in three positions (e.g. 0.5D, 1D and 2D, where D is the tunnel diameter). Its vertical distances (vd) are (0.5D, 1D, and −1D). See Figure 2.

Table 2. Soil properties.

Properties	Unit	Value
γ_{unsat}	[kN/m^3]	15
γ_{sat}	[kN/m^3]	16
k_x	[m/day]	7.36E−05
k_y	[m/day]	3.68E−05
k_z	[m/day]	7.36E−05
$\lambda*$	[–]	0.1
$\kappa*$	[–]	0.02
μ	[–]	0.005
ν_{ur}	[–]	0.15
K_0^{NC}	[–]	0.669124
M	[–]	1.117471
c_{ref}	[kN/m^2]	5
ϕ	[°]	28
ψ	[°]	0
R_{inter}	[–]	1

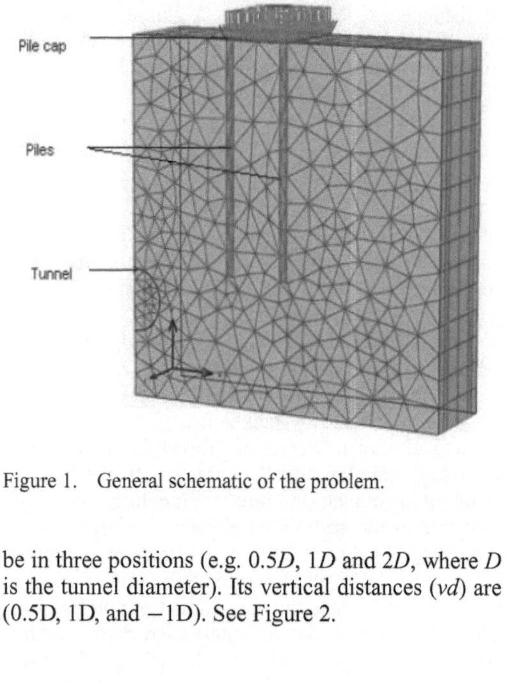

Figure 1. General schematic of the problem.

4.1 *Effect of horizontal distance of pile tip to the tunnel axis*

Figures 3 and 4 show the vertical displacements of pile A and B at different pile positions. From these figures it can be seen clearly that the piles settle uniformly along their lengths. This is because of their high axial stiffness. The vertical displacement increases with decreasing the distance between the pile tip and the tunnel center. The vertical displacements of the nearest pile A are larger than pile B. In Figures 5 and 6, it is shown that the lateral displacements of piles A and B are the maximum for $hd = 1D$. Also the displacement in pile A is larger than that in pile B.

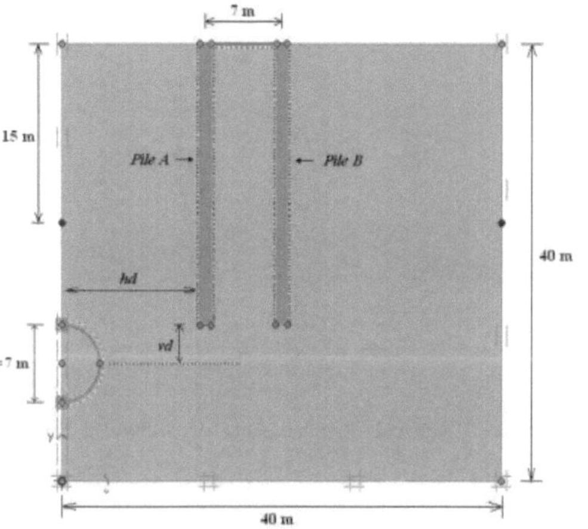

Figure 2. Location of piles relative to the tunnel.

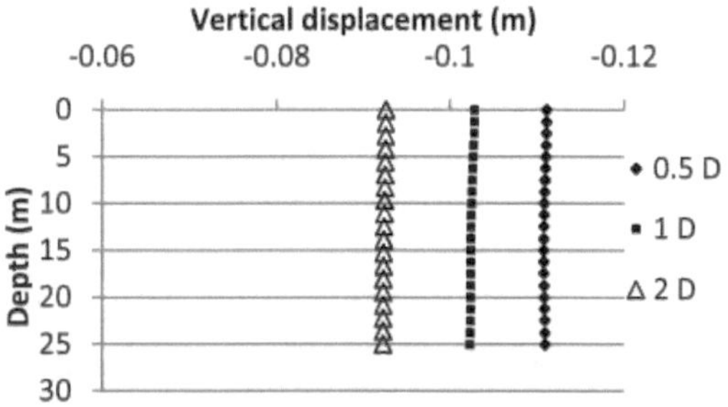

Figure 3. Vertical displacement of pile A at different horizontal distances.

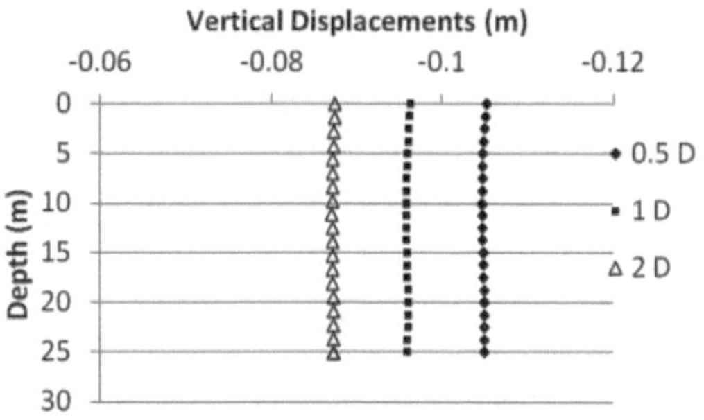

Figure 4. Vertical displacement of pile B at different horizontal distances.

4.2 *Effect of vertical distance of pile tip to the tunnel axis*

To study the effect of changing the vertical position of the pile tip, the later is located at four different positions from the tunnel center (e.g. $vd = 0.5D$, 1D and $-1D$), meanwhile the horizontal distance (*hd*) is the same for all cases. The results show for the cases where the pile tip is located above the tunnel centerline, the vertical displacements increased with decreasing vertical distance between the pile tip and the tunnel center

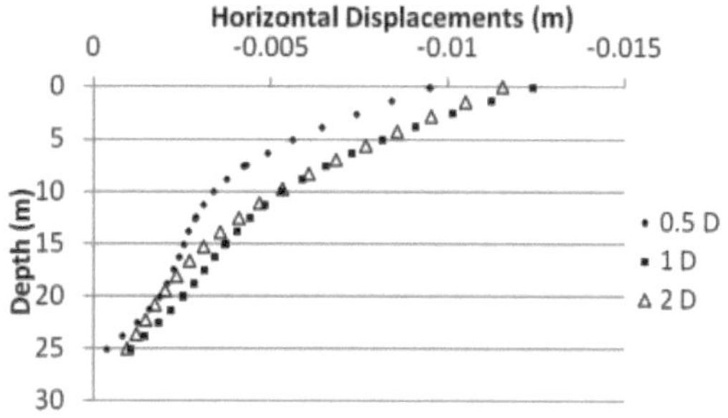

Figure 5. Horizontal displacements of pile A at different horizontal distances.

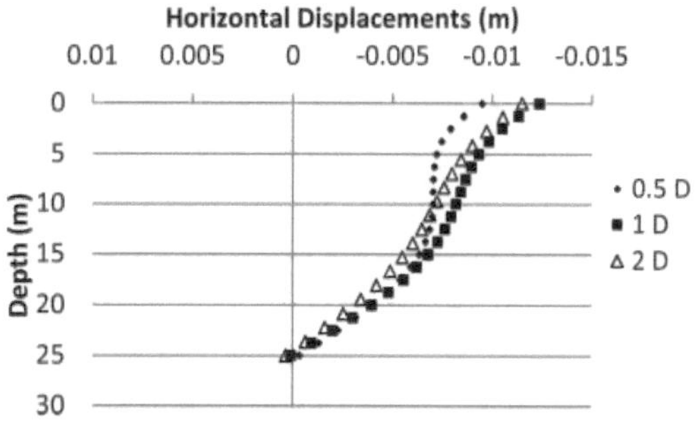

Figure 6. Horizontal displacements of pile B at different horizontal distances.

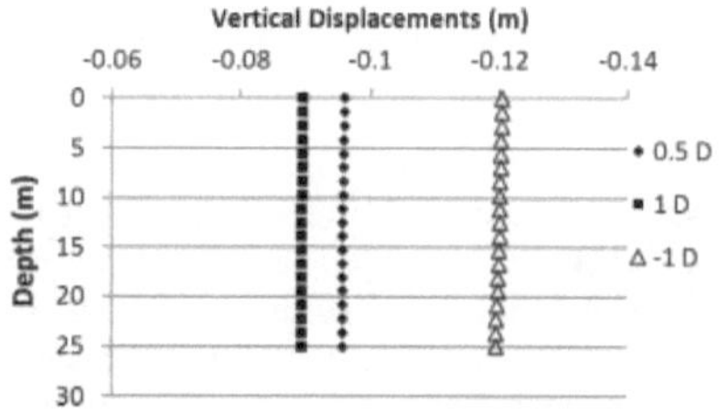

Figure 7. Vertical Displacement of pile A at different vertical distances.

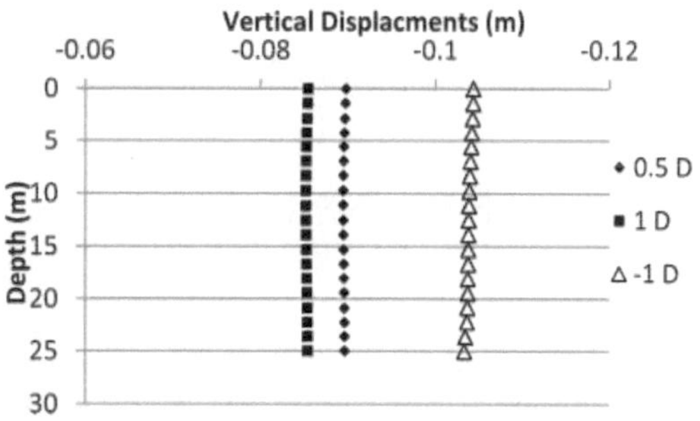

Figure 8. Vertical displacement of pile B at different vertical distances.

(Figures 7 and 8). On the other hand, when the pile tip is located below the tunnel center line ($-1D$), its vertical displacement will be higher as shown in Figures 7 and 8.

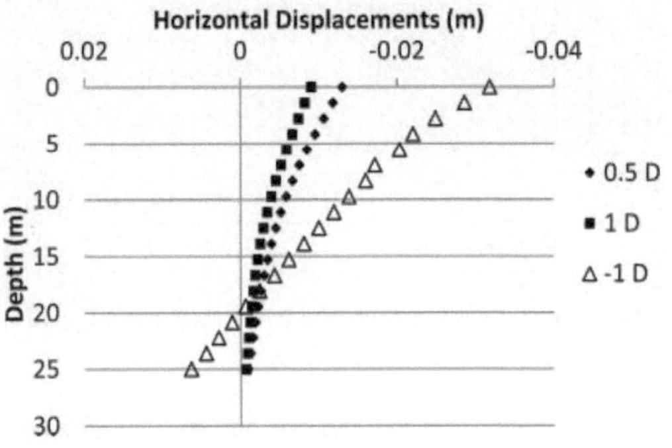

Figure 9. Horizontal displacements of pile A at different vertical distances.

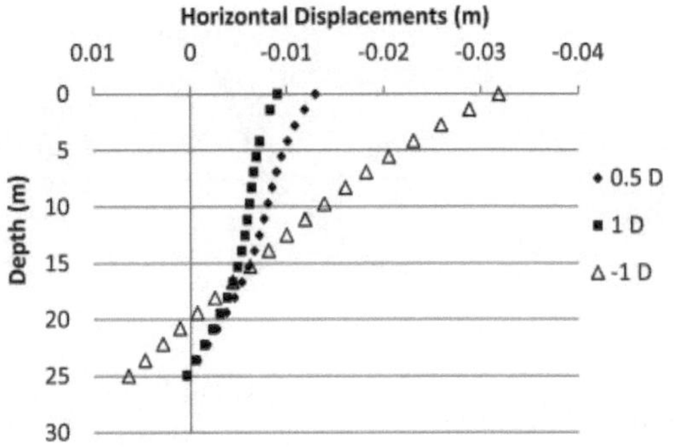

Figure 10. Horizontal displacements of pile B at different vertical distances.

Figures 9 and 10 show the horizontal displacements of pile A and B, respectively. The horizontal displacement of the pile increases with decreasing pile tip distance from the tunnel center, for the cases when the pile tip is located above the tunnel centerline. For the case of the pile tip is below the tunnel center line, its vertical displacement is higher. Both the vertical and horizontal displacements of pile A are larger than of those of pile B, that is because the pile A is like a buffer which absorbs large amount of the stresses caused by tunneling operations.

5 THE SENSITIVITY ANALYSES OF THE SOFT-SOIL CREEP MODEL PARAMETERS (c, ϕ, λ^*, κ^*)

In this paper, a parametric study has been carried out to investigate the effects of soil properties on the tunnel-soil-pile interaction. These parameters are: failure parameters (e.g. cohesion intercept (c) and friction angle (ϕ)) and stiffness parameters (e.g. modified compression index (λ^*) and modified swelling index (κ^*)). Three different values for each parameter were compared in the analysis with fixing the values of the remaining parameters. The pile A is located 12 m from the tunnel center.

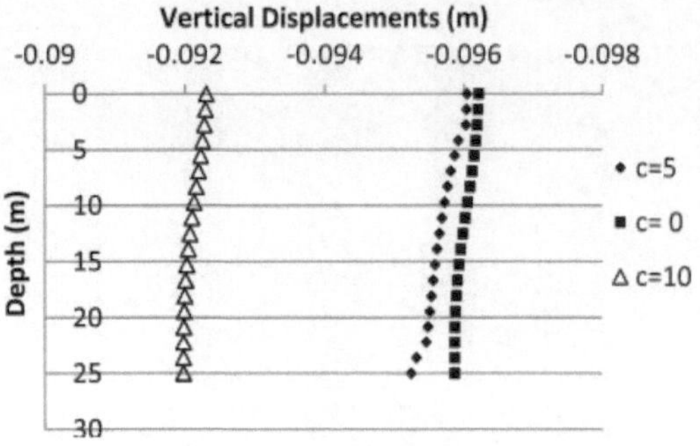

Figure 11. Vertical displacements of pile A at different c values.

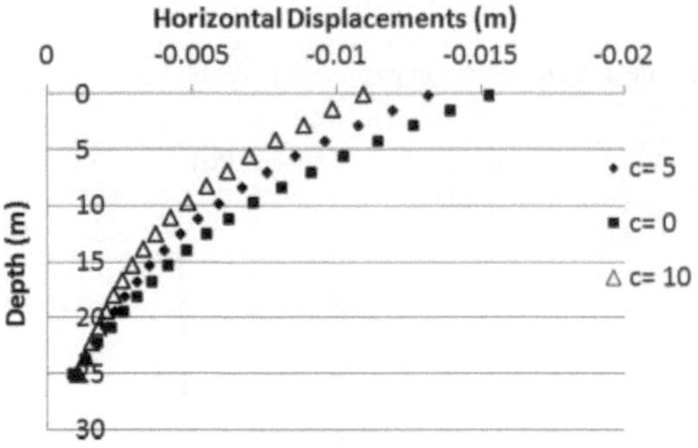

Figure 12. Horizontal displacements of pile A at different c values.

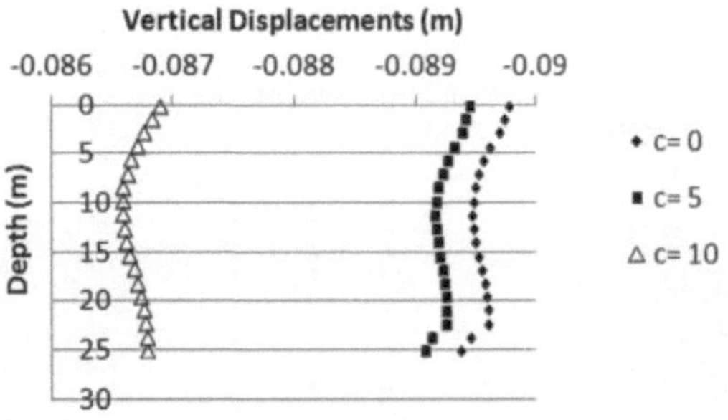

Figure 13. Vertical displacements of pile B at different c values.

5.1 *Effect of c*

Three different values of cohesion parameter c (e.g. 0, 5, and 10 kN/m^2) were used in the analysis. Figures 11 and 13 show that the vertical displacements of the piles A and B, respectively, decrease with increasing the value of c. This observation coincide with Mohr-Coulomb failure criterion that the soil strength decreases with decreasing cohesion intercept c. On the other hand, the lateral displacements in piles A and B with these different values of c are approximately converging together as shown in Figures 12 and 14. Both the vertical and lateral displacements in pile A are larger than in pile B.

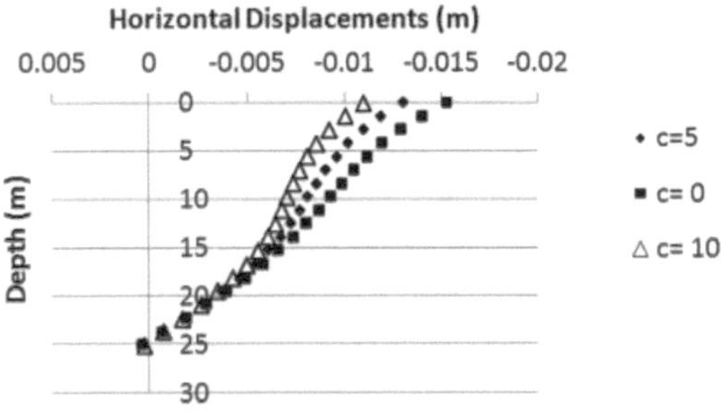

Figure 14. Horizontal displacements of pile B at different c values.

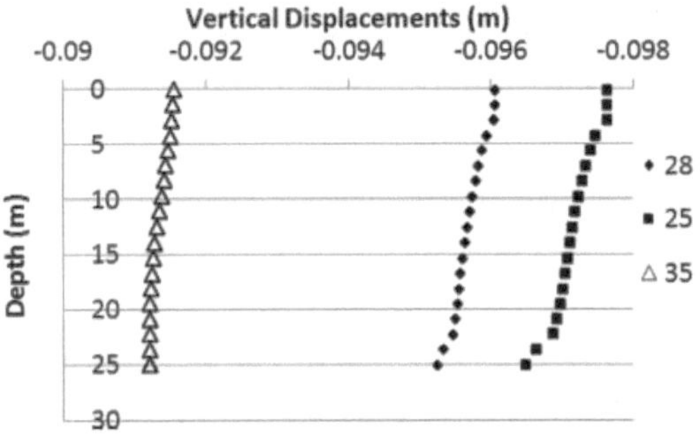

Figure 15. Vertical displacements of pile A at different ϕ values.

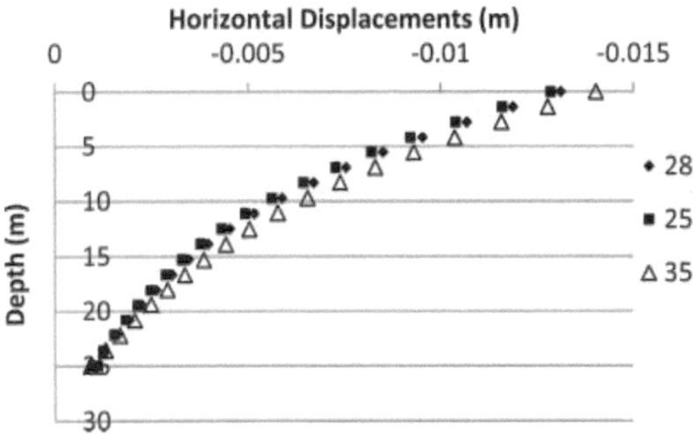

Figure 16. Horizontal displacements of pile A at different ϕ values.

5.2 *Effect of* ϕ

To study the sensitivity of changing the internal friction angle (ϕ), three different values of ϕ (e.g. 25°, 28° and 35°) have been studied. Figures 15–18 show the vertical and horizontal displacements of piles A and B with different values of internal friction angle (ϕ). It is shown that the vertical and lateral displacements decrease with increasing the value of ϕ. The difference between lateral displacements for the different values are slight as shown in Figures 16 and 18. Also these figures show that the displacements in pile A are larger than in pile B.

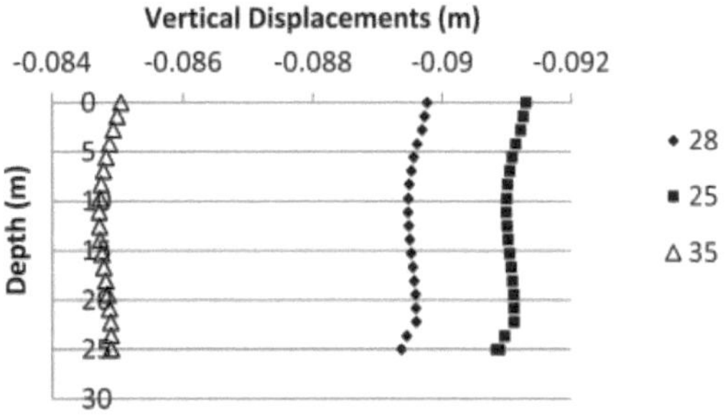

Figure 17. Vertical displacements of pile B at different ϕ values.

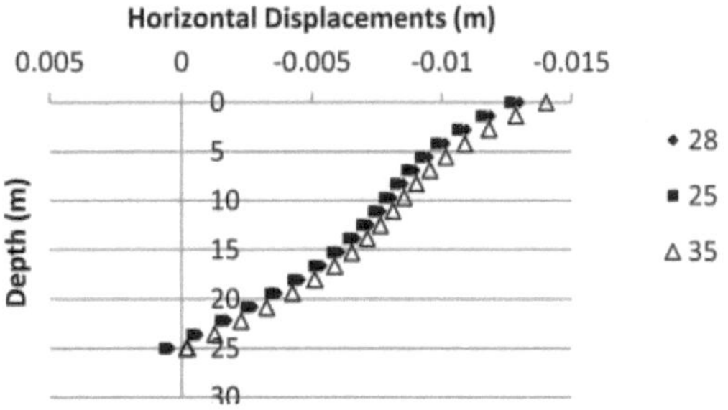

Figure 18. Horizontal displacements of pile B at different ϕ values.

5.3 *Effect of* λ^*

The modified compression index (λ^*) is obtained from the slope of the normal consolidation line from isotropic compression test and oedometer test. In this study three different values of λ^* (0.10, 0.11, and 0.12) are considered to study their effects. This parameter has a significant effect on the behaviour of piles in soft clays due to tunneling operations. It is seen that the lateral and vertical displacements of the piles increase with increasing the value of λ^* (see Figures 19–22). This is due to the volumetric strain of the soil which increased with increasing λ^* as shown in Eq. 2. Figures 19 and 21 show that the settlement of the near pile A is larger than that in the rear pile B. In addition, pile A experienced high lateral displacements compared to pile B as shown in Figures 20 and 22.

5.4 *Effect of* κ^*

This parameter can be obtained from an isotropic compression test and an oedometer test. The slope of the unloading (swelling) line is then used to compute the modified swelling index κ^*. Three different values of κ^* (e.g. 0.020, 0.022, and 0.024) were used to check the sensitivity of changing this parameter on the tunnel-soil-pile interaction. Figures 23–26 show the vertical and lateral responses of the piles A and B in different κ^* values.

It is shown that the lateral and horizontal displacements of piles A and B increase with increasing κ^*

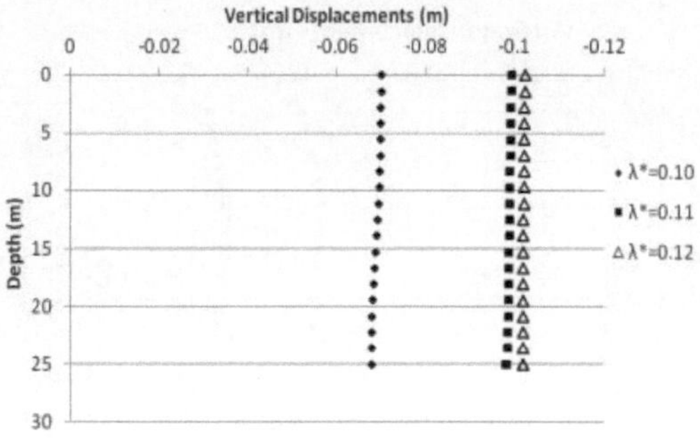

Figure 19. Vertical displacements of pile A at different λ* values.

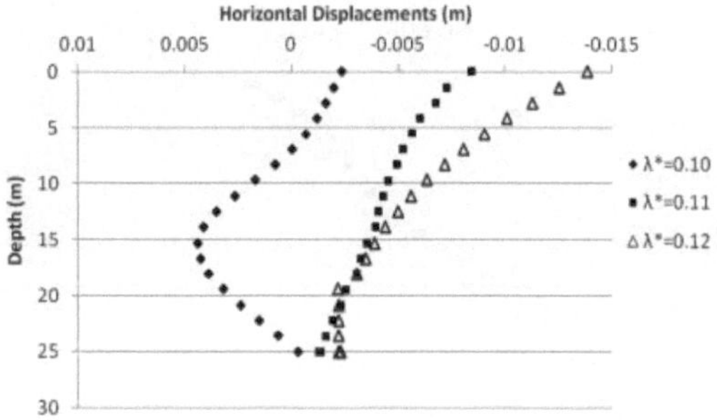

Figure 20. Horizontal displacements of pile A at different λ* values.

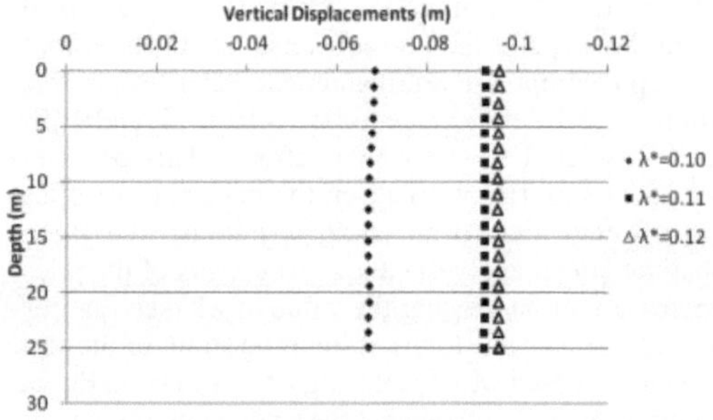

Figure 21. Vertical displacements of pile B at different λ* values.

value, because the elastic volumetric strain increases with increasing κ* value of the soil as shown in Eq. 3. It is clearly seen in Figures 23 to 26 that the nearest pile to the tunnel, i.e. pile A suffers from high settlement and lateral displacements more than the rear pile, B. The effects of tunneling on adjacent piles embedded in soft clays are more pronounced when changing the modified swelling index (κ*). The change in κ* value has a considerable effect on the analyses. This change does not affect the amount of displacements only, but it changes the behaviour of pile's lateral displacement curve quite significantly. Thus it is important to get its value accurately.

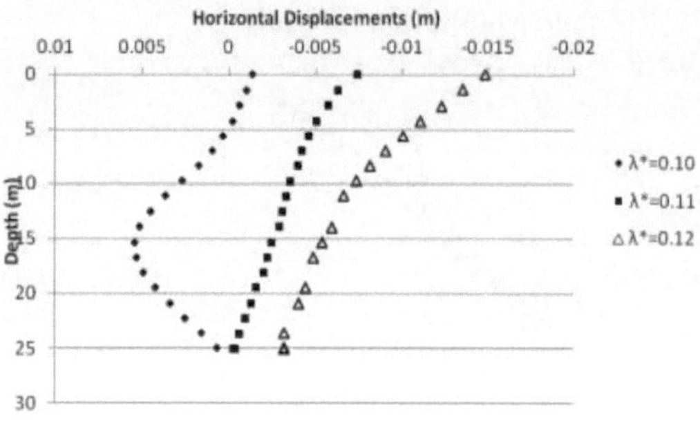

Figure 22. Horizontal displacements of pile B at different λ* values.

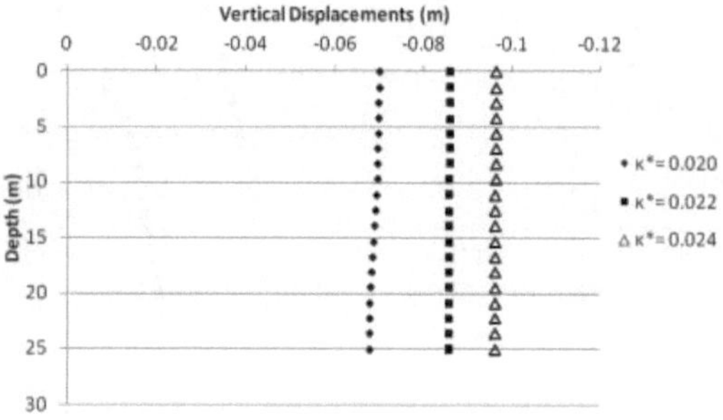

Figure 23. Vertical displacements of pile A at different κ* values.

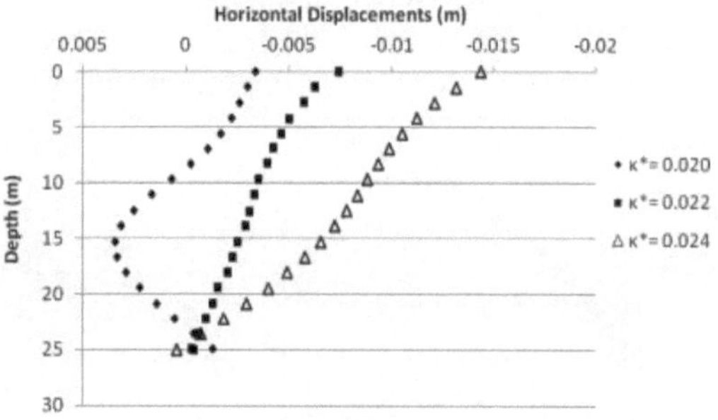

Figure 24. Horizontal displacements of pile A at different κ* values.

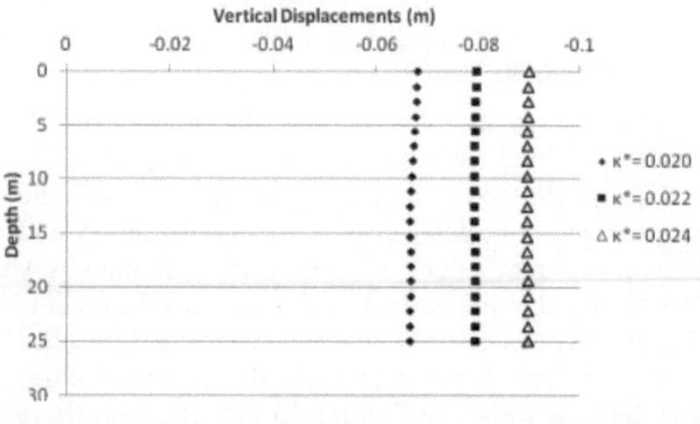

Figure 25. Vertical displacements of pile B at different κ* values.

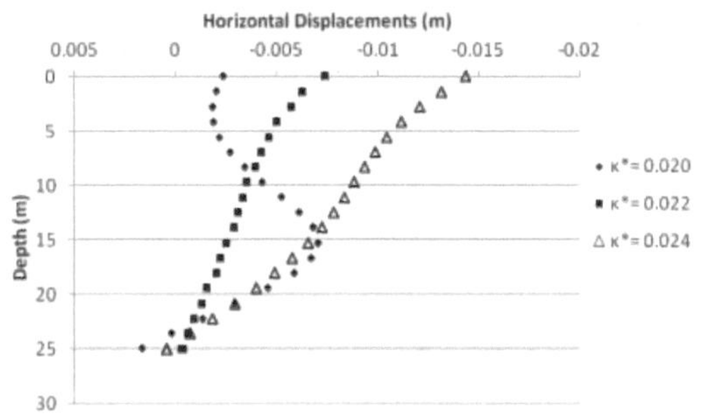

Figure 26. Horizontal displacements of pile B at different κ^* values.

5.5 *Discussion*

From studying the sensitivity analysis of the Soft Soil Creep model (SSC) parameters it is clearly seen that the changing in values of the parameters c and ϕ have minor effects on piles lateral displacements, but a major impact on vertical displacements. On the other hand, the changing in λ^* and κ^* have a major effects on vertical and lateral displacements. These effects are larger than those of c and ϕ. Thus it is important to obtain the values of these parameters carefully in order to get realistic results. The other important point that has to take into consideration is the difference in vertical settlement between the near and the rear pile. This differential settlement may increase and cause damage to the building.

6 CONCLUSIONS

This paper deals with the behavior of piles under effects of nearby tunnel construction in soft clay and focuses on obtaining the displacements of the piles. Based on the numerical study, the following remarks can be seen:

1. With regard to the position of the tunnel, the nearest pile experiences higher lateral and vertical displacements than the rear pile. Thus, the tunneling processes will cause uneven settlements of piles that lead to differential settlement of the building.
2. Vertical displacements of the pile increase with decreasing horizontal distance between pile tip to the tunnel for the same vertical distance.
3. The horizontal displacement of the pile increases with decreasing the pile tip distance from the tunnel center, for the cases when the pile tip is located above the tunnel centerline.
4. The piles settle uniformly along their lengths, because of their high axial stiffness.
5. The vertical and lateral displacements of piles increase with decreasing the values of c and ϕ.
6. The pile suffers high lateral and vertical displacements with increasing the values of modified compression index λ^* and modified swelling index κ^*.
7. The changing in λ^* and κ^* values has a significant and great effect on the analysis especially on the lateral displacements of the piles, while the changing in c and ϕ has a minor effect on the lateral displacements.

REFERENCES

Bjerrum, L. 1967. Engineering geology of Norwegian normally-consolidated marine clays as related to settlements of buildings. *Géotechnique*, Vol. 17, No. 2: 81–118.

Buisman, K. 1936. Results of long duration settlement tests, *Proceedings 1st International Conference on Soil Mechanics and Foundation Engineering*, Cambridge, Massachusetts, USA, vol. 1: 103–107.

Garlanger, J.E. 1972. The consolidation of soils exhibiting creep under constant effective stress, *Géotechnique* 22: 71–78.

Huang, M., Zhang, C., Li, Z. 2009. A simplified analysis method for the influence of tunneling on grouped piles. *Tunneling and Underground Space Technology* 24: 410–422.

Lee, Y.J. & Bassett, R.H. 2007. Influence zones for 2D pile–soil-tunnelling interaction based on model test and numerical analysis. *Tunnelling and Underground Space Technology* 22: 325–342.

Lee, Y., & Yoo, C. 2006. Behaviour of a Bored Tunnel adjacent to a Line of Loaded Piles, *Tunneling and Underground Space Technology* Volume 21, Issues 3–4, page 370.

Meguid, M.A & Mattar, J. 2009. Investigation of Tunnel-Soil-Pile Interaction in Cohesive Soils. *Journal of Geotechnical and Geoenvironmental Engineering, ASCE*: 973–979.

Mroueh, H. & Shahrour, I. 2003. A full 3-D finite element analysis of tunneling–adjacent structures interaction. *Computers and Geotechnics* 30: 245–253.

Standing, J.R. & Leung, W.Y.M.T. 2005. Investigating stresses around tunnels and piles using photo-elasticity techniques. *Geotechnical Aspects of Underground Construction is Soft Ground, Proceedings of the 5th International Symposium TC28, Amsterdam, Netherland.*

Šuklje, L. 1957. The Analysis of the Consolidation Process by the Isotaches Method. *Proceedings of the 4th International Conference on Soil Mechanics and Foundation Engineering*, London, United Kingdom: 200–206.

Surjadinata, J., Hull, T.S., Carter, J.P., and Poulos, H.G. 2006. Combined Finite- and Boundary-Element Analysis of the Effects of Tunneling on Single Piles. *International Journal of Geomechanics:* 374–377.

Vermeer, P.A. & Neher, H.P. 1999. A soft soil model that accounts for creep. *Proceedings of the Plaxis Symposium on Beyond 2000 in Computational Geotechnics*, Amsterdam, the Netherlands: 249–262.

Baltic Piling – Mets & Raudsepp (Eds)
© 2013 Taylor & Francis Group, London, ISBN 978-0-415-64334-4

Numerical investigation of the installation process of full displacement piles

T. Pucker & J. Grabe
Institute of Geotechnical Engineering and Construction Management, Hamburg University of Technology, Hamburg, Germany

ABSTRACT: Full displacement piles (FDP) are commonly used deep foundations in geotechnical engineering. These piles are designed to achieve high bearing capacity. During the drilling process the soil is densified, thus no soil material is unearthed. The advantages of this pile type are high productivity, high bearing capacity and small impact on the environment, like noise and vibration emissions. The installation process of piles influences the soil and some of its properties. Several investigations have been performed to comprehend the influence of installation effects on the pile bearing capacity. The developing stress state around the pile during the pile installation processes has been investigated by several authors. It is obvious, that the pile installation process changes the stress state and the void ratio in the surrounding soil. Due to this reason, differences between the results of cone penetration tests before and after pile installation occur. A method to determine the changes of the soil state variables is presented in this paper. A Coupled-Eulerian-Lagrangian (CEL) approach is applied in this paper to simulate the drilling process of a FD-Pile. The simulations are used to explain influences on the soil state variables and measured at a test area in Hamburg. The CEL-method is briefly introduced and a numerical model to simulate the installation process of a FD-Pile is presented. The results of the numerical simulations are discussed.

1 INTRODUCTION

The installation process of piles influences the surrounding soil depending on the installation method. Several investigations have been performed to comprehend the influence of installation effects on pile bearing capacity, see for example (De Beer 1988, Peiffer & Van Impe 1993, Hartung 1994, Schmitt 2004). (Peiffer & Van Impe 1993) and (Peiffer 2009) investigated the development of the stress state around the pile during the pile installation processes. It is obvious, that the installation process changes the stress state and the void ratio in the surrounding soil. Therefore, there are differences between the results of cone penetration tests done before and after pile installation, see e.g. (De Beer 1988, Van Weele 1988, Peiffer & Van Impe 1993) and (Busch, Grabe, Gerressen, & Ulrich 2010).

The piles can be driven, jacked, vibrated or drilled into the ground. The effects inside the soil are still not completely investigated. Therefore, most methods to predict the pile behavior or its bearing capacity are based on empirical approaches. Using numerical methods, the influences of the installation process on the surrounding soil have been investigated numerically for driven, jacked and vibrated piles e.g. by Mabsout and Tassoulas (1994), Cudmani (2001), Mahutka (2008), Henke (2008) and Henke and Grabe (2009b).

In this article, the numerical simulation of the installation process of drilled full displacement piles is presented. Using this numerical model, new insight can be gained in understanding effects in the soil nearby installed piles. First results and detailed information about the numerical method and model can be found in Pucker and Grabe (2012).

To simulate the installation of a full displacement pile, the full displacement pile with lost bit (FDP), invented by BAUER (2009), is used. The installation process is divided into six steps schematically presented in Figure 1 (Busch et al. 2010). In the first step the drilling tool is taken into position. During the second and the third step, the drilling tool is drilled to the purposed depth. During the drilling process, the soil is transported to the displacement body where the soil is compacted. The reinforcement basket is installed in step four. The drilling tool is removed in step five and the pile is concreted simultaneously. The concrete hardens in step six and the installation is completed. The presented numerical study simulates only the steps two and three. A detail of the drilling tool is illustrated in Figure 2 (Pucker and Grabe 2012).

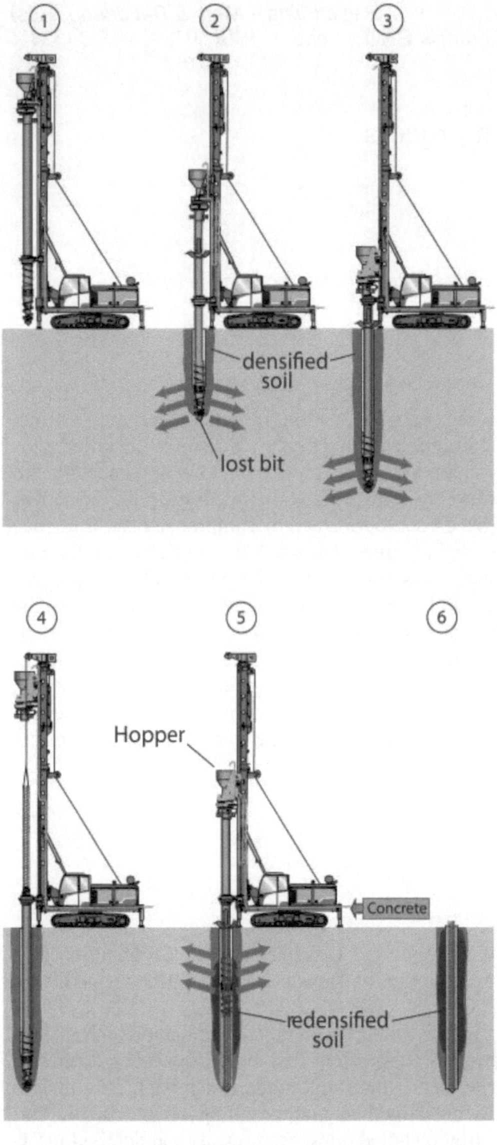

Figure 1. Schematic illustration of the installation process of full displacment piles with lost bit.

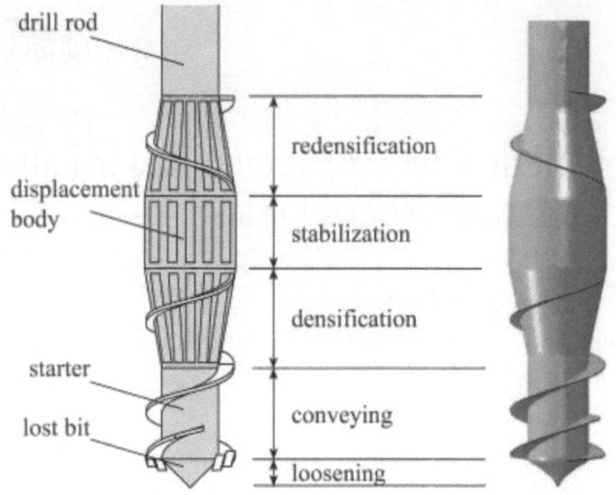

Figure 2. Scheme of the FDP drilling tool invented by BAUER (left); numerical discretization of the drilling tool (right).

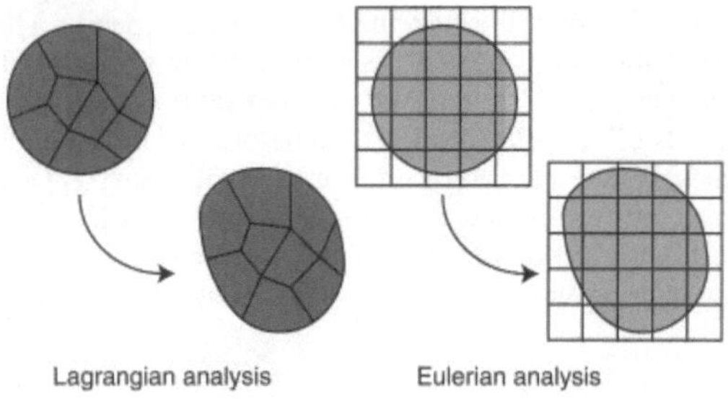

Figure 3. Deformation of a continuum in a Lagrangian (left) and a Eulerian analysis (right).

2 NUMERCIAL METHOD

2.1 *Coupled Eulerian-Lagrangian-Method*

The numerical simulations are performed using the Coupled Eulerian-Lagrangian-Method (CEL-Method). The method is briefly introduced in this section. Detail information can be found in Dassault Systéemes (2010). The numerical simulations in Mabsout and Tassoulas (1994), Cudmani (2001), Mahutka (2008), Henke (2008) and Henke and Grabe (2009b) are restricted to vertically installed piles with non-changing cross sections, because the Lagrange formulation of the Finite-Element-Method (FEM) is applied. During the simulated installation process large deformations in the soil body occur that possibly go along with mesh distortions. Additionally the rotation of the drilling tool also causes mesh distortion problems. Therefore, the massively changing cross section of the FDP drilling tool (see Figure 2) cannot be installed into a soil body using the classical Lagrangian FEM formulation.

The CEL-Method is able to overcome mesh distortion problems in simulation of pile installation processes. It combines the advantages of the Lagrangian formulation with the advantages of the Eulerian formulation and uses an explicit time integration scheme. Both formulations differ in their description of the movement of small volumetric elements as a function of time.

The Lagrangian formulation describes the movement of a continuum as a function of material coordinates and time. Each node of the Lagrangian mesh moves together with the material during the simulation, see Figure 3 (Qiu et al. 2009). Thus, the elements are deformed during the simulation and problems due to mesh distortion can occur. The surface of the continuum is precisely specified within this formulation. Furthermore, each Lagrangian element is always associated with one type of material. The Lagrangian formulation is used in classical Finite-Element analyses and often applied in solid mechanics.

In contrast the Eulerian formulation describes the movement of a continuum as a function of spatial coordinate and time. The nodes of an Eulerian mesh are fixed during the simulation, so that the elements cannot deform. To realize the movement of a continuum, the material flows through the Eulerian mesh. Therefore a Eulerian element is not only dedicated to one type of material. Not even the element has to be filled with any material at all. Concerning this formulation, no mesh distortion occurs and the simulation of large deformations is possible.

The Coupled Eulerian-Lagrangian-Method allows an interaction between Lagrangian elements and Eulerian material by using an Eulerian-Lagrangian contact formulation. The Eulerian time integration is realized applying the "Lagrange-plus-remap" formulation. First of all, a traditional Lagrangian phase is calculated in each time increment and the nodes of the Eulerian mesh are temporarily fixed within the material. Therefore the Eulerian elements can deform temporarily in the Lagrangian phase. Second, in the Eulerian phase, the so called transport phase, elements are tested for significant deformation. These elements are automatically remapped and the material flow through these elements is calculated. The timestep has to be sufficiently small, so no element distortions occur during one timestep.

The application of the CEL-Method on geomechanical problems involving large deformations has been approved e.g. by Qiu et al. (2009), Qiu et al. (2010), Qiu and Grabe (2011), Henke et al. (2010), Bienen et al. (2011), Pucker and Grabe (2012), Pichler et al. (2012).

2.2 *Constitutive model*

This study investigates the penetration into non-cohesive, granular materials like sand. The soil is described using the hypoplastic constitutive model after Gudehus (1996) and von Wolffersdorff (1996). Hypoplasticity has been shown to describe the non-linear and anelastic behavior of granular materials well (Niemunis 2003, Kelm 2004). The constitutive law is consistent with mechanical soil properties like dilatancy, contractancy, dependency of stiffness on stress state and void ratio and different stiffness for loading and unloading.

Due to the different stiffness for loading and unloading typical effects like different wave velocities of the loading or unloading front for the numerical simulation of wave propagation can be simulated with the hypoplastic constitutive model. This characteristic has been investigated by Kelm (2004) and Henke and Grabe (2009a) who studied the wave propagation in hypoplastic media and found typical effects like shock fronts to occur. The hypoplastic constitutive model describes accurately compaction processes. Especially, the dependency on void ratio in the constitutive model allows for a realistic simulation of compaction processes. Due to this dependency wave energy can be dissipated into compaction of the surrounding soil (Kelm 2004, Mahutka 2008) during the analyses.

Table 1. Hypoplastic material parameters for Karlsruher Sand.

Symbol	Value	Description
φ_c	30	critical state friction angle (°)
h_s	5800	granular hardness (MPa)
n	0.28	exponent
e_{d0}	0.53	minimum void ratio
e_{i0}	0.84	critical void ratio
e_{c0}	1.00	maximum void ratio
α	0.13	exponent
β	1.05	exponent
R	1×10^{-4}	maximum value of intergranular strain
m_R	5.00	stiffness ratio at a change of direction of 180°
m_T	2.00	stiffness ratio at a change of direction of 90°
β_R	0.50	exponent
χ	6.00	exponent

The formulation depends on the current stress state **T** and the void ratio e, whereby the bulk density is taken into account. In order to model the accumulation effects and the hysteretical material behavior under cyclic loading Niemunis and Herle proposed the concept of intergranular strain (Niemunis and Herle 1998). This enhancement is used in the present study to take into account small strain behaviour of soils. Pore pressure effects have not been taken into account.

The hypoplastic parameters chosen for the present numerical study are those of Karlsruher Sand (Herle 1997), see Table 1.

2.3 *Calculation of the cone penetration resistance profile*

Hsu and Huang (1999), Jamiolkowski et al. (2003), Grabe and König (2004) and Grabe et al. (2010) showed that cone penetration resistance q_c is correlated with the stress state σ_p' and the bulk density D_R of the soil. The correlation can be specified using Equation 1. Three soil dependant parameters C_0, C_1 and C_2 are required to calculate the cone penetration resistance. The calibration of C_0, C_1 and C_2 in Karlsruher Sand results in Equation 3. Using this correlation, the cone penetration resistance profile before and after the simulated installation process can be directly estimated from the numerical model.

$$q_c = C_0 \cdot p_a \left(\frac{\sigma_p'}{p_a}\right)^{C_1} \cdot e^{(C_2 \cdot D_R)} \tag{1}$$

$$q_{c,KS} = 610 \cdot p_a \left(\frac{\sigma_p'\,[\mathrm{kPa}]}{p_a}\right)^{0.45} \cdot e^{(0.90 \cdot D_R)} \tag{2}$$

$$D_R = \frac{e_{max} - e}{e_{max} - e_{min}}, \tag{3}$$

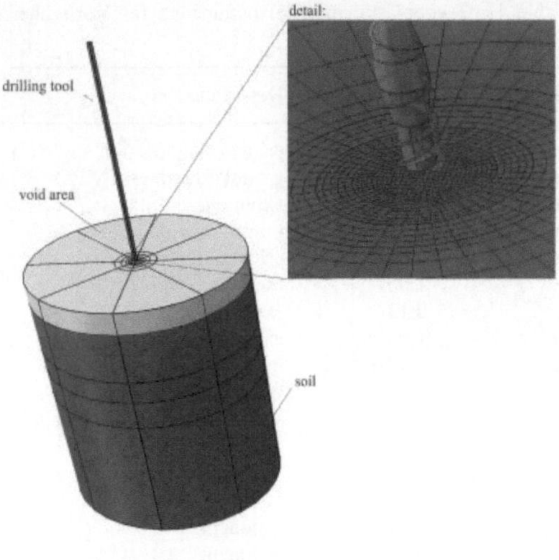

Figure 4. Numerical model with void area and mesh of the soil in detail.

σ'_P as

$$\sigma'_P = \frac{\sigma'_1 + \sigma'_2 + \sigma'_3}{3}\,. \quad (4)$$

and the reference pressure p_a is chosen to be 1 kPa. In Equation 3, e_{max} is the maximum, e_{min} is the minimum and e the actual void ratio of the soil; in Equation 4, σ'_1, σ'_2 and σ'_3 are the principal effective stresses in the soil.

3 SIMULATION OF THE INSTALLATION PROCESS

3.1 *Numerical model*

Due to the asymmetric drilling tool and the applied CEL-Method, a three dimensional model has to be used. It is assumed, that the deformations of the drilling tool a neglectable according to the large deformations of the soil, so the drilling tool is modelled as a rigid body. The displacement body has a diameter about 51 cm and the drill rod about 32 cm. The diameter of the auger elements relate to the displacement body.

The soil is modelled in an Eulerian area. The shape is cylindrical with a diameter about 16 m and a height of 20 m. Thereby, the distance between the boundaries and the drilling tool is more than 15 D (15 times of the pile diameter D), so that no influences of the boundary conditions on the results are expected. Above the soil, a 2 m void area is provided, so that the soil can avoid into this free space during the drilling process. The whole Eulerian area is discretized with approximately 163,000 eight-noded elements with reduced integration. The discretized model is shown in Figure 4. The drilling tool penetrates the soil velocity controlled. Assuming, that the response of the soil model does not depend on the velocity, the penetration velocity is taken to 1 m/s. The influence of the ratio between the

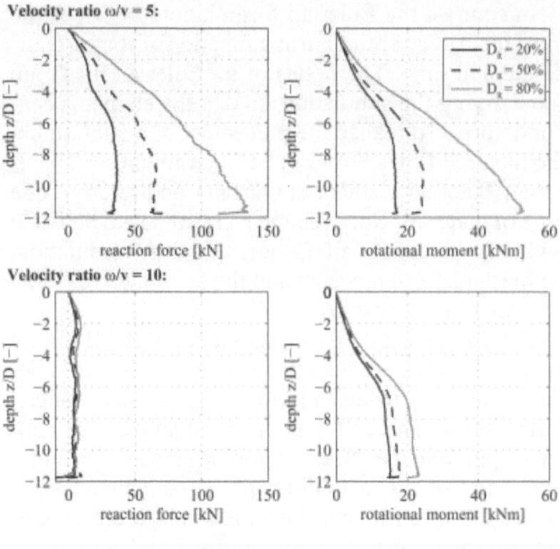

Figure 5. Karlsruher Sand: Reaction forces and rotational moments of the drilling tool versus normalized penetration depth for different velocity ratios ω/v.

penetration velocity and the rotation velocity is investigated. Therefore the rotational velocity is chosen to 5 and to 10 U/s (rotations per meter). The reaction forces and moments are recorded during the simulation.

3.2 *Results*

3.2.1 *Reaction force and rotational moment of the drilling tool*

The evaluated reaction forces and rotational moments of the drilling tool are plotted versus the normalized penetration depth z/D in Figure 5 The penetration depth z is normalized with the outer diameter of the drilling tool D. The results for different bulk densities D_R are evaluated and shown for two velocity ratios in Figure 5. The reaction force and the rotational moments ascend with increasing bulk density D_R. The small ratio of $\omega/v = 5$ between the rotational velocity ω and the penetration velocity v causes a significantly higher reaction force compared to penetration with a ratio of $\omega/v = 10$. The reaction force varies from 30 kN up to 130 kN at a ratio of $\omega/v = 5$ and from 1 kN up to 10 kN at a ratio of $\omega/v = 10$. The rotational moment varies from 18 kNm up to 52 kNm at a ratio of $\omega/v = 5$ and from 15 kNm up to 25 kNm at a ratio of $\omega/v = 10$. At the ratio of $\omega/v = 10$, the reaction forces close to zero are caused by a screwing effect due to the fact that the distance between the auger elements fits the penetration per rotation. Therefore it can be stated, that, regarding the investigated soil, the optimum drilling performance is rather influenced by the combination of the ratio ω/v and the geometry of the drilling tool than by the soil properties.

3.2.2 *Changes of void ratio*

The changes of the void ratio distribution caused by the drilling process are shown in Figure 6. In Figure 6, left, the void ratio distribution is plotted for different initial bulk densities along horizontal paths at depths

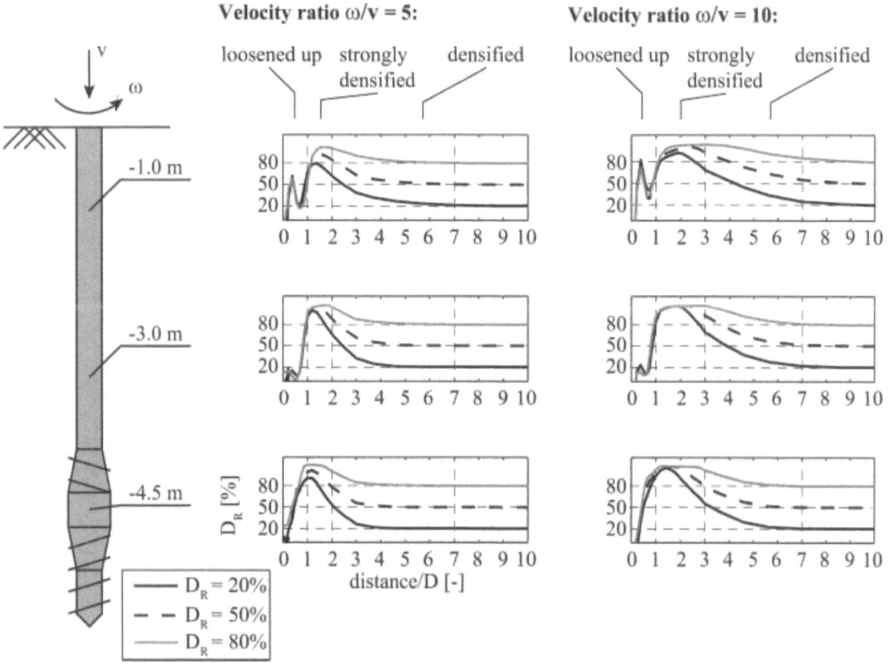

Figure 6. Void ratio along horizontal paths at depths of 1 m, 3 m and 4.5 m after completion of the drilling process for different drilling ratios.

of 1 m, 3 m and 4.5 m below the ground level versus the distance to the drilling tool for a velocity ratio $\omega/v = 5$. Close to the drilling tool the soil is loosened up to a distance of approximately 1D. Accordingly, the soil is strongly densified up to a distance of approximately 2D. Moreover, the densification influence ranges from 6D for the dense sand up to 8D for the loose bulk density. A higher velocity ratio $\omega/v = 10$ results in an expansion of the strong densification area up to 3D.

3.2.3 *Changes of cone penetration resistance*

Changes of the cone penetration resistance profile nearby the pile after the installation process compared to the initial state has been determined in several measurements, see e.g. De Beer (1988), Van Weele (1988), Peiffer and Van Impe (1993) and Busch et al. (2010). The numerically estimated cone penetration test results are shown in Figure 7 for a velocity ratio of $\omega/v = 10$. The cone penetration resistance q_c is plotted versus the normalized penetration depth z. The cone penetration resistance mainly increases at a distance of approximately 1 D at the depth of the displacement body.

In Figure 7, middle, the vertical distribution of the void ratio is plotted versus the normalized penetration depth z at the initial state and after the installation process. Nearby the displacement body, the void ratio decreases independently from initial bulk density to the minimum possible void ratio at the distance of 1 D. The bulk density can reach values higher than 100%, because the minimum and maximum void ratios are determined in a normalized test and can be exceeded in reality. Additionally the void ratio and its limits are defined stress dependent in the hypoplastic constitutive formulation, see (Bauer 1996). The calculation of bulk density D_R is related to the limits at zero stress. At a distance of 3D the decrease of void ratio depends on the bulk density.

The distribution of horizontal stresses σ_h is plotted in Figure 7, right. The horizontal stresses mainly increase at a distance of 1D at the depth of the displacement body, therefore the cone penetration resistance also increases. Above the displacement body the horizontal stresses reach a nearly constant value at the distance of 3D and are below the initial state, so the cone penetration resistance profile in this range is quite similar compared to the initial values although the void ratio is decreased.

Regarding the investigated soil, the change of cone penetration resistance measured in-situ before and after FDP installation is mainly caused by decreasing the void ratio.

4 SUMMARY

The presented numerical results prove that the installation process of drilled piles can be simulated using the CEL-Method. This method is able to overcome mesh distortion problems due to complex loading conditions along with large deformations. The applied hypoplastic constitutive model is able to reproduce the soil behaviour in this case. The influences on the stress

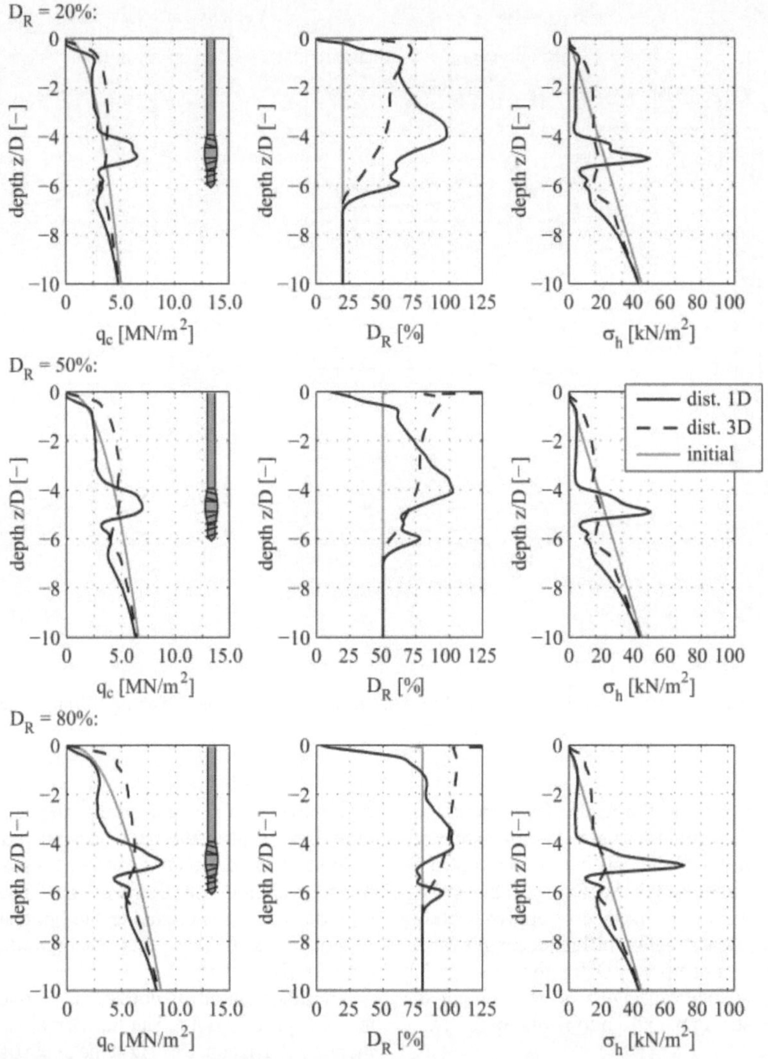

Figure 7. Karlsruher Sand: Influence of the drilling process on the soil states for a velocity ratio of $\omega/v = 10$ at two distances after 6 m of penetration: Cone penetration resistance before and after the drilling process (left); void ratio (middle) and horizontal stresses (right).

state and the soil density can be investigated and visualized. The effects inside the soil and the movement of the soil can be explored more precisely. It can be stated for the investigated soil, that changes of cone penetration resistance after the drilling process are mainly caused by decreasing void ratio. Additionally it was figured out, that the optimal drilling parameters are rather influenced by the geometry of the drilling tool than by the soil attributes.

In conclusion, it is possible to simulate the installation process using the CEL-Method as well as the influences that are associated with the application of different pile types and drilling methods. The numerical analyses also provides a deeper insight into the mechanics, that occur inside the soil. In future work, the installation parameters for the drilling rig can be determined and parametric studies can be executed in order to optimize the drilling process.

REFERENCES

BAUER MASCHINEN GMBH, FDP – full displacement pile system, process and equipment. *Process description 3/2009. Schrobenhausen, Germany.*

Bauer, E. (1996). Calibration of a comprehensive hypoplastic model for granular materials. *Soils and Foundation 36, (1)*, pp. 13–26.

Bienen, B., S. Henke, and T. Pucker (2011). Numerical study of the bearing behaviour of circular footings penetration into sand. *Proc. of 13th International Conference of International Association for Computer Methods and Advances in Geomechanics (IACMAG)*, pp. 939–944.

Busch, P., J. Grabe, F. Gerressen, and G. Ulrich (2010). Use of displacement piles for reinforcement of existing piles. *Proc. of DFI and EFFC 11th Int. Conf. in the DFI series: Geotechnical Challenges in Urban Regeneration in London/UK*, pp. 113–119.

Cudmani, R. (2001). *Statische, alternierende und dynamische Penetration in nichtbindigen Bden*. Veröffentlichungsreihe des Instituts für Bodenmechanik und Felsbau der Universität Fridericiana in Karlsruhe, Heft 152.

Dassault Systémes (2010). *Abaqus User Manual Version 6.10*.

De Beer, E. (1988). Different behavior of bored and driven piles. *Proceedings of the 1st international geotechnical Seminar on Deep Foundations on bored and auger Piles (BAP I), Rotterdam, Balkema*, S. 47–82.

Grabe, J., S. Henke, S. Kinzler, and T. Pucker (2010). Inverse determination of soil density and stress state using dispersion wave measurements and cone penetration tests in a non-layered soil. *Soil Dynamics and Earthquake Engineering, 30(6):481–489*.

Grabe, J. and F. König (2004). Zur aushubbedingten Reduktion des Drucksondierwiderstandes. *Bautechnik 81, (7)*, S. 569–577.

Gudehus, G. (1996). A comprehensive constitutive equation for granular materials. *Soils and Foundations 36*(11), 1–12.

Einflüsse der Herstellung auf die Pfahltragfähigkeit. Dissertation, Mitteilung des Instituts für Grundbau und Bodenmechanik der Technischen Universität Braunschweig, Heft 45.

Henke, S. (2008). *Herstellungseinflüsse aus Pfahlrammung im Kaimauerbau*. Dissertation, Schriftenreihe des Instituts für Geotechnik und Baubetrieb der Technischen Universität Hamburg-Harburg, Band 18.

Henke, S. and J. Grabe (2009a). Elastische und anelastische wellenausbreitungsph nomene in der boden-bauwerks-interaktion. *VDI-Berichte Nr. 2063 (Baudynamik)*, S. 521–542.

Henke, S. and J. Grabe (2009b). Numerical modeling of pile installation. *Proc. of the 17th International Conference on Soil Mechanics and Geotechnical Engineering, Alexandria, Egypt*, pp. 1321–1324.

Henke, S., G. Qiu, and J. Grabe (2010). A coupled eulerian-lagrangian approach to solve geotechnical problems involving large deformations. *Proc. of 7th European Conference on Numerical Methods in Geotechnical Engineering (NUMGE) in Trondheim/Norway*, pp. 233–238.

Herle, I. (1997). Hypoplastizität und granulometrie einfacher korngrüste. *Heft 142 Institut für Bodenmechanik und Felsmechanik der Universität Fridericana in Karlsruhe*.

Hsu, H.-H. and A.-B. Huang (1999). Calibration of cone penetration test in sand. *Proc. Of National Science Council, 23*, S. 579–590.

Jamiolkowski, M., D. Lo Presti, and M. Manassero (2003). Evaluation of relative density and shear strength of sands from cone penetration test (cpt) and flat dilatometer (dmt). *Soil Behaviour and Soft Ground Construction, Eds. J.T. Germain, T.C. Sheahan and R.V. Whitman, ASCE, GSP 119*, 201–238.

Kelm, M. (2004). *Simulation der Verdichtung rolliger Böden mittels Vibrationswalzen*. Ph. D. thesis, Institut für Geotechnik und Baubetrieb, Hamburg University of Technology.

Mabsout, M. and J. Tassoulas (1994). A finite element model for the simulation of pile driving. *International Journal for Numerical an Analytical Methods in Geomechanics* 37, pp. 257–278.

Mahutka, K.-P. (2008). *Zur Verdichtung von rolligen Böden infolge dynamischer Pfahleinbringung und durch OberflÃ¤chenrüttler*. Ph. D. thesis, Institut für Geotechnik und Baubetrieb, Hamburg University of Technology.

Niemunis, A. (2003). *Extended hypoplastic models for soils*. Verffentlichungsreihe des Instituts fr Grundbau und Bodenmechanik der Ruhr-Universitt Bochum.

Niemunis, A. and I. Herle (1998). Hypoplastic model for cohesionless soils with elastic strain range. *Mechanics of Frictional and Cohesive Materials*, 2, 279–299.

Peiffer, H. (2009). The dmt as tool for the monitoring of the effect of pile installation on the stress state in the soil. *Proceedings of the 5th international geotechnical Seminar on Deep Foundations on bored and auger Piles (BAP V), Ghent, CRC Press*, S. 135–142.

Peiffer, H. and W. Van Impe (1993). Evaluation of pile performance based on soil stress measurements – field test program. *Proceedings of the 2nd international geotechnical Seminar on Deep Foundations on bored and auger Piles (BAP II), Rotterdam, Balkema*, S. 385–389.

Pichler, T., T. Pucker, T. Hamann, S. Henke, and G. Qiu (2012). High-performance abaqus simulations in soil mechanics reloaded – chances and frontiers. *Proc. of International Simulia Community Conference in Providense, Rhode Island/USA*, pp. 237–266.

Pucker, T. and J. Grabe (2012). Numerical simulation of the installation process of full displacement piles. *Computers and Geotechnics*, 45, pp. 93–106, DOI: 10.1016/j.compgeo.2012.05.006.

Qiu, G. and J. Grabe (2011). Explicit modeling of cone and strip footing penetration under drained and undrained conditions using a visco-hypoplastic model. *Geotechnik 34*(3), 205–271.

Qiu, G., S. Henke, and J. Grabe (2009). Applications of coupled eulerian lagrangian method to geotechnical problems with large deformations. *Proc of SIMULIA Customer Conference 2009 in London*, pp. 420–435.

Qiu, G., S. Henke, and J. Grabe (2010). Application of a coupled eulerian-lagrangian approch on geomechanical problems involving large deformation. *Computers and Geotechnics*, DOI:10.1016/j.compgeo.2010.09.002.

Schmitt, A. (2004). *Experimentelle und numerische Untersuchungen zum Tragverhalten von Ortbetonpfählen mit variabler Bodenverdrängung*. Dissertation, Mitteilungen des Instituts und der Versuchsanstalt für Geotechnik der Technischen Universität Darmstadt, Heft 70.

Van Weele, A. (1988). Cast-in-situ piles – installation methods, soil disturbance and resulting pile behaviour. *Proceedings of the 1st international geotechnical Seminar on Deep Foundations on bored and auger Piles (BAP I)*, Rotterdam, Balkema, S. 219–226.

von Wolffersdorff, P.-A. (1996). A hypoplastic relation for granular materials with a predefined limit state surface. *Mechanics of Frictional and Cohesive Materials*, 1, 251–271.

Baltic Piling – Mets & Raudsepp (Eds)
© 2013 Taylor & Francis Group, London, ISBN 978-0-415-64334-4

Design of screw piles

S. Efthymiou
National Technical University of Athens, Athens, Greece

J.R. Standing
Imperial College London, London, UK

M. Kavvadas
National Technical University of Athens, Athens, Greece

ABSTRACT: The paper compares conventional pile design methods with the ScrewFast methodology for estimating the capacity of screw piles. The ScrewFast method is based on the conventional design method, but uses several pile-specific assumptions which improve its accuracy and reliability. The predictions of the conventional and the ScrewFast methodology were correlated with actual torque measurements during pile installation and with several pile load tests on 16 screw piles (13 tests in compression and 3 tests in tension) in both cohesive and granular soils. As only two of the pile load tests reached failure, Chin's extrapolation method was applied in the other tests for the estimation of the failure load. The analysis showed that pile designs based on the empirical torque-to-load relationship (where torque is the minimum installation torque) can provide the capacity of screw piles, at least for the range of soil types examined.

1 INTRODUCTION

Screw pile foundations are an original, and at the same time, modern solution for small scale foundation requirements. A screw pile can be best characterized as a large self-tapping screw that is placed into the ground. Single or group pile calculations are based on standard formulae for end bearing piles, with skin friction not being a major factor under vertical and horizontal loading or bending.

Screw piles consist of a 2 m-long lead section with one or more helices, followed by additional lengths with multiple helices. The additional length(s) of the pile are usually 2 m-long steel shaft sections (sometimes with additional helices).

The diameter and location of the helices, along with the length and diameter of the shaft are designed on the basis of the soil investigation information and the required loading capacity. The lead pile section, which consists of the helical plates welded to the shaft is connected to the power head and rotated into the ground. The helical plates enable the lead section to be pulled into the ground. This section is then extended by an extension section which is connected to the lead pile by high tensile bolts. Extensions are added until the required design depth is achieved.

Screw piles are screwed into the ground using a hydraulic rotator attached to earth-moving equipment, mini excavators, bobcats or large excavators depending on the size and capacity of the screw pile or the accessibility of the site. The plant operator continually monitors the torque during the installation process and at the completion of each installation, to ensure the correct capacities are achieved. This monitoring process provides an indication of the ground profile and founding soil capacity.

2 SCREW PILE DESIGN METHODS

2.1 *General considerations*

The general idea of a screw pile foundation is not so much in the manufacturing the product but giving a best quality specific engineering design. In simplistic terms the foundation is designed as a solution to a particular requirement or condition, subsequently the solution becomes the product. Most screw pile companies use standard end bearing and tension anchor piling formulae.

Screw piles do not fit easily under any of the pile categories covered in the Highways Agency Specification Series 1600 or the ICE Specification for Piles and Embedded Retaining Walls (Thomas Telford 1996).

The design of a screw pile is a rather complicated procedure as there are a lot of factors which have to be considered. The central shaft will transmit torque to the helical plates. Design depth is achieved by bolting on additional shaft sections. Once installed onto or into the bearing stratum the shaft transfers the axial load to the helical plate(s). Each helical plate on any single pile will be of the exact same pitch which mitigates soil disturbance whilst installing, unlike conventional piling techniques. The diameter of these plates is designed by structurally calculating all the applied loads. Loads are transferred either above or below the helical plates therefore they are designed to maximize the bearing capacity of the soil. The actual working mechanism involves a complex plate to soil interaction (Legrand 2001, Legrand & Poorteman 2003).

2.2 *Modes of failure*

There are two basic modes of failure.

- Helix width shaft failure. In this case the helices have been placed relatively close to each other, and the ratio of shaft diameter to helix diameter is inadequate to resist the loads applied to the pile because of poor soil conditions.
- Multiple helix base failure. The helices are spaced far enough from each in order to avoid any possible influence of stress zones.

In most cases screw piles are designed so that each helix works independently; so failure would occur in mode 2.

2.3 *Conventional geotechnical screw pile design method*

In clayey soils, in the basic equation for the base load capacity of a single pile assuming an undrained analysis, φ equals to zero and c equals the undrained strength, S_u. So the pile base resistance given by a total stress analysis can be obtained from the equation:

$$Q_b = A_b.\ S_u.\ N_c \quad (1)$$

where, N_c is a dimensionless parameter typically taken as $N_c = 9$.

In granular materials, the helix bearing capacity Q_b is computed as follows:

$$Q_b = \sigma'_v.\ N_q.\ A_b \quad (2)$$

where, σ'_v = effective overburden pressure at the pile base; and N_q = a dimensionless bearing capacity factor.

A wide range of values are reported for N_q but the most widely used are those derived by Berezantev. N_q is very sensitive to the friction angle, particularly for friction angles greater than 36°. Because of the sensitivity to φ', the most sensible value to choose in design is the ultimate (critical state) value, φ'_{cs} (Peck, Hanson & Thorburn 1974). But it is important to note that it is not easy to determine the critical state of sands.

2.4 *The ScrewFast design method*

This method is based on the conventional method of calculating the helix and shaft capacities of a screw pile but relying on only very limited information about the soil profile. As a consequence it can be characterized as an 'over-design' technique as it relies on conservative factors of safety.

This approach has been adopted by Screwfast Foundations Ltd in order to review the available data and methodology of the design of screw piles for Arup Geotechnics, UK and produce a technical basis for their existing design.

Screw piles are often installed in locations difficult to access for a standard, comprehensive ground investigation (e.g. railway / motorway embankments). The main soil 'parameter' used is the Standard Penetration Test (SPT) N-value. Where access for a drilling rig is particularly difficult, a dynamic probe kit is used and the results are correlated with SPT N-values.

$$\text{Helix Capacity} = 5.\ S_f.\ A_r.\ N.\ \pi.\ (\text{zone of influence base diameter x } \sin 30^\circ)^2\ .f \quad (3)$$

where S_f = Soil factor, which is usually taken as 1.0 but there are cases where it varies from 0.6–1.0 depending on soil conditions; $A_r = (\pi\ (0.5d)^2 - \text{shaft area})/\pi(0.5d)^2$; d = diameter of helix plate; N = SPT N value for the helix founding stratum; f = reduction factor, equal to 1.0 for dry soil and 0.5 for saturated soil. For fine grained soils, ScrewFast's reduction factor is equal to 0.5 and 1.0 for coarse grained soils. There is a standard pile (shaft and helix as a unity) spacing of 1.5d–2d when using the ScrewFast formula.

The formula and calculation of bearing capacity is based on the assumption that the helix plates fail independently of each other and the whole screw pile does not suffer by 'shaft' type failure and the friction is negligible.

The ScrewFast formula is mostly based on ground information, in particular SPT N-values or dynamic probe results. The formulation is therefore the same for both cohesive and granular soil types (i.e. there is no differentiation).

Relatively high factors of safety are applied in each case. A reasonable factor of safety is always determined from available subsurface and surface information, variability of the soil, soil layering and strengths, type and importance of the structure and past experience The FoS will typically be between 2 and 4 (for tension is equal to 4.0 and for compression equal to 3.0).

ScrewFast's design methodology states that it is reasonable to reduce the spacing of the helix plates to less than 3d, provided that the factor of safety is increased to take account of their closer spacing. This is a reasonable approach to ensure against loss in capacity in the case where the failure mechanism is not one of bearing capacity failure of the plates.

2.5 *Installation and predicted torque correlations*

As part of a screw pile design, minimum and maximum torque is calculated. In simple terms minimum torque is dictated by the design loads, the soil conditions and it is generally a function of the sum of chosen helix ultimate capacities. The minimum torque is calculated using the ultimate load applied on the design helices and this value is then divided by 18.0, which is a value empirically used by ScrewFast from experience and field testing. On the other hand the maximum torque is related to the breaking point of pile and its fabric, i.e. bolts, joints of pile shaft – British Standard BS 5950 (2000).

The installation torque should be within a certain limit of the value measured in field as the screw pile is installed. The installation torque in order to be adequate for the overall design should be within the margins of maximum and minimum torque values.

2.6 *Determination of ultimate load capacity using Chin's method*

Chin's method assumes a hyperbolic relation between the pile head settlement and the mobilized soil resistance expressed by the applied load. It then provides a method by which the ultimate load of a pile can be obtained from the results of a load test without having to load the pile to failure. Definitions of this sort are capable of interpretation for single piles, but when applied to pile groups the question immediately arises as to what comparative order of settlement should then be used.

There are different ways to extrapolate the load-settlement curves to estimate the limit load. Chin's method has been experimented with to try to estimate the limit load capacity of screw piles. The method (Chin 1970) is based on the assumption that the load settlement relation is hyperbolic.

$$s/Q = C_1 s + C_2 \tag{4}$$

where Q = load applied to the pile; s = settlement corresponding to the load Q; C_1 and C_2 = slope and intercept of the load-settlement curve in s/Q vs. s space. The limit load capacity = $1/C_1$.

3 CASE STUDIES

The data analyzed for this paper have been taken from 13 compression and 3 tension tests. The information was taken from sites spread across the UK and kindly provided by ScrewFast Ltd. A variety of pile load tests have been conducted, at different geological profiles, type of loading procedure; i.e. cyclic or incremental, design load and maximum applied load for the completion of test as well as the torque values applied for every pile. All pile load tests have been divided into nine projects according to the site where they took place.

3.1 *Compression pile load test in cohesive soil*

Project 1: The borehole revealed the presence of 5.30 m of very stiff laminated dark grey clay, underlying 0.30 m of moderately weak grey mudstone, underlying 6.40 m of firmly thinly laminated dark clay with occasional pockets of medium sand-sized gypsum crystals and occasional shell fragments underlying 2.45 m of soft to firm grey brown clay with occasional fine gravel sized calcareous nodules underlying 0.55 m of made ground.

Table 1. Comparison of helix capacity using conventional and ScrewFast approaches.

Comparison of Helix Capacity q_f (kN)			
Depth (m)	Helix Diameter (m)	q_f (kN): General Method	q_f (kN): ScrewFast Method
2,5	0,60	78,34	74,82
3,5	0,60	111,91	106,89
4,5	0,45	88,13	80,40
5,5	0,45	88,13	80,40

A compression test was performed with cyclic loading up to 300% of design load of 250 kN of a 6.0 m screw pile with a variation of helical plate diameters; test data show an initial good behaviour of pile movement when 50% of the design has been applied with a magnitude of 2.5 mm while at 100% design load the pile deflection was 8.7 mm. At 150% design load the pile settled 12.7 mm from its initial position and as more load was been applied, the pile showed a stiffening behaviour with a displacement value of 16 mm at 300% of design load and at the release of the last compression load increment, the residual displacement reduced to 12 mm.

The test pile was installed with a torque of 37 kNm and there was a good correlation between the bearing capacity of pile using a conventional geotechnical approach and the Screwfast formula as shown in Table 1. The difference was within a margin of 8 kN for all the helical plates.

Project 2: The borehole revealed the presence of 2.30 m of very stiff dark clay with occasional fine gravel sized shell fragments, underlying a thin layer of 0.20 m of moderately weak light grey siltstone recovered as fine to coarse angular gravel, underlying 5.0 m of stiff dark grey clay, with occasional fine gravel sized shell fragments, underlying 3.30 m of firm dark clay, underlying 1.10 m stiff very closely fissured light grey mottled brown slightly sandy clay with occasional fine angular sand sized gypsum fragments, underlying 2.50 m soft fissures slightly gravelly clay, underlying 0.80 m of made ground.

Project 2 refers to a compression test similar to Project 1 but it is more consistent with a better pile behaviour at the recovery stage. It was also a cyclic loading test with a maximum applied load 300% of the 250 kN design load. The length of pile was 6.0 m with 450 mm and 600 mm helical plates. The initial displacement reached up to 3.3 mm after the application of 100% design load and it increased to 5.3 mm at 150% design load showing a softening behavior. At 200% design load the deflection reached 8.3 mm and at the release of this load increment it recovered up to 2 mm, and there is a consistent increase and decrease

Table 2. Comparison of helix capacity using the conventional and ScrewFast approaches.

Comparison of Helix Capacity q_f (kN)			
Depth (m)	Helix Diameter (m)	q_f (kN): General Method	q_f (kN): ScrewFast Method
2,5	0,60	179,06	171,02
3,5	0,60	134,30	128,27
4,5	0,45	62,95	57,43
5,5	0,45	88,13	80,40

Table 3. Comparison of helix capacity using the conventional and ScrewFast approaches.

Comparison of Helix Capacity q_f (kN)			
Depth (m)	Helix Diameter (m)	q_f (kN): General Method	q_f (kN): Screwfast Method
4,5	0,60	100,72	65,50
5,5	0,60	179,06	133,08
6,5	0,45	100,72	69,78
7,5	0,35	60,93	47,02

of the magnitude of settlement within this margin until the end of the test.

The test pile installation torque was 43 kNm and there was a good relationship between the conventional geotechnical approach and the ScrewFast formula with a maximum difference in helix bearing capacity between the two techniques of 8 kN (Table 2). The SPT N-values were the similar in both sites for projects 1 and 2.

Project 3: The borehole revealed the presence of 11.0 m of very stiff very closely fissured grey clay with occasional silty laminations (London Clay) underlying 3.0 m of very stiff mottled grey and brown clay, underlying 6.0 m of brown gravelly clay, underlying 1.0 m of slightly sandy gravelly clay, underlying 1.0 m of greyish brown gravelly clay, underlying 3.0 m of brown slightly sandy clay, underlying 1.0 m of made ground, concrete and asphalt.

A compression test was performed with incremental loading and unloading up to 150% of design load. It was an 8 m helical pile with a design load of 73 kN. Initially the pile shows a stiffening behaviour as the load is applied. It reached up to 2.3 mm of displacement at the first 100% of design load and in the unloading stage it recovers reasonably well. Then new load increments were applied and the deflection is 5.4 mm at 150% design load. The tested pile held the load for the required six hours period at both 100% and 150% without any significant creep. The displaced pile did not have detrimental effect on pile capacity and deflection under load which are well within the allowable deflection criteria.

The installed torque of 52 kNm is a very good indicator of pile capacity and this can be observed clearly in the results calculated. On the other hand Table 3 shows that there is an inconsistency in the bearing capacities calculated with the two design approaches as they have a difference of around 30kN which is unacceptable compared with the other study cases. This is because the SPT N-value in the design excel sheet and the borehole log had a small difference; the design sheet has a more uniform flow of the N-values.

3.2 *Tension pile load test in cohesive soil*

Project 4: The borehole revealed the presence of 6.5m of soft light brown/orange clay with grey veins, underlying 0.8 m of soft light brown mottled clay with occasional fine to medium angular gravel, underlying 0.2 m of made ground.

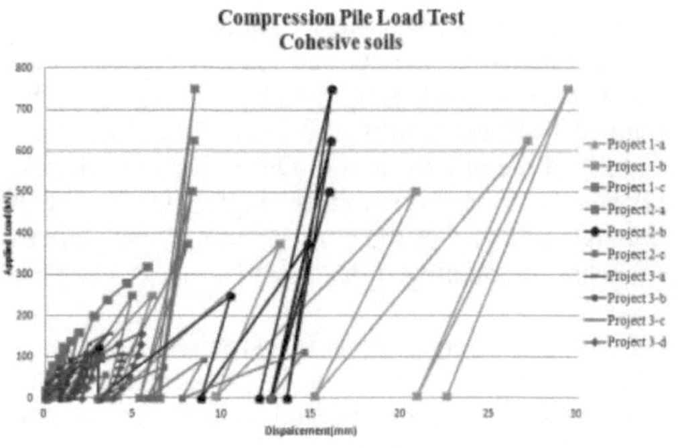

Figure 1. Summary pile load – settlement results for compression pile loading in cohesive soils.

A tension test using incremental loading was performed up to 150% of design load of 128 kN. The length of pile is 5 metres with a variety of sizes of helical plates closely placed together compared to the usual screw piles. The pile moved 6.2 mm at 100% design load reaching a value of 12.7 mm at 150% with an increasing softening of the tested pile. This could be due to surrounding soil disturbance or movement of dial gauge as the load has been applied. The pile was recovered well at the end of the test reaching a value of 5.6 mm as its final deflection.

The pile was installed to 7 m to an average torque of 28 kNm which is close to the value of minimum design torque. The calculation of the bearing capacity of helices using both, Conventional and ScrewFast techniques showed a good correlation as shown in Table 4 although there was a difference of 20 kN but this due to different SPT N-values between borehole logs and design sheet.

Project 5: The borehole revealed the presence of 3.7 m of very dense to grey very thinly bedded silty firm to medium sand and sandy clay, underlying 0.9 m very dense dark grey rounded fine to coarse gravel, underlying 0.2 m of very stiff grey very sandy clay (London clay), underlying 12.0 m of very stiff grey brown sandy to very sandy clay (London clay), underlying 4.0 m very stiff grey brown thinly laminated

Table 4. Comparison of helix capacity using the conventional and ScrewFast approaches.

Comparison of Helix Capacity q_f (kN)			
Depth (m)	Helix Diameter (m)	q_f (kN): General Method	q_f (kN): ScrewFast Method
4,5	0,60	279,78	240,50
5,0	0,45	201,44	165,40
5,5	0,45	239,21	196,41

Table 5. Comparison of helix capacity using the conventional and ScrewFast approaches.

Comparison of Helix Capacity q_f (kN)			
Depth (m)	Helix Diameter (m)	q_f (kN): General Method	q_f (kN): ScrewFast Method
4,5	0,60	123,10	117,57
5,5	0,45	56,66	51,69
6,5	0,45	50,36	45,95
7,5	0,35	53,31	45,23

extremely closely to and very closely fissured clay (London clay), underlying 6.0 m of stiff becoming very stiff by 11.0 m grey brown clay (London clay), underlying 4.0 m loose becoming medium dense brown angular to subrounded fine to coarse flint gravel (Thames gravel), underlying 1.0 m soft dark grey sandy clay, underlying 0.9 m of soft grey locally thinly laminated clay pockets, underlying 1.6 m of loose dark brown and mottled brown clay.

The pile testing involved a tension test and a shear test to 150% design load of 110 kN and 30 kN respectively. As in this study only compression and tension tests are analyzed, the shear test will be ignored for this analysis. The pile length was 8 m with a variety of helix diameters from 600 mm to 350 mm. The cyclic loading test showed a sensible behaviour as the load was increased, reaching a displacement of 4.7 mm at 100% design load (74 kN) and 5.7 mm at 150% design load. At the recovery period the pile moved back to 4.6 mm. Between the 50% design load and 100% design load the pile showed a relative big settlement.

The average installation torque was 17 kNm, relative to the worst case pile design (11 kNm). The two design techniques of estimating the bearing capacity showed a good correlation of a difference 8 kN (Table 5) as well as the SPT N-values from the site investigation and Screwfast's design sheet looked similar.

Project 6: There are records for 12 Boreholes made at this site but in this study only one will be described. This screw pile test has been done in sand but as it is the only tension load test conducted in non cohesive soil, it is good to be described in this category.

Table 6. Comparison of helix capacity using the conventional and ScrewFast approaches

Comparison of Helix Capacity q_f (kN)			
Depth (m)	Helix Diameter(m)	q_f (kN): General Method	q_f (kN): ScrewFast Method
1,5	0,80	457,59	535,36
2,5	0,80	616,75	825,11
3,0	0,80	835,60	901,87
3,8	0,60	559,55	519,83

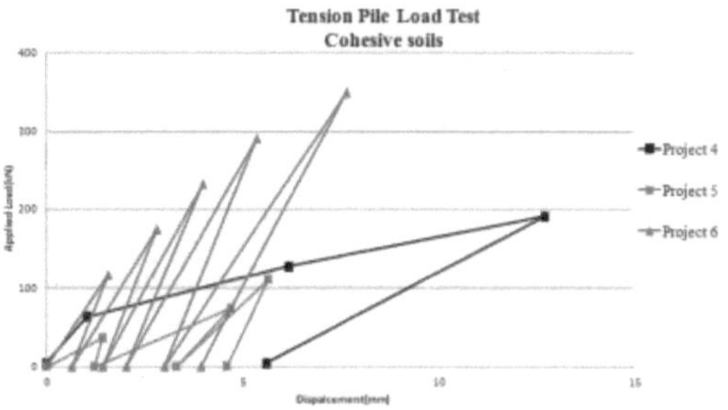

Figure 2. Summary pile load – settlement results for tension pile loading in cohesive soils.

The borehole revealed the presence of 1.5 m of very dense red fine sand, underlying 1.5 m of medium dense becoming dense, red laminated fine to medium sand, underlying 1 m of medium dense, light brown becoming red fine sand, underlying 0.5 m of grass over loose gravelly fine sand topsoil.

The test performed was a tension test to 150% design load with load increments of 25% as per the data sheet in the appendix. Two concrete blocks either side of the test pile acted as reaction platforms for the pile test. The design load for the tested pile was 460 kN shared equally between the two concrete blocks each designed for a working load of 233 kN in tension. The pile deflected to 3.1 mm for the 100% design load increment and 4.9 mm for 150% design load.

The test pile was installed to the design depth with an average installation torque of 114 kNm. The pile performed well and deflections were well within the defined performance criterion. The above test result is a very good indication of the pile capacity of the installed piles on the project across the site and proves the pile design and the torque to bearing capacity relationship. Although because of the large number of boreholes a probable misreading on the data was made and the values of the bearing capacity between the two methods were relative different; as shown in Table 6. ScrewFast bearing capacity was much greater that the conventional method, around 70 kN for all the helices. The SPT N-values defined from the 12 borehole logs and the SPT N-values estimated from the ScrewFast design sheet were not consistent, this could lead to misinterpretation of the soil strength.

3.3 *Compression pile load test in granular soil*

Project 7: The borehole revealed the presence of 0.4 m of dense light grey silty sand with fine to medium gravel, underlying 5 m of dense becoming medium dense with depth grey becoming sand with occasional thin bands of stiff known clay, underlying 3.5 m of medium dense becoming very loose light brown sand, underlying 2.3 m of soft grey sandy clay made ground, underlying 0.7 m of soft grey made ground, underlying 0.5 m light brown tarmac over sand made ground. The borehole was extended to 12.5 m below the existing ground level and terminated in material with an SPT N-value of 29.

For this compression pile test some data were possibly misread and some correction was required for better analysis. The test pile was designed for a working load of 80 kN and the pile was made of a 7.0 m long pile with 350 mm diameter helix and two of 450 mm diameter helices spaced below ground level. The deflection of the pile as the load was increased was large, causing a displacement of 11.8 mm at 100% design load and then it was relatively stabilized compared to the initial constant increase of deflection. It was recorded 13.2 mm displacement at 150% design load and 15.5 mm at 200% design load. After the unloading period the pile has been displaced 23.3 mm at 300% design load and 25.77 mm at 350% design load; this showed that something was wrong as well as the fact that the pile has achieved its designed factor of safety 3.0. After careful observation, at 350% design load the pressure in the recently calibrated pump had dropped. This caused the pile displacement to stabilize. This drop in pressure may be assumed from decay in the 'elasticity' of the soil. The reaction piles were not creeping significantly.

The test pile was installed approximately 500 mm above ground surface level and the reaction piles 900 mm above the ground surface. These piles have been installed in poor quality made ground/fill type material and maybe they were not able to restrain any horizontal movement brought about by eccentrically installed piles. Once this initial horizontal movement was taken up, the average displacement per load increment on the test pile has been reduced considerably, with the profile flattening out.

The installation torque for this test was 19 kNm, showing a good correlation with the designed bearing capacity of the screw pile. Table 7 shows a major difference in the bearing capacities calculated with the two design approaches in granular soils.

Project 8: As there is only borehole record made for this project in granular soils and there is no proper indication for which test it is related to, it is assumed that provides a soil profile indication for all three pile tests conducted at this site.

The pile details of the second pile of Project 8 case test were an 8.0 m long pile with a variation of helical plate diameters. The design load was 80 kN. At 100% design load a displacement of 3.4 mm was recorded with an increase of 2 mm approximately as the design was increased 50% reaching 5.1 mm of deflection.

Table 7. Comparison of helix capacity using the conventional and ScrewFast approaches.

Comparison of Helix Capacity q_f (kN)			
Depth (m)	Helix Diameter (m)	q_f (kN): General Method	q_f (kN): ScrewFast Method
4,5	0,45	1087,30	104,20
5,5	0,45	358,46	119,10
6,5	0,35	500,05	104,20

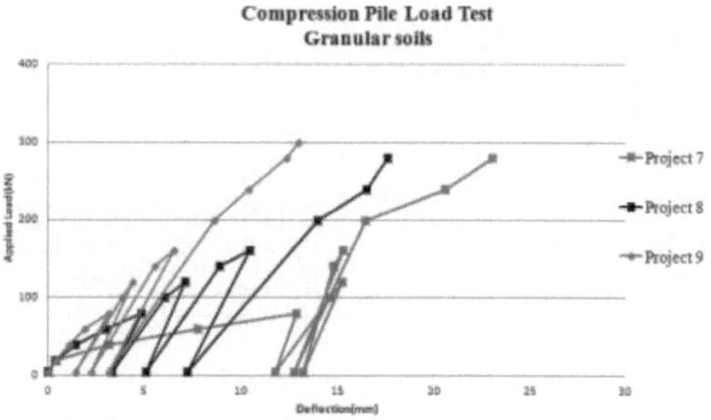

Figure 3. Summary pile load – settlement results for tension pile loading in granular soils.

The increase of deflection was relatively stable; at 200% design load it was 7.3 mm. At 300% design load and 350% design the pile displacement stabilized, the pile reached its failure point. The pile settled 16.5 mm at 300% design load and Z17.6 mm of settlement at 350% design load.

For this test the pile has been re-installed and tested again at ground level and it performed much better this time. The installation torque was measured at 18 kNm for this project.

Project 9: The pile load test for this project was relocated about 1.5 m north of the projects 7 and 8. For this compression test the design load was 80 kN and the displacements of the tested pile were relatively small. At 100% design load the piled had a deflection of 1.4 mm, 2.3 mm of displacement at 150% design load and 3.2 mm for 200% design load. This test reached failure too, at 300% design load, it settled 12.3 mm and kept increasing infinitely as the pile achieved its designed factor of safety of 3.0. The pile was loaded up to 375% which was the limit of the testing equipment.

During installation, the required minimum torque was not achieved. Therefore extra steel sections have been added to the designed pile until the minimum required torque was achieved at a depth of 11.0 m.

4 APPRAISAL OF DESIGN METHODS

4.1 *Ultimate load and failure*

Despite the variety of design loads and maximum applied loads for all the study cases only two tests have actually reached failure considering the fact that few of the pile tests have been taken to 300% of design load. For most of the others, the behaviour of tests pile under incremental loading indicated that the ultimate load capacity was considerably more than the maximum load applied. As loads increased closer to ultimate levels, it was noted that the magnitude of movement increased. This indicates the effects of overlaps of bulbs of pressure in the soils from the individual helix plates.

4.2 *Relationship between load capacity and installation torque*

The relationship between ultimate load capacity and installation torque is very important and it has to be carefully considered for every screw pile design. It provides a good estimation of the long lasting life of the pile as well as the actual strength of the ground.

In nearly all of the cases that have been studied, there was a good correlation between these two values showing the consistency of the ScrewFast technique in designing the right dimensions of a screw pile in order to avoid the wrong estimation of the bearing capacity of the pile. The predicted working load based on the torque factors is greater than the required design load, sometimes by quite a large margin. For the case where the torque was below the allowable limits, it is not known why the predicted torque was not achieved, but during the test this pile did not perform significantly worse than the other tests.

As it is mostly a technique tested and qualified on site the estimated soil strength is lower or higher than the actual conditions and this leads to the successful installation of the pile even though the design parameters are different. But even in the cases where the installation torque is below the design limits more pile sections can be added to the designed pile or other changes of the design pile can lead a safe and long lasting foundation.

4.3 *Estimation of bearing capacity of a helix using conventional geotechnical approaches and the ScrewFast formulation*

The ScrewFast design approach is based on conventional theory, but various assumptions were part of forming the formula for all the study cases to be analyzed.

This study aimed to test the consistency of both approaches with each other. It could be easily noticed from the calculations of bearing capacity in cohesive soils that there is a good correlation between the two methods, giving little margins of difference. Equation (1) is based mainly on the SPT N-value as it calculates the undrained shear strength, S_u ($=4.5 \times N$) as well as the equation (3) uses the SPT N-value in order to simplify the methodology for the site investigation required. The necessary SPT N-values can be determined using probe devices rather than conventional borehole drilling techniques. Both equations do not involve the depth of helical plate; they just concentrate on the helix diameter as an individual circular foundation. This is the main reason for obtaining similar results.

On the other hand, as the Screwfast equation has been developed, one of the assumptions was that the depth of a helix was a lost term and it was not considered in the estimation of the bearing capacity of a helical plate. This assumption, though, influences the consistency of equations (2) and (3). In order to estimate the bearing capacity of a helix using the conventional method, the vertical effective stress value is needed and that includes the depth of required helix, which is a lost term in Screwfast equation. This causes large errors when comparing the two bearing capacities.

5 CONCLUSIONS

5.1 *Summary*

The aim of this study was to compare the conventional geotechnical technique with the ScrewFast methodology for estimating the capacity of a pile. ScrewFast's design formula is based on conventional theory, but using several assumptions which simplify the calculation method. As has been discussed earlier the comparison of the two methods for cohesive soils showed good correlation, but for granular soils the differences are large and this is due to the lost 'depth' term, which reduces confidence in ScrewFast's design approach.

The torque measurements are a valuable parameter in estimating the expected capacity of a screw pile, but the best method of improving the certainty of estimated pile capacities is the knowledge of true failure capacity derived from load tests. In this way the calculation method of ScrewFast as well as any relationship with torque will be improved.

The analysis carried out showed that by using the results of ground investigation at each study case, pile designs based on the empirical torque to load relationship which outputs a minimum installation torque required, can be relied upon to provide the foundation capacity required, for the range of soil types within which ScrewFast Foundations Limited state that the method is valid.

It is clear that true failure is not an issue according to the ScrewFast design approach as it is referring to small-scale structures and high factors of safety applied for a highly conservative design. The parameter that is important for this technique is the expected degree of movement under load which may affect the safe and reliable design of the screw pile. Failure could be characterized by creep which is not catastrophic.

It is recommended that that more emphasis has to be given to the design procedure using better soil profile information and additional independent checking of the pile designs. The recorded installation torque can be used as substantial measure of pile performance, in addition to competent pile design and design checking, which can be considered to prevent the need for extensive testing.

It is important that further research of the ScrewFast approach is required as it is inconsistent with the conventional design approach for granular soils. The 'depth' term used in the general formula for calculating the capacity of a pile is a very important parameter as it reflects the effective stress applied at the base of the piles and in this case the base of helical plate. ScrewFast's approach eliminates this term and replaces it with the 'A_r' term which is a ratio of helix area and pile shaft area and it is usually less than 1.0. This is unconservative for granular soils as it underestimates the capacity of the helix.

ACKNOWLEDGEMENTS

The authors would like to acknowledge the contribution of ScrewFast Foundations Ltd, who kindly provided the test data and agreed to its publication, and especially wish to thank Messrs Charles Wark and Rajen Raghwani for all their help during this study. Thanks are also due to Arup Geotechnics, in particular Mr Tim Chapman, for their input into the study.

REFERENCES

Berezantev, K. & Golubkov. 1961. Load Bearing Capacity and Deformation of Piles Foundations. *Proceedings of 5th International Conference on Soil Mechanics and Foundation Engineering.* Volume 1: 11.

BS 5950-1, 2000. *Structural use of steelwork in building. Code of practice for design. Rolled and welded sections.* BSI, London.

Chin, F. K. 1970. Estimation of the ultimate load of piles not carried to failure. In *Proceedings of the 2nd Southeast Asian Conference on Soil Engineering*, Singapore, pages 81–90.

ICE. 1996. Specification for piling and embedded retaining walls. Thomas Telford Publishing, London.

Ingram P., Glazebrook A. & Chapman T. 2005. Design of Screw Piles – Assessment of pile design methodology. *Tech Rep, ArupGeotechnics,* UK: London.

Legrand, C. 2001. Deep foundations and the need for research on screw piles in Belgium. *Proceedings of the Symposium Screw Piles – Installation and Design in Stiff Clay.* Brussels: Belgium, 3–9.

Legrand, C. & Poorteman, F. 2003. Deep foundations and the need for research on screw piles in Belgium, *Proceedings of the Symposium on screw piles in sand – design & recent developments*, Brussels: Belgium, 3–10.

Peck, R.B., Hanson, W.E., Thornburn, T.H. 1974. *Foundation Engineering.* Wiley International edition (2nd ed.).

Baltic Piling – Mets & Raudsepp (Eds)
© 2013 Taylor & Francis Group, London, ISBN 978-0-415-64334-4

Lida brewery pile foundations

M. Mets & V. Leppik
Institute of Forestry and Rural Engineering, Estonian University of Life Sciences, Tartu, Estonia

T. Ruben
Geoengineering OÜ, Tallinn, Estonia

ABSTRACT: Intaraction of piles and grill

1 INTRODUCTION

At present, one of the oldest breweries in Belarus is expanding. In this article, we will be looking at the pile foundation problems presented in establishing a new bottling section to the brewery [1].

The brewery is located in Belarus, on the banks of the river running through the Lida city. The area is covered with fill and flood plain sediments (peat and mud). The flood plain sediments cover the alluvial fluvioglacial sands, located on top of a layer of till. Under the till there lie very dense fluvioglacial sands from the previous Glacial period.

Pile foundation design was based on the geological cut [2], that is shown in Figure 1; under fill (1) there is located a weak layer of till (2), a strong layer of sand (3), medium strength till (4), and a very strong layer of till (5). Their characteristics are shown in Table 1.

Based on the geological structure, the suggestion was to use the CFA piles which had to be driven through high density till, their diameters were 630, 550 and 400 mm. To calculate the bearing capacity of the pile SNiP (building norms and rights in Russia) tables were used. The ultimate bearing capacity of the pile end was 10000 kN/m^2 and side friction was 50 kN/m^2. The partial factor of bearing capacity that was used was 1.5, which is much higher than the recommendation on SNiP. The load-bearing capacities of piles is given according to diameter: 2330 kN (630 mm), 1763 kN (550 mm) and 709 kN (400 mm) [3].

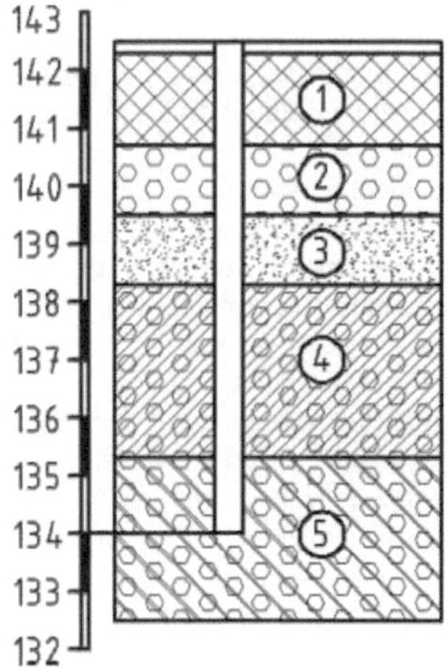

Figure 1. Layers of soil.

The piles were driven into strong till and after two months 4 static load tests were made. The choice of piles was made by the local building inspector and the load test made by the Belarus company of "STROI complex". For processing the pile test results, the authors used the method of characteristic points [4]. This method allows evaluating the side friction of the pile, the bearing capacity of the pile end and the creep limit. The pile test results are shown in Table 2.

Piles no. 92 and no. 56 got the ultimate load. For piles no. 167 and no. 68 the ultimate load was not reached because the bearing capacity of the anchorage pile system was too small [5]. The pile testing quality was unsatisfactory. Anchor piles were very close to the testing piles, as were also the reference marks. The

Table 1. Soil characteristics.

Classification of soil	Density of soil kN/m^3	Angle of internal friction degrees	Cohesivness kPa	Module of deformation MPa
Sand	20	36	0	27
Low density till	21	23	15	7
Medium density till	22	24	25	25
High density till	22.5	26	42	35
Fluvioglacial sand	20	41	2	55

Table 2. Pile load test results.

Pile number and diameter	Side friction N_a kN	Average side friction τ_a kN/m²	Creep limit N_y kN	Pressure to pile end at creep limit σ_y kN/m²	Ultimate load N_f kN	Pressure to pile end at ultimate load σ_f kN/m²	Maximum load to pile N_{max} kN	Maximum pressure to pile end σ_{max} kN/m²
92/550	500	29	1000	2173	1400	3913		
56/630	600	39	1000	1290	1800	3870		
68/ 630	500	40	1200	2258			1500	3225
167/500	500	40	1000	2173			1400	3913

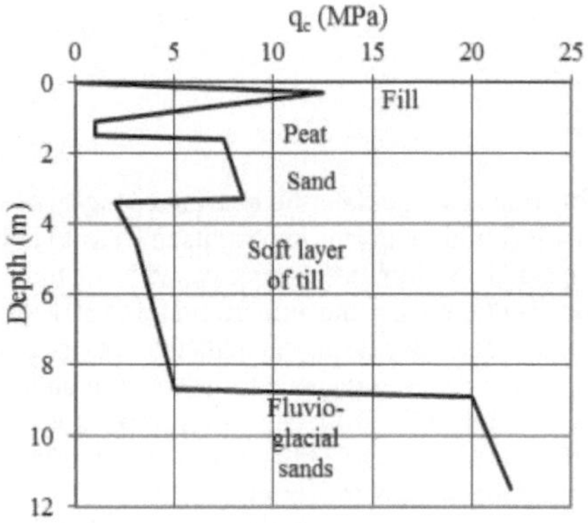

Figure 2. Layers of soil.

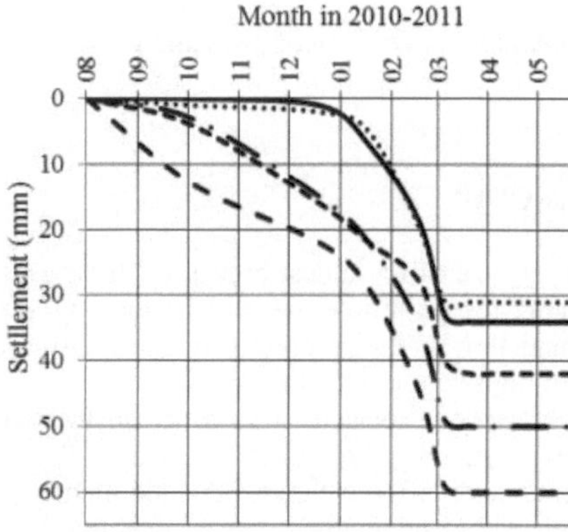

Figure 3. The settling time.

testing system was working normally up to a settlement of 10 mm. The pile test results show that the ultimate bearing capacity of the tested piles was significantly lower than of the designed piles.

The stress under the bottom of piles was less than 40% estimated in ultimate load. In view of this situation UP Geoservice made additional geotechnical studies, with 18 CPT tests with the depth of 15 meters. Figure 2 shows the penetration test number 7 results, with the resistance changing in the depth of the cone.

These tests showed that all of the till layer is of low density and the pile ends are located in it. On SniP tables, this pile end resistance was 3000–3500 kN/m², which is below the test result.

2 DISCUSSION

The usual practice in similar situations has been to take into account the cooperation between the pile and the grill and share the load with the grill. There have been two ways to solve the situation. The first is the strengthening of piles with injection piles, which have been driven into fluvioglacial sands, and that would have taken 60% of the grill's load. Another solution is to remove the filling and the floodplain, replacing them with sand that should be compressed to the density factor 0.95. In that case, parts of the load would have been on the piles and the rest on the grill. The side friction in piles was designed to be 40 kN/m² and the compression stress under the pile end 3200 kN/m². The designed grill surface was increased by 15% to make sure that the pressure under the grill did not exceed 250 kN/m².

The displacement of the new designed foundation is based on the following calculation. The design load was 6600 kN, which was divided between piles 4500 kN and grill 2100 kN. The grill measurements were 2.9 × 2.7 m and the pressure under the grill was 277 kN/m². The calculated total displacement was 8,6 cm.

It was decided to install the reference marks in all of the building foundations and start the geotechnical control in parallel with building construction. The displacement measurements took place in August 2010–July 2011 and by the time the building was fitted with all the equipment and the foundations had been targeted at the designed load. During this period, estimations were performed 11 times.

Figure 3 shows five of the following reference marks to be settling over time. The maximum building settlement was 6 cm and minimum 2.5 cm. Characteristic of these foundation settlements was the higher settlement rate of the soil in the melting period in spring. It seems that in the beginning the pile side takes the load, and then the load gets transferred to the end of pile. The soil under the pile end reaches the ultimate state and propels a relatively rapid settlement of the foundation, which is accelerated by the weather. The acceleration takes place in the snow melting periods. The soil under the grill thickens and the grill starts to work. Due to the pin effect the settlement process stops.

3 SUMMARY

The Lida brewery foundation solution showed that using a combination of piles and the grill could reduce the number of piles under the building and increase the efficiency of the foundations. If under the grill the soil has high compressibility ($E < 5$ MPa) this solution is not viable.

REFERENCES

[1] Insinööritoimisto Savolainen OY, Design project of Lida brewery warehouse. 2009

[2] Инженерно-геологическое изыскание. ГОИИ УП "Геосервис". Объект No. 1230/09-11, 2009.

[3] Lidskoe pivo brewery warehouse, Geoengineering OÜ, 2009/013.

[4] M. Mets Iselomulike punktide meetod, Ehitusgeoloogia kogumik nr. 5. Tallinn 1991, p. 42–63.

[5] Stroi complex, work number 876, Pile tests.

Baltic Piling – Mets & Raudsepp (Eds)
© 2013 Taylor & Francis Group, London, ISBN 978-0-415-64334-4

Assessment of time dependent influence on the lateral response of single pile subjected to lateral load end embedded into elasto-plastic soil

Jasim M. Abbas
Department of Civil Engineering, Diyala University, Baquba, Iraq

Mohd Raihan Taha & Zamri H. Chik
Department of Civil and Structural Engineering, Universiti Kebangsaan Malaysia, Bangi, Selangor D H, Malaysia

Qassun S.M. Shafiqu
Department of Civil Engineering, Nahrain University, Baghdad, Iraq

ABSTRACT: The time dependent response of the geotechnical structures can be categorized as one of the important elements that should be take during into account its analysis and design. From the literatures, there is very little knowledge reported on the finite element technique of time dependent behavior of laterally loaded piles embedded in two types soils. There appear to be no published solutions for laterally loaded pile foundations in consolidating elasto-plastic soil to determine the design parameter such as maximum lateral pile displacement, ultimate lateral soil resistance and *p-y* relationship. Therefore, this study investigates the effect of soil type with different intensities of loading on the lateral pile deformation and lateral soil pressure with time. Finite element analysis is carried out to evaluate the lateral pile response embedded in cohesionless and cohesive soil subjected to pure lateral load. The simulation include linear elastic model to represent the pile structural material and Mohr-Coulomb elasto-plastic model to represent the surrounding soil. Biot's equation of consolidation is used to govern the elasto-plastic material. The complete model of the whole geotechnical system are used to assess the lateral pile displacement and lateral soil pressure developed at pile face of 15 m and 1 m pile length and diameter, respectively. It is shown that higher lateral pile displacements and lateral soil pressure were observed with time due to consolidation process.

Keywords: single pile, consolidation, lateral response, axial load intensity, finite element method.

1 INTRODUCTION

In fact, pile foundation is one of the underground structures, therefore many geotechnical problems predicted during the development of the Megacities. One of these problems is the time dependent behavior of the pile foundation which possibly caused by the lateral load. The lateral load usually results from the near new building or any large near civil engineering projects as well as the other normal sources of lateral load. Time dependent analysis of the structure has been limited because of complexity of the time dependent interaction between the soil and structure (Taiebat & Carter, 2001; Small & Liu, 2008). In cities where high rise buildings are closed together, lateral load must be considered as the most important factor in analysis and design of pile foundation.

Piles design to carry either vertical load or horizontal load. In case of piles subject to lateral loading, the failure mechanisms of short piles under lateral loads are different with long piles case (Poulos & Davis 1980). The approaches for analysis of piles are divided into two categories depending on the direction of the applied loads (Karthigeyan et al. 2007). First category includes the axial loaded pile, while the second category includes the analysis of pile under pure lateral load. To asses the lateral pile response, four methods are available: (a) limit state method (b) subgrade reaction method or *p-y* method, (c) elastic continuum method and (d) the finite element method. The brief review of historical used of the finite element technique for the analysis of lateral pile response was firstly developed by (Desai 1974, Muqtadir & Desai, 1986, Trochanis et al. 1991, Abbas et al. 2008 and Abbas et al. 2009).

The modeling of consolidation was studied and solve early by Biot (1941). Generally essential to alternative to a numerical simulation (e.g. FEM) to solve time-dependent problems because it is complex to solve analytically. Very few examples observed to solve such problems, i.e. Carter & Booker (1984) and Taiebat & Carter (2001) analyze lateral loaded piles using two-dimensional finite element approach include efficient formula based on semi-analytical

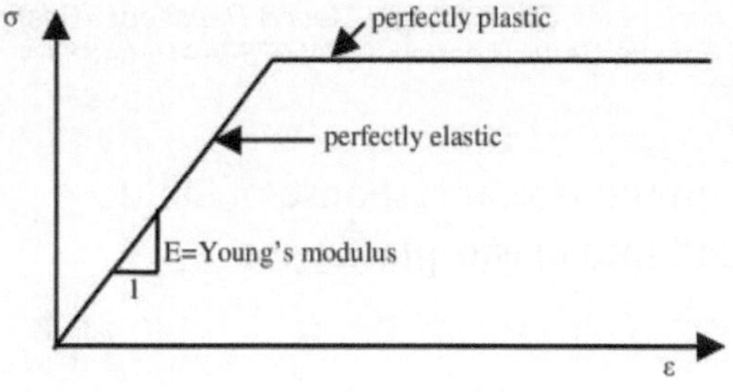

Figure 1. Stress–strain curve (Johnson et al., 2006).

finite element method. These studies limited to predict the lateral pile response subjected to pure lateral load embedded in the cohesionless soil.

The present paper focuses on the study of time-dependent behavior of piles subjected pure lateral loads through finite-element analyses. The details of the numerical models, the finite element formulation, and results from parametric studies are discussed in the paper.

2 MATERIAL AND METHOD

Pile Model (linear-elastic model): This model used represents Hooke's law of isotropic linear elasticity used for modeling the stress-strain relationship of the pile material as shown in Fig. 1. The model involves two elastic stiffness parameters, namely Young's modulus, E, and Poisson's ratio, ν. It is primarily used for modeling of stiff structural member for example piles in the soil (Abbas et al. 2009).

According to (abbas et al. 2008), the soil was model as Mohr-Coulomb Model. This elasto-plastic model is based on soil parameters that are known in most practical situations. The model involves two main parameters, namely the cohesion intercept, c' and the friction angle, ϕ'. In addition three parameters namely Young's modulus, E', Poisson's ratio, ν', and the dilatancy angle, ψ' are needed to calculate the complete $\sigma{-}\varepsilon$ behavior. Mohr-Coulomb's failure surface criterion is shown in Fig. 2 (Potts & Zdravkovic 1999). The failure envelope as referred by Johnson et al. (2006) only depend on the principal stresses (σ_1, σ_3), and is independent of the intermediate principle stress (σ_2).

Transient Formulation: An incremental formulation was used in the current work producing the matrix version of the Biot (1941) equation at the element level presented below (Smith & Griffiths 2004)

$$\begin{bmatrix} K & L \\ L^T & S+\overline{\alpha}H\Delta t_k \end{bmatrix} \begin{Bmatrix} \overline{u} \\ \overline{p} \end{Bmatrix} = \begin{bmatrix} K & L \\ L^T & S-(1-\overline{\alpha})H\Delta t_k \end{bmatrix} \begin{Bmatrix} \overline{u} \\ \overline{p} \end{Bmatrix} + \begin{Bmatrix} dF/dt+C \\ \overline{F} \end{Bmatrix} \quad (1)$$

where: K = element solid stiffness matrix, L = element coupling matrix, H = element fluid stiffness matrix, u = change in nodal displacements, p change in nodal excess pore-pressures, S = the compressibility matrix, $\overline{F}$ = load vector, Δt = calculation time step, $\overline{\alpha}$ = time stepping parameter (equal to one in this work), dF/dt = change in nodal forces.

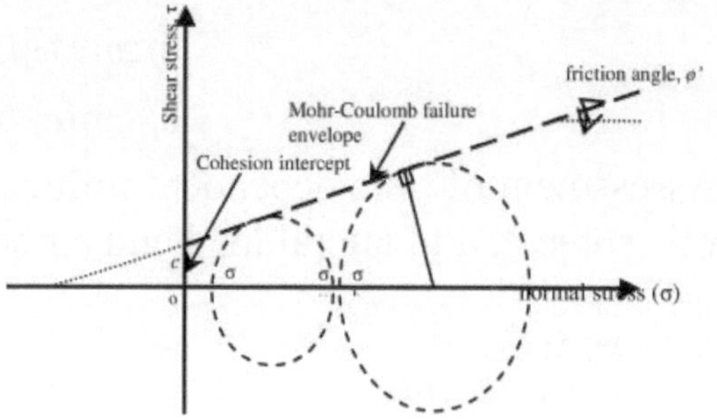

Figure 2. Mohr-Coulomb's failure surface (Potts & Zdravkovic, 1999).

3 FINITE ELEMENT MODEL

The finite element program with two-dimensional approach was developed and applied in case of time-dependent behavior of laterally loaded single isolated piles. In order to cover all the issues of this problem, it is supported by a pre-processor to develop 2-dimensional meshes include both rectangular type prismatic elements and 8 node quadrilateral elements. The developed program has the ability to plot the 2-D mesh as illustrated in Fig. 3. The pile and the surrounding soil are modeled using 8-node quadratics elements. Analysis was performed with several trail meshes with increasing refinement until the displacement did not change with more refinement. The aspect ratio of elements used in the mesh range from small closed to the pile body and when near to the pile head and base and increase refinement to wide spacing far from the pile body. All the nodes of the lateral boundary are restrained from moving in the normal direction to the respective surface representing rigid, smooth lateral boundary. The nodes at the bottom surface are restrained an all the two direction representing rough, rigid bottom surface.

4 NON-DIMENSIONAL TIME FACTOR

In order to examine the time dependent consolidation behavior of the pile, it is convenient to introduce a non-dimensional time factor T, defined as (Carter & Booker 1984, Taiebat & Carter 2001, Small & Lui 2008 and Abbas et al. 2009)

$$T = \frac{c_v t}{D^2}, \quad \text{where} \quad (2)$$

$$c_v = \frac{k(1-\upsilon'_s)E'_s}{\gamma_w(1-2\upsilon'_s)(1+\upsilon'_s)}$$

then:

$$T = \frac{k(1-\upsilon'_s)E'_s t}{\gamma_w(1-2\upsilon'_s)(1+\upsilon'_s)D^2} \quad (3)$$

where the coefficient of consolidation c_v is defined in term of the permeability k, the drained modulus E',

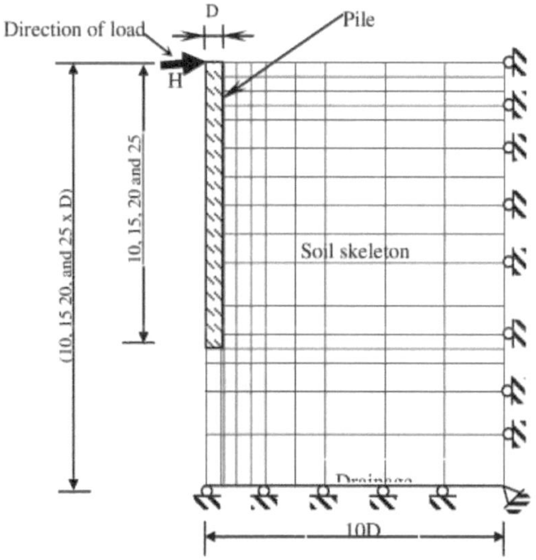

Figure 3. 2-Dimensional finite element mesh.

and Poison's ratio ν', the unit weight of water γ_w and the diameter of pile D.

5 COMPARISON WITH THE EXISTING RESEARCHES

The analysis of the behavior of a vertical pile embedded in a saturated elasto plastic soil and subjected to a lateral load was studied by Carter and Booker (1984) and Taiebat and Carter (2001) with elastic and elasto-plastic skeleton, respectively.

According to Taiebat & Carter (2001) a pile studied with diameter D is embedded in a layer of saturated cohesionless soil which obeys the Mohr-Coulomb failure criterion. The friction angle of the soil is assumed to be $\phi' = 30°$. The soil is also assumed to have a submerged unit weight of $\gamma_{sub} = 0.7\gamma_w$, where γ_w is the unit weight of pore water, a Young's modulus for fully drained conditions given by $E'_s = 3000\gamma_w$ and a Poisson's ratio $\nu' = 0.30$. The initial value of the coefficient of lateral earth pressure is $K_0 = 0.5$. The Young's modulus of the pile material is $E_p = 1000E'_s$. The problem was analyzed by assuming elastic and elasto-plastic models for the soil. All elasto-plastic analyses have been carried out using 8-node quadrilateral finite elements on the other hand the same sequence of loading. Good comparisons were obtained between the published case results of Taiebat & Carter (2001) and the present simulation model at lateral load intensity of $15\gamma_w \times D^3$ as shown in Fig. 4. This loading was maintained constant with time and the analyses were continued, allowing excess pore pressures to dissipate, and thus for the soil to consolidate during a total time of (T = 0.0001).

The predicted load-displacement curves for the pile head, for cases where the pile deforms under fully

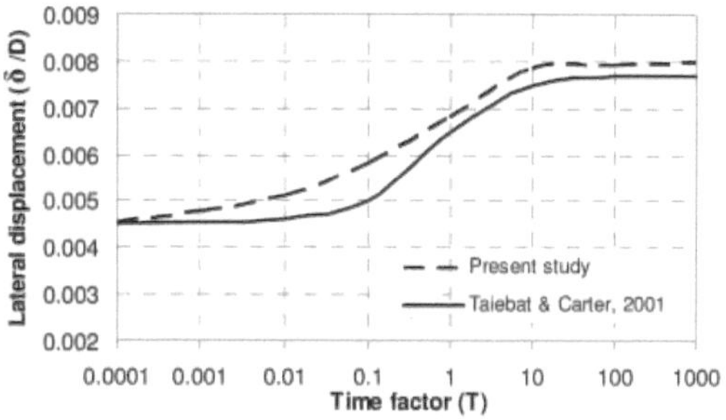

Figure 4. Comparison of the lateral displacements of the pile head in elasto-plastic soils.

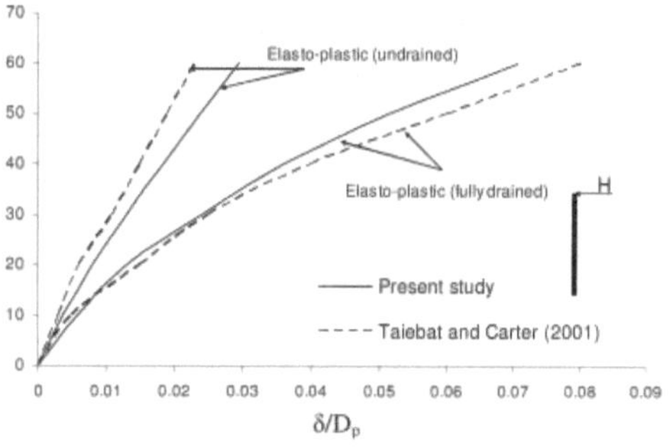

Figure 5. Lateral displacement relationships for laterally loaded piles under drained and undrained conditions.

drained state and rapid loading (i.e., undrained) conditions, are presented in Fig. 5. Case is plotted for the Mohr-Coulomb soil model. The response of the pile during rapid loading is almost linear and close to the elastic response with head displacement about twice that of elastic analysis. Again good agreement was observed between present study and Taiebat & Carter (2001) results.

6 ANALYSIS LAYOUT

To assess all numerical analysis in this investigation the finite element program, which has the feature of modeling two-dimensional (plane strain and axisymmetric) geotechnical problems such as consolidation is developed. The finite element model of the whole geotechnical structure developed was verified based on the results published cases. The analysis of the behavior of a vertical pile embedded in a saturated elasto-plastic soil and subjected to a lateral load was studied by Carter & Booker 1984 and Taiebat and Carter 2001 with elastic and elasto-plastic skeleton, respectively. This study include: (1) the load intensity which taken from low value of 50 kN and increased reached to 450 kN. (2) time factor is taken from 0.0001 which means rapid load (short term) to 1.0 for long time after loading (long term), and (3) two type of soil are considered (i.e., cohesionless and cohesive soil). The main advantage of this study is to increase

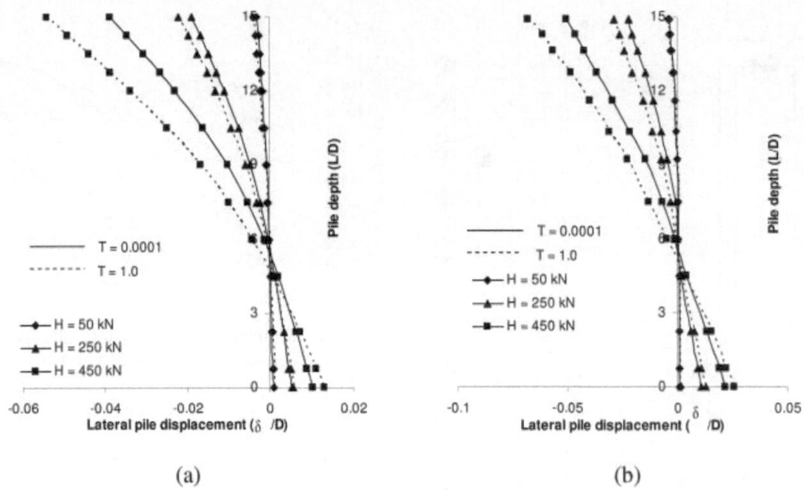

Figure 6. Lateral pile displacement with depth, (a) cohesionless soil, (b) cohesive soil.

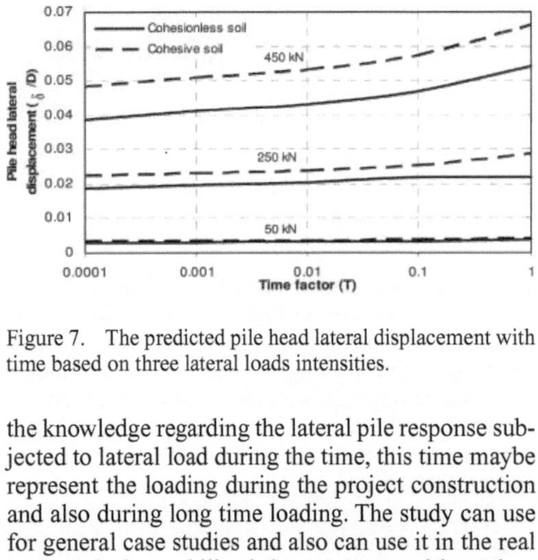

Figure 7. The predicted pile head lateral displacement with time based on three lateral loads intensities.

the knowledge regarding the lateral pile response subjected to lateral load during the time, this time maybe represent the loading during the project construction and also during long time loading. The study can use for general case studies and also can use it in the real case study by mobilized the program and boundary condition according to new cases. From this simulation can assess the lateral pile displacement with depth, lateral soil resistance with depth, and finally can estimate *p-y* curve with is very important key when design the pile under lateral loading.

7 RESULTS AND DISCUSSION

In order to analysis and design of laterally loaded pile, it is important to calculate both the maximum lateral pile displacement as well as the ultimate lateral soil resistance. In this study, the maximum lateral pile displacement has been draw with time factor. In addition, the ultimate lateral soil resistance has been developed with depth. Besides that, in order to understand the lateral soil distribution that help to know the position of the ultimate lateral load that take in the *p-y* design curves, this study also includes the lateral soil pressure which is developed with depth under time dependent condition.

7.1 *Development of lateral pile displacement*

The lateral pile displacement that developed with depth is illustrated in Fig. 6(a and b) for two types of soil. For the load intensity of ($5\gamma_w \times D^3$), small differences in the lateral settlement can be observed. Whereas the lateral pile displacement increased after increase the load intensity to reach the maximum value of ($45\gamma_w \times D^3$). This is possibly due to dissipation of pure water pressure. In addition, the figure shows small lateral displacement in case of rapid (instantaneous) and large deformation due to consolidation (long time). In this case, when large lateral pile displacement occur the lateral pile capacity reduced to minimum value with time. The main deflection of the pile occurred near to the surface with significant negative deflection appeared in the opposite pile face and below the rotation point which is between 5–7.5 D. the maximum negative deflection occurred at the toe of pile (closed to pile base).

In case of cohesionless soil, 78.6% from the total settlement occurred in the initial stage (rapid load) and 21.4% for the long term loading. This means that when low intensity if loading of ($5\gamma_w D^3$) is applied the pile is less resist in the first stage of load and being strong with time after dissipation of pore water pressure. While, when the pile carry large amount of loading (i.e. $45\gamma_w D^3$) the 29% from the total lateral settlement is carried in long term. This means that more effect of the long term loading in case of high lateral load. In the other hand, for the pile embedded on the cohesive soil, the long team settlements are 16.3% and 26.8% for pile under low and high load intensities, respectively. Overall, the pile in cohesionless soil resist more in rapid load and resist less in case of long term loading.

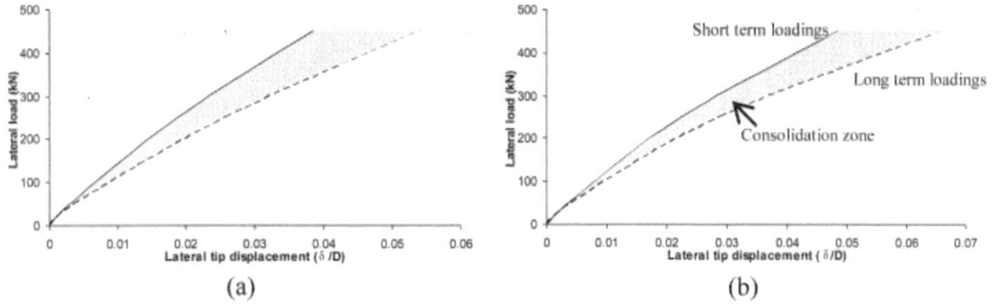

Figure 8. Lateral pile displacement with depth, (a) cohesionless soil, (b) cohesive soil.

Table 1. Percentage of lateral displacement increment

	δ% (%)	
Load intensity (kN)	Cohesionless soil	Cohesive soil
50	21.4	16.3
250	25.2	22.7
450	28.9	26.8

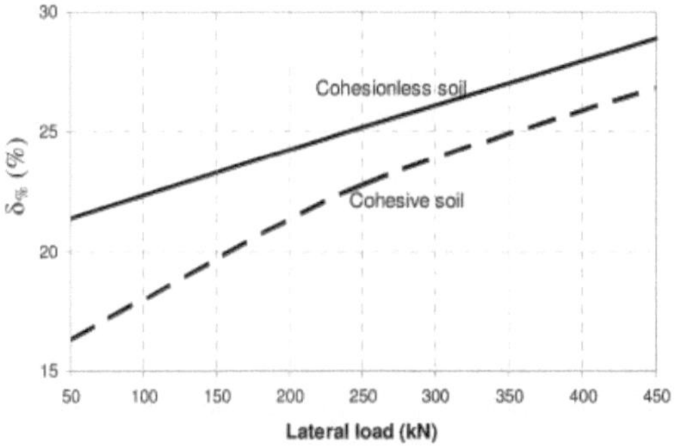

Figure 9. Lateral pile displacement increment with time fore three load intensities.

Normally, the maximum lateral pile deflection occurred in the tip of pile and this due to the pile is the free-headed pile. Therefore, the lateral pile displacement with time at the point at pile head in both cohesionless and cohesion soil is shown in Fig. 7. The figure draw in both of different displacements with time factor (T = 0.0001 – 1.0) as well as different load magnitude. The predicted load-displacement relationship under rapid and long time loading is presented in Fig. 8(a&b) for two types of soil. These figures used to predict the lateral pile displacement in certain load. These values limited by shadowed area for long and short time loadings.

The percentage of lateral displacement increment between T = 0.0001 and T = 1.0 is large for cohesionless soil which calculated using Equation below. This means the pile in cohesionless soil can resist more at the rapid loading and resist less when long time loading compared with the pile in cohesive soil as illustrated in Table 1. This is due to the fact that cohesionless soil has more void due to the higher permeability value compared to cohesive soils, thus in the case of rapid loading causes the pore water to carry more part of the applied load. Therefore the pore pressure increased the pile resistance by reducing the lateral displacement. However, in long term more water dissipated that caused decreased in lateral pile resistance and as results large displacement occurred.

$$\delta_{\%} = \frac{\delta_{T=1.0} - \delta_{T=0.0001}}{\delta_{T=1.0}} \times 100$$

where: $\delta_{\%}$ = Percentage of lateral displacement increment between T = 0.0001 and T = 1.0; $\delta_T = 1.0$ = the lateral displacement δ/D at T = 1.0 (effect of consolidation time); $\delta_T = 0{,}0001$ = the lateral displacement δ/D at T = 0.0001 (rapid load)

This increment in lateral load with time is important to study and increase understanding regarding the real pile behavior with time. From this study can notice that the pile carried more than 70% of the total lateral pile capacity in the first time of loading for both type of soils. This gives indication that the pile being more risky in the first time of loading. Proposed design curve for lateral pile displacement increment with time can see in Figure 9. The figure include the comparison between cohesionless and cohesive soil under pure lateral load (i.e. low intensity, H = 50 kN, intermediate intensity, H = 250 kN, and high intensity, H = 450 kN).

7.2 *Development of lateral soil resistance*

Lateral soil pressures p in soil resulting from the lateral loads is shown in Figs. 10. It can be seen that the pressure redistributed with time. Higher values of lateral pressure occurred at L/D between 1 and 3 scaled from pile head for all amount of loads. In case cohesionless soil, the maximum lateral pressure occured at L/D = 2.5. In addition at L/D = 1.5 in case of cohesive soil case from pile tip, this also recommended Broms (1964a,b) with depth of 1.5D.

It can seem that the load intensity is significantly effect on the front lateral load resistance distribution. The soil resistance starts from the small value near

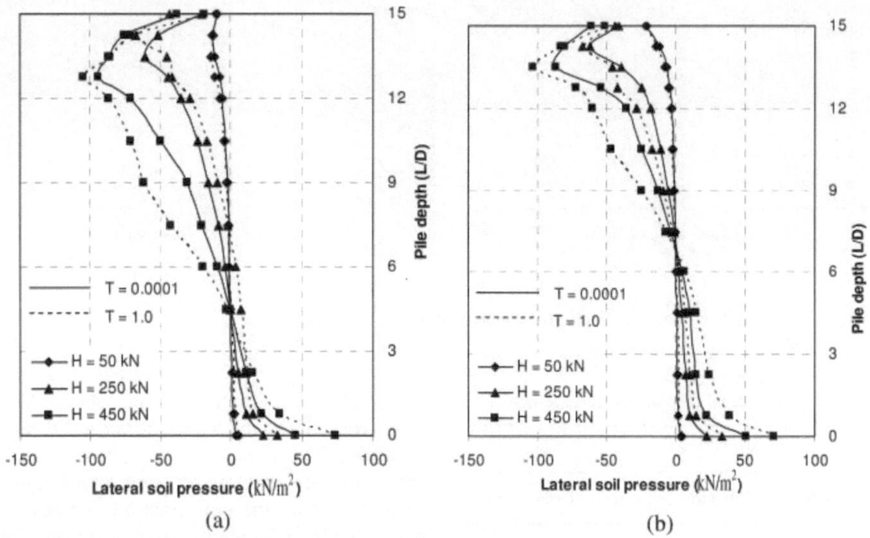

Figure 10. Lateral soil resistance with depth, (a) cohesionless soil, (b) cohesive soil.

to the surface and be the maximum in case of low loading. While in case of intermediate and high loading, can see the maximum value occurred not on the surface but in depth and it means that the soil near surface failed when increase the lateral load. Because of this, recommended to take the result from certain depth into account for design parameters. In addition, the negative lateral soil pressure occurred at the lower part of the opposite pile face. The value of the maximum negative pressure appeared at the pile toe. This values started from zero at the point of rotation and then increased to reached the maximum magnitude near the base (at pile toe) for both cases (i.e. cohesionless and cohesive soil) which also recommended by Broms (1964a,b).

7.3 *Prediction of p-y curve*

The computed *p-y* curves at the tip of pile (at surface) and at the depth of 1.5D and 3D for pile embedded in two types of soil are shown n Fig. 11. The FE results indicate that the *p-y* curve sensitive to the type of soil and the calculated level. Thus, this paper conducted herein supports the assumption that the effect of the soil type on *p-y* curve,

It can seem that, clear differences appeared when calculate *p-y* curve based on the depth from the surface. At the point near to the surface, can see evident change of the results came from piles in cohesionless and cohesive soil. For the same amount of lateral settlement 10% and 20% of pile diameter, the result obtained from cohesionless soil gives fewer amounts than these results with cohesive soil. It is may be due to early collapse of surface soil mass in case of cohesionless soil. Also it can see that the effect of time dependent factor on p-y curve, the response with cohesionless soil more sensitive with long time loading.

When predict p-y curve deep and close with maximum ultimate lateral soil pressure it can see convergent performance of the two types of soil. The long term loading is affected and same time give good increase in developed lateral soil pressure. The figure in 1.5D is more accurate and can used in the design because it is give a great value of lateral soil pressure which is more critical and also gives a significant large amount of lateral pile displacement.

8 CONCLUSIONS

The finite element approach coupled with consolidation equation is used to assessing the lateral pile response when subjected to pure lateral load. Based on the results the following conclusions can be drawn:

1. The lateral pile response includes both lateral pile displacement and lateral soil pressure is affected by lateral load intensity and also change mainly by the long time after applied load.
2. The pile in cohesionless soil is more resist in the first stage of load (rapid load) and being weak with time after dissipation of pore water pressure (long term loading) compared with the pile embedded in cohesive soil.
3. The front lateral soil pressure distribution is changes mostly in the upper part of pile and reach the maximum value at 1.25D this mean the final stage of loading. While in the first stage of low load intensity, the ultimate lateral load intensity occurred much closed with surface in both types soil. In addition, maximum negative lateral soil pressure occurred at the pile toe.
4. For the same amount of lateral deflection 10% and 20% of pile diameter, the result obtained from

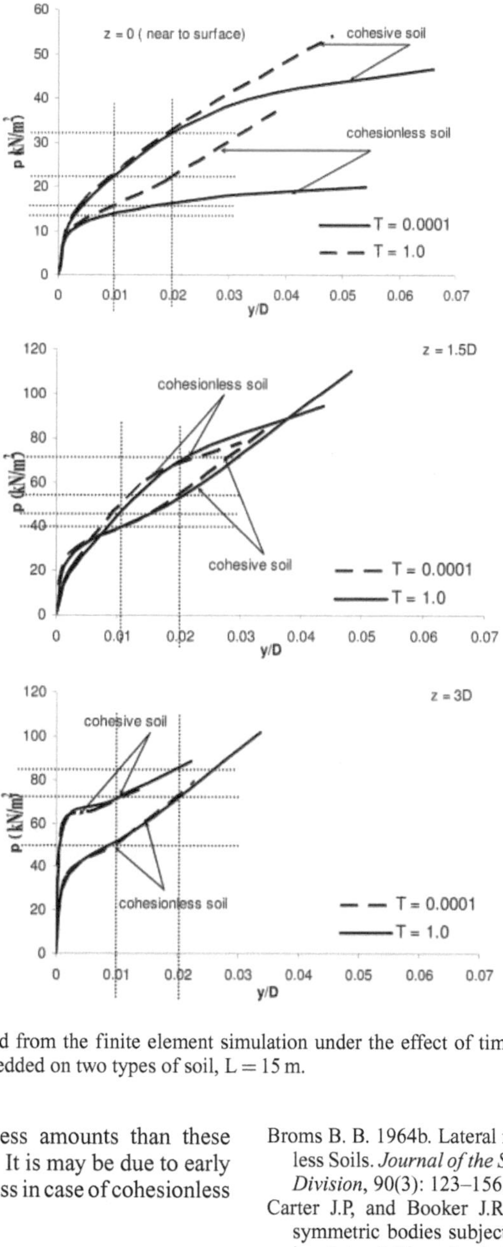

Figure 11. *p-y* curves predicted from the finite element simulation under the effect of time-dependent and with different depth below pile tip of pile embedded on two types of soil, L = 15 m.

cohesionless soil gives less amounts than these results with cohesive soil. It is may be due to early collapse of surface soil mass in case of cohesionless soil.

REFERENCES

Abbas J M, Chik Z, Taha M R. 2008. Single pile simulation and analysis subjected to lateral load. Electronic Journal of Geotechnical Engineering, SCOPUS. 13 Bundle (E).: 1–15.

Abbas J M, Chik Z, Taha M R, Shafiqu Q S M. 2009. Analysis of Time Dependent Laterally Loaded Pile in Clay. *Electronic Journal of Geotechnical Engineering*. 14 Bundle (N): 1–10.

Biot, M.A. 1941. General theory of three-dimensional consolidation. J. App. Phys., (12):155–164.

Broms B. B. 1964a. Lateral resistance of Piles in Cohesive Soils. *Journal of the Soil Mechanics and Foundations Division*, 90(2): 27–63.

Broms B. B. 1964b. Lateral resistance of Piles in Cohesionless Soils. *Journal of the Soil Mechanics and Foundations Division*, 90(3): 123–156.

Carter J.P, and Booker J.R. 1983. Consolidation of axi-symmetric bodies subjected to non-axi-symmetric loading. Inter. J. Num. Analy. Meth. Geomech, 7: 273–281.

Desai C. S. and Appel G. C. 1976. 3-D Analysis of Laterally Loaded Structures. *Proceedings of the 2nd Int. Conf. Num. Methods in Geomechanics, Blacksburg, Virginia, ASCE*, (1).

Johnson, K., Lemcke, P., Karunasena, W., Sivakugan, N. 2006. Modelling the load – deformation response of deep foundation under oblique load. *Envir. Model. Soft.*, 21:1375–1380.

Karthigeyan, S., Ramakrishna, V. V. G. S. T., and Rajagopal K. 2006. Influence of vertical load on the lateral response of piles in sand. *Comp. Geotech.*, 33: 121–131.

Karthigeyan, S., Ramakrishna, V. V. G. S. T., and Rajagopal K. 2007. Numerical Investigation of the Effect of Vertical Load on the Lateral Response of Piles. *J. Geotech. Geoenvir. Eng.* 133(5): 512–521.

Mroueh H. and Shahrour I. 2009. Numerical analysis of the response of battered piles to inclined pullout loads. *Int. J. Numer. Anal. Meth. Geomech.* 33: 1277–1288

Muqtadir, A., and Desai, C. S. 1986. Three-dimensional analysis of a pile-group foundation. *Int. J. Num. Anal. Meth. Geomech.* 10: 41–58.

Poulos, H.G., Davis, E.H. 1980. *Pile Foundation Analysis and Design*. John Wiley & Sons, Inc, United States.

Potts, D. M., and Zdravkovic, L. 1999. *Finite element analysis in geotechnical engineering: theory*. Thomas Telford. Heron Quay, London.

Small, J. C., and Lui, H. L. S. 2008. Time-settlement behavior of piled raft foundation using infinite elements. *Comp. Geotech.* 35: 187–195.

Smith, I.M., and Griffiths, D.V. 2004. *Programming Finite Element Method*. 4th Ed., John Wiley and Sons, Chisester, U.K.

Taiebat HT and Carter JP. 2001. A semi-analytical finite element method for three-dimensional consolidation analysis. *Comp. and Geotech.*, 28: 55–78.

Trochanis, A.M. Bielak J., and Christiano, P. 1991. Three-dimensional nonlinear study of piles. *J. Geotech. Engrg.*, 117(3): 429–447.

Baltic Piling – Mets & Raudsepp (Eds)
© 2013 Taylor & Francis Group, London, ISBN 978-0-415-64334-4

Influence of the installation process of full displacement bored piles on the subsoil

P. Busch & J. Grabe
Institute of Geotechnical Engineering and Construction Management, Hamburg University of Technology, Germany

F.-W. Gerressen
Bauer Maschinen, Schrobenhausen, Germany

ABSTRACT: Penetration and bearing behaviour of full displacement bored piles is mostly unexplored, due to large variations in the bearing behaviour among the different types of screw cast-in-place piles. Therefore a test area in Hamburg, Germany, has been set up, at which the penetration and bearing behaviour was analysed. In this paper the influence of the installation process of full displacement bored piles on the subsoil is investigated. For this purpose inclinometer measurements as well as cone penetration tests have been conducted to identify the changes in the soil. The results of the cone penetration test are used to estimate the change of the void ratio and the stress state. At the end of the paper a laboratory test rig is shortly presented, which allows you to measure directly the horizontal stresses during pile installation.

1 INTRODUCTION

In past years several full displacement bored piles, also called screwed cast-in-place piles, have been developed in various countries, especially in the Benelux Countries. Prezzi (2005) provides an overview of the conventional well-established systems. Screwed cast-in-place piles are installed into the ground by a combination of torque and vertical feed and displace the soil primarily to the sides. Thus they combine the advantages of driven piles, such as high productivity and good bearing capacity, while simultaneously avoiding environmental problems like noise and vibrations. Further information about the different pile systems can be found in the recommendations of the working committee "piles" (EA-Pfähle 2012).

The penetration and bearing behaviour of full displacement bored piles is mostly unexplored, due to large variations in displacement and bearing behaviour among the different types of screwed cast-in-place piles (Vermeer et al. 2008). No verified findings for changes in the ground during pile installation exist yet.

For this reason BAUER in collaboration with the Hamburg University of Technology set up a test area within the port area of Hamburg, at which the penetration and bearing behaviour was analysed. Inclinometer measurements to determine the horizontal displacement in the soil as well as cone penetration tests before, during and after the installation of the pile have been conducted to identify the changes in the soil.

2 EMPLOYED PILE SYSTEM

BAUER full displacement piles with lost bit (FDP lost bit) were installed in a test area in Hamburg and compared with partial displacement and normal bored piles. The assembly of the 2 m long drilling tool is shown in Figure 1. An auger extension of variable

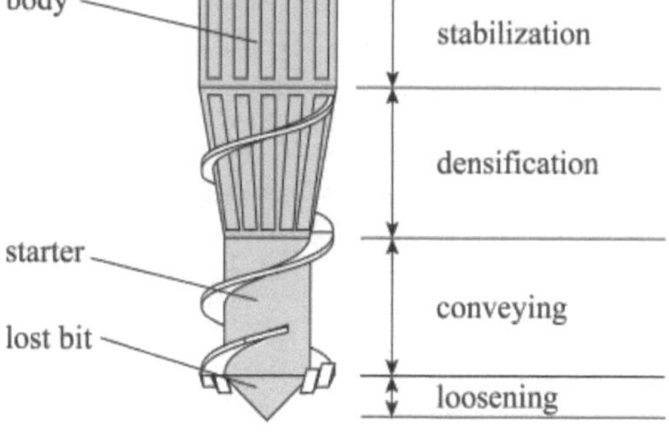

Figure 1. Assembly of the FDP drilling tool by BAUER.

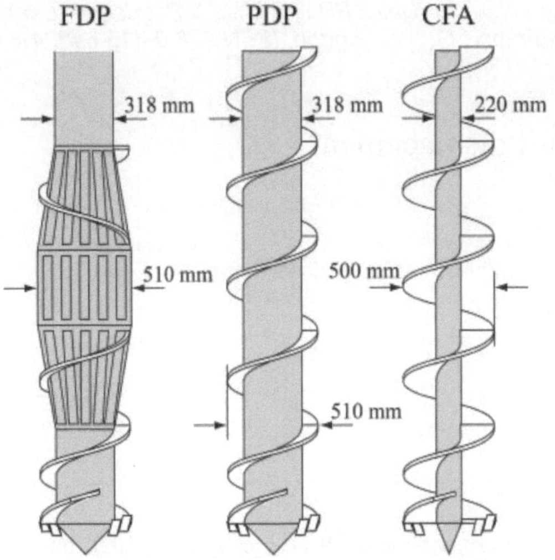

Figure 2. Comparison between used drilling tools at the test area: full displacement pile (FDP), partial displacement pile (PDP), continuous flight auger pile (CFA).

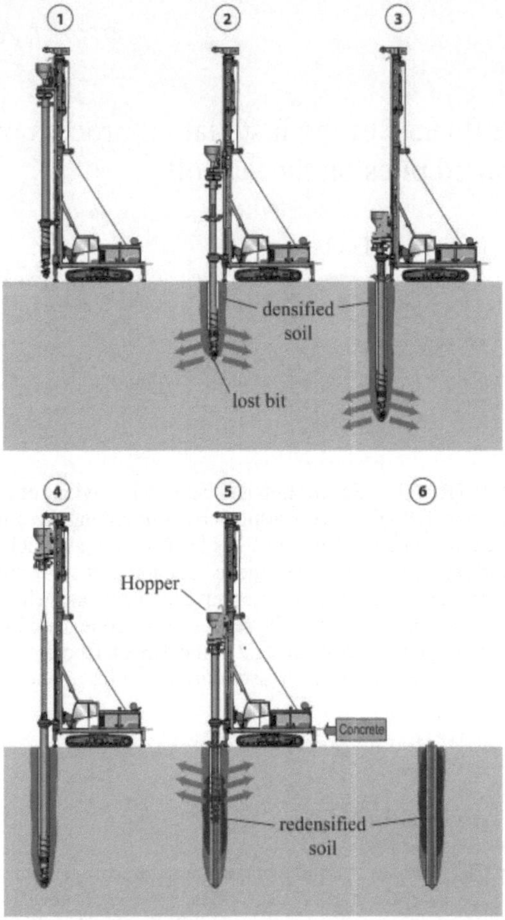

Figure 3. Installation process of an FD-pile (Bauer 2009).

length may be installed between displacement body and starter to penetrate non-displaceable deposits. The auger extension generally extends the length of the pile by approximately its length. At the test area two drilling tools with diameters of 440 and 510 mm were employed.

For the purpose of comparison partial displacement bored piles (PDP) and continuous flight auger piles (CFA) with diameters of approximately 500 mm were installed at the test area, see Figure 2. The PD-piles have a mandrel diameter of 318 mm, identical in construction to the auger extension. The CFA-piles have a mandrel diameter of 220 mm.

The installation process of an FD-pile is shown in Figure 3. After positioning and setting up the drilling rig over the pile position, a sacrificial drill bit is attached (1). Then the displacement tool is drilled into the ground by rotating and pushing of the tool (2). The soil is loosened by the lost bit, conveyed upwards by the starter and then pushed laterally into the surrounding soil by the displacement body. The displacement tool is drilled down to the final depth (3). The drilling depth can for example be extended by the addition of jointed drill string sections.

On reaching the final depth the rotary drive is disconnected from the hollow drill string and a full length reinforcement cage is inserted into the hollow drill stem (4). During extraction of the displacement tool, concrete is simultaneously discharged by the concrete hopper and placed "unpressurised" in the pile through the hollow drill stem (5). The displacement tool is rotated in the drilling direction during extraction and redensifies any soil material that may have fallen into the borehole or any loosened borehole wall areas with the counter-rotating flight section above the displacement body. The last picture (6) shows the installed pile.

3 TEST AREA IN HAMBURG

The test area is located in the harbour of Hamburg on the isle "Hohe Schaar" in the river Elbe. The stratification of the subsoil is affected by the glacial valley of the Elbe. The 3.5 to 5 m thick fill consists of sand with silty fractions. Beneath the fill lies an approximately 2 m thick layer of organic silt, followed by medium dense sand. The transition between these layers can be gradual. At a depth of approximately 16 m the gravel fraction increases. The ground profile is shown in Figure 4 together with the corresponding CPT results. The thin curves are the arithmetic mean from eight CPT's carried out on the test area before pile installation. The thick lines show the average over one meter.

In spite of the small extend of the test area with 900 m^2 the ground conditions and depths of the layers differ relatively strongly. The pictured ground profiles therefore show simplified and averaged ground conditions.

Overall, 21 piles were installed on the test area: 16 FDP, three PDP and two CFA-piles. The FD-piles were installed to the maximum depth, which is limited

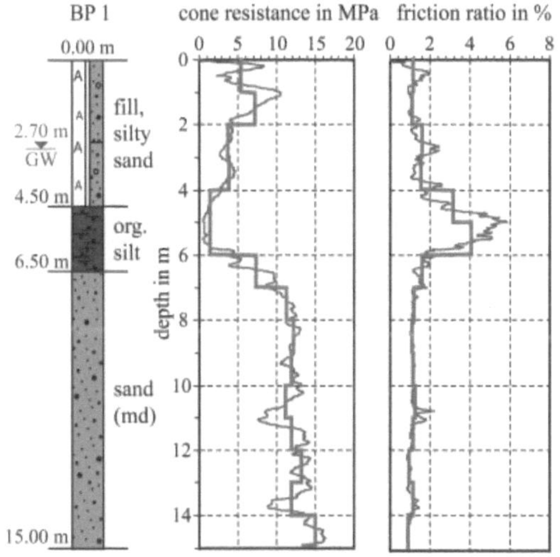

Figure 4. Ground profile and CPT results.

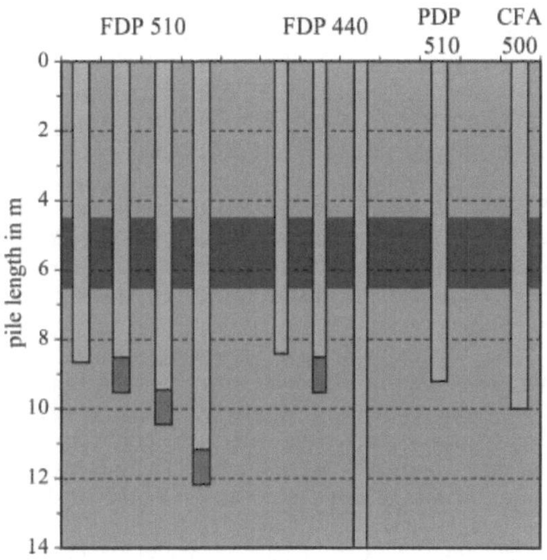

Figure 5. Achieved pile lengths of the different pile types.

by the maximum torque of 200 kNm and the maximum crowd force of 108 kN of the drilling rig. Pile lengths of 8.4 to 14 m with embedment lengths of 1.5 to 7 m in medium dense sands have been achieved. Figure 5 indicates some of the reached pile lengths for the different pile types. Through this comparison the additional embedment length due to the use of an auger extension of 1 m in length becomes apparent.

The lengths of the PD-piles and the CFA-piles with 9 to 10 m were chosen in such a way as to ensure the comparison to the FD-piles.

4 FIELD MEASUREMENTS

To investigate the changes in the ground due to the installation process of bored piles, inclinometer

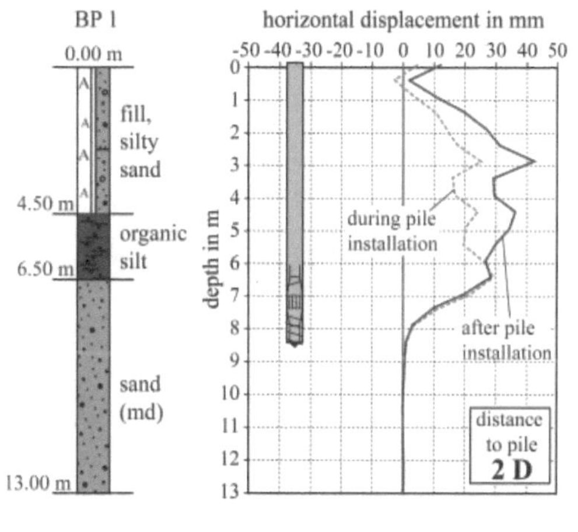

Figure 6. Horizontal soil deformation due to the installation of an FDP (Busch 2009).

measurements and cone penetration tests (CPT) were carried out in various distances to the piles before, during and after the pile installation.

4.1 *Inclinometer measurements*

With the help of inclinometer measurements the horizontal displacement in the soil due to the installation process are determined. The results of the inclinometer measurements are shown in Figure 6. The dotted curve marks the profile of horizontal displacements with depth at a distance of twice the pile diameter from pile centre after installation. For a better understanding the schematic of a finished pile and the displacement tool are shown within the diagram. The soil body above the displacement tool relaxes as the drilling rod does not support the borehole walls. The resulting loosening of the soil body is rectified by redensification and concreting during extraction. The solid curve shows displacements with depth at a distance of twice the pile diameter after extraction and concreting. Up to 40 mm of horizontal displacement were measured. Near surface deformations were influenced by operational loads due to the piling activities and are not representative of the displacement behaviour of FD-piles.

Figure 7 shows the horizontal displacements after FD-pile installation in a distance of twice and three times the pile diameter. At a distance of three times the pile diameter plastic deformations reached 20 mm on average after pile installation. The influence of the pile installation decreases with the distance to the pile, see Figure 7 and 8. It can be assumed that the influence domain of an FDP is up to three times the pile diameter.

The horizontal soil deformation during pile installation is shown in Figure 8. The yellow profiles were measured at a distance of 1D and 3D while boring was stopped at a depth of 6.5 m. The green dotted profile was measured while the displacement tool reached the final depth. The schematic of the displacement tools are shown in the given depths within the diagram.

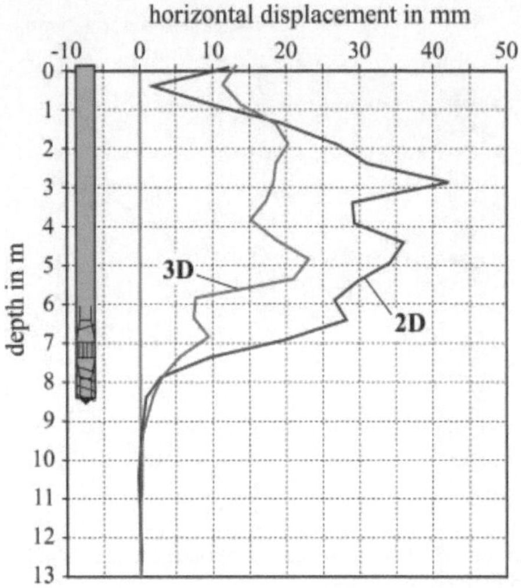

Figure 7. Horizontal soil deformation after FDP installation in different distances to pile.

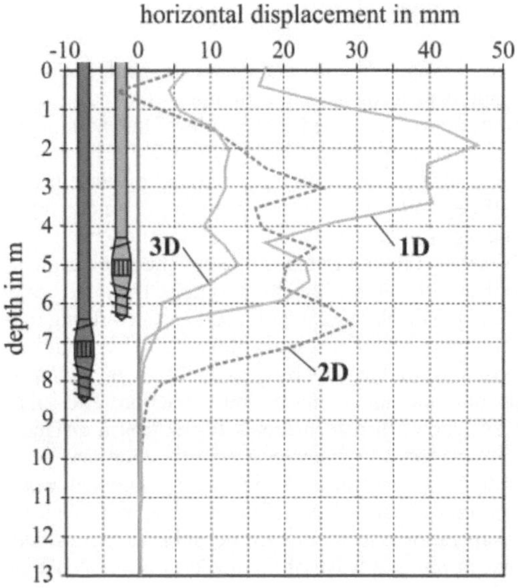

Figure 8. Horizontal soil deformation during FDP installation.

The colours of the displacement tools correspond with the colours of the profiles. The horizontal displacements during pile installation range from 46 mm at a distance of 1D to 10 mm at a distance of 3D.

Unfortunately, it was not possible to measure the maximum horizontal displacement at a distance of one pile diameter after pile installation. The extraction and concreting causes such big deformations in the subsoil, that the inclinometer tube was damaged and no more passable for further measurements.

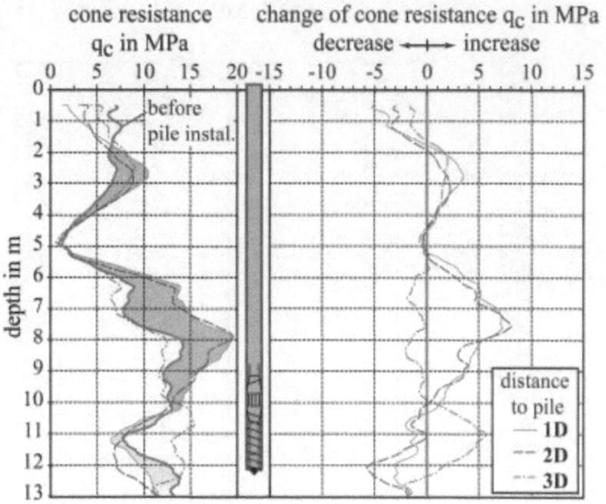

Figure 9. CPT results before and after pile installation with various distances to the FDP.

4.2 *Cone penetration tests*

Changes in the ground can be indicated by CPT's conducted before and after pile installation (see Figure 9, left). After pile installation CPT's in different distances to the pile (1D, 2D and 3D) were carried out. On the right-hand side of Figure 9 the changes of cone resistance before and after pile installation are shown. The changes in cone resistance near surface were also influenced by operational loads due to the piling activities like the inclinometer measurements and are not representative of the displacement behaviour of the FD-piles.

The effects of the pile installation are less pronounced within the fill and the soft organic silt deposits than in the medium dense sands. Within the sand deposits approximately down to the depth of the displacement body the cone resistance locally shows the largest increases. The CPT results in Figure 10 show for instance an increase of up to 12 MPa due to the pile installation. Below the displacement body and generally down to 0.5 to 1 m below the drilling tool the cone resistance decreases slightly, see yellow marking in Figure 9 and 10, left.

The length of the decrease depends on the length of the auger below the displacement body. The longer the auger extension, the deeper reaches the decrease in cone resistance. The behaviour of the decrease is similar to the one of a PD-pile because of the same auger geometry. If no auger extension is used, there no changes in cone resistance before and after pile installation are detected. One meter below the drilling tool, the profiles of cone resistance vs. depth are thought to be the result of ground inhomogeneity.

The changes of the cone resistance reduce with increasing pile distance. The maximum changes are achieved in a distance of one or two times the pile diameter, see Figure 9, right. At a distance of three times the pile diameter the changes of the cone resistance are very small. They cannot be confirmed through measurements, because the influences of the

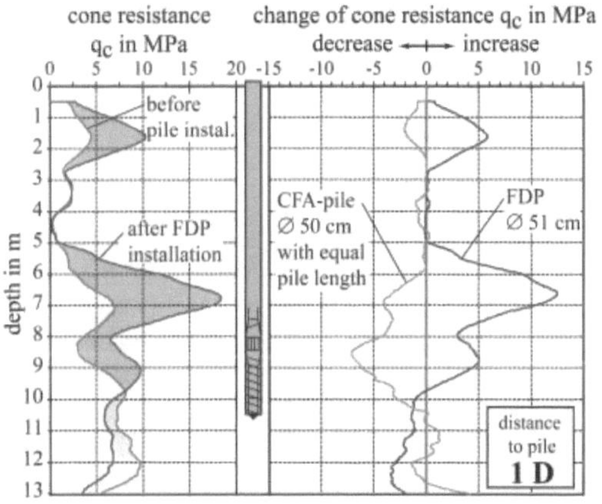

Figure 10. CPT results before and after pile installation with a distance of 1D to the piles.

ground inhomogeneity are bigger than the influence of the pile installation. The measured cone resistance shows the changes of the ground inhomogeneity and not the changes of the pile installation.

The CPT results corroborate the assumptions from the inclinometer measurements. The influence domain of an FDP is up to the three times the pile diameter for the existing ground conditions.

For comparison, the right side of Figure 10 shows the changes in cone resistance before and after installation of an FD-pile and for a CFA-pile of the same length. Because of the extraction of soil during the installation of a CFA-pile, the density as well as the stress state of the surrounding soil body decreases, resulting in a decrease in measured cone resistance, see Hannink & von Tol (2005) and compare with section 5.1.

The results of PD-piles are between the two previously discussed pile types. For loose deposits ($q_c < 7.5$ MPa) the cone resistance increases slightly while it decreases for medium dense to dense deposits ($q_c > 10$ MPa).

5 EFFECTS OF THE PILE INSTALLATION

The pile installation process influences the state variables of the soil nearby the installed pile as well as the bearing capacity. As shown in Figure 10 the installation process of bored piles can have different influences on the subsoil. Especially the installation of continuous flight auger piles can cause loosening of the soil, see Hird et al. (2008). The loosening effect can be significant when boring proceeds in an erroneous manner, see Fleming (1995). An optimal ratio of boring velocity and revolution rate is required to reduce these effects to the extend shown in this paper, see Viggiani (1993) and Van Impe (1997). In contrast, the installation process of FD-piles can increase the bulk density and the stress state, see Busch et al. (2010).

5.1 *Effects on the soil behaviour*

The results of the cone penetration tests before and after the pile installation are used to estimate simplified the changes in soil parameter values due to the installation process. The presented estimation is based on a limit value consideration and the results should show, whether the void ratio or the stress state changes due to the pile installation.

Jamiolkowski et al. (1988) showed that the resistance of a cone penetration test correlates to the stress state and the bulk density of the soil. Hence they developed equation (1) depending on the cone resistance q_c, a reference pressure p_a, the mean stress σ'_p, the bulk density D_R and three soil dependent parameters a, b and c.

$$q_c = a \cdot p_a \left(\frac{\sigma'_p}{p_a} \right)^b \cdot \exp(c \cdot D_R) \tag{1}$$

Grabe and König (2004) estimated the parameters a, b and c for a typical sand in the area of Hamburg, Germany, see equation (2).

$$q_c = 795 \cdot p_a \left(\frac{\sigma'_p}{p_a} \right)^{0.42} \cdot \exp(1.4 \cdot D_R) \tag{2}$$

using

$$\sigma'_p = \frac{\sigma'_v + 2 \cdot \sigma'_h}{3}$$

and

$$p_a = 1\,\text{kPa}$$

Busch et al. (2010) analyzed the results of cone penetration tests before and after the installation process of CFA and FD-piles. Some results are shown in Figure 11 and the differences between the cone resistance in the sand layer are brought into focus. After the installation process of the continuous flight auger pile, the cone penetration resistance decreases approximately 4.5 MPa (yellow area). In contrast, the cone penetration resistance increases approximately 5 MPa (brown area) after the installation of the full displacement pile.

According to Grabe et al. (2010) it is assumed, that the installation process mainly changes the stress state. Therefore, the horizontal stress state is varied between the active and the passive earth pressure, trying to match the cone resistance in-situ with the calculated resistance via equation (2). If the change of the horizontal stress does not suffice to explain the changes of the cone penetration resistance, the bulk density D_R is also changed.

The initial state is characterized by the friction angle $\varphi' = 30°$, the coefficient of the earth pressure at rest $K_0 = 0.5$, the vertical stress $\sigma'_v = 96.25$ kPa, the horizontal stress $\sigma'_h = 64.2$ kPa and the bulk density $D_R = 0.6$.

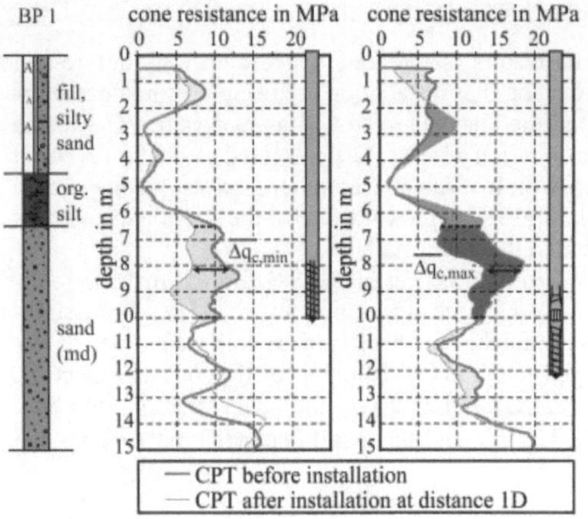

Figure 11. Measured cone penetration resistance before and after the pile installation of a CFA-pile and FDP.

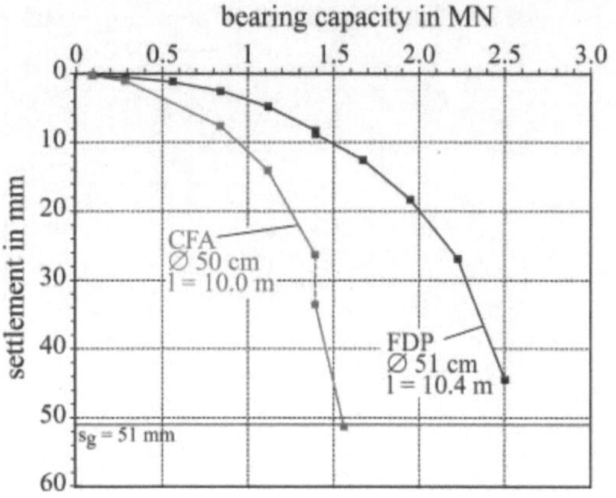

Figure 12. Comparison of the bearing capacities of an CFA and an FDP-pile.

The decrease of the cone penetration resistance caused by the installation of the CFA-pile can be back calculated with a decrease of the horizontal stress up to the active earth pressure at an earth pressure coefficient of $K_a = 0.3$ and with an decrease of the bulk density up to $D_R = 0.26$. Hence, it can be inferred that approximately 20% of the change of the cone penetration resistance is induced by the decreasing horizontal stresses and approximately 80% of the change is induced by the loosening of the soil.

The installation of an FD-pile increases the cone penetration resistance. The horizontal stress is increased up to $\sigma'_h = 178.6\,\text{kPa}$ which correlates to a passive earth pressure coefficient of approximately $K_p = 2.28$. Hence, a change of the bulk density does not have to be occurred.

This shows, that the estimation of the changes of soil parameters used here is oversimplified, because a change of the bulk density due to the installation process of a full displacement bored pile is obvious. Nevertheless, this estimation shows the potential influences from pile installation effects on the subsoil.

5.2 *Effect on the bearing capacity*

The changes of the soil parameters due to different installation methods are not only theoretical but also can be seen in the bearing capacities of the installed piles. In Figure 12 the load settlement curves of an CFA and an FD-pile are compared. For the same soil structure, the same pile diameter and the same pile length a FD-pile achieves an approximately 1 MN increased bearing capacity than the CFA-pile. This difference results from the changes of the state variables of the soil due to the pile installation process, compare Figure 11.

The measured ultimate bearing capacity of $R_{CFA,meas} = 1.55\,\text{MN}$ is in spite of the reduction of the cone resistance due to the installation process higher than the calculated design load capacity from lower empirical values for CFA-piles $R_{CFA,cal,min} = 1.0\,\text{MN}$. With the upper empirical values for CFA-piles the ultimate bearing capacity is 1.4 MN. The fluctuation range of the empirical values results from the error-prone installation process of CFA-piles.

The CFA-piles achieve despite the reduction of the cone resistance due to the installation process the calculated ultimate bearing capacities. Hence, for the bearing capacity the decrease of the cone resistance after the pile installation is classified as non-critical. This is confirmed by investigations on the pile foundation of the Elbe Philharmonic hall in Hamburg, at which the installed CFA-piles achieved their calculated minimum bearing capacity in spite of significant cone resistance reductions of up to 20 MPa (Franke & Woldt, 2009).

6 LABORATORY TEST RIG

To identify the changes of soil parameters due to the pile installation process the Institute of Geotechnical Engineering and Construction Management of Hamburg University of Technology is constructing a laboratory test rig. In this test rig the horizontal and vertical stresses within the tested soil profile can be measured during the pile installation. It is possible to control the crowd speed and the rotation speed or to control the crowd force and the torque of the drilling tool. Hence, tests with any combination of drilling parameters can be conducted, e.g. the installation of piles at a certain construction or test site can be reconstructed. The soil sample has a variable diameter of 0.3 to 1 m with a height of 1 m and it is possible to produce different initial states in terms of stress state and bulk density for non-cohesive soils.

With this test rig the influence of the different drilling tools on the subsoil can be investigated with variable installation parameters and variable initial soil conditions.

7 CONCLUSIONS

This paper shows the displacement effects of full displacement bored piles by means of cone resistance and horizontal displacement measurements in the ground. The information gained clarifies differences between partial displacement and continuous flight auger piles. The measurement results show that the cone resistances due to the FD-pile installation are increased down to the depth of the displacement body. The slight decreases below the displacement body have no negative effect on the bearing capacity of the pile.

The changes of the cone resistances can cause a change in density or in the stress state of the ground. To rate the changes plausibility calculations with the help of the relationship between cone resistance and void ratio and stress state were done. These calculations show that there are only minor changes in density and primarily changes in the horizontal stresses of the ground. The influence domain of an FDP is approximately up to three times the pile diameter.

The gained results should be confirmed through other in-situ measurements. But it is very difficult and expensive to measure the horizontal stress state and the void ratio in-situ. Therefore a new laboratory test rig was developed, with whose help it is possible to measure the mentioned soil parameters directly during the pile installation. The innovative characteristics of the test rig are the path or power controlled rotary drive and the possibility to reconstitute representative initial states for the ground sample (bulk density and stress state).

ACKNOWLEDGEMENTS

At this point the input from the BAUER Maschinen GmbH is gratefully acknowledged. Without their help the project would not have been accomplished.

REFERENCES

Bauer Maschinen GmbH (2009). FDP – Full displacement pile system, process and equipment. *Process description 3/2009*. Schrobenhausen, Germany.

Busch, P. (2009). Research and development of displacement piles. Proceedings of the 4th International *Young Geotechnical Engineers Conference,* Alexandria, Egypt. 201–204.

Busch, P., Grabe, J., Gerressen, F.W. & Ulrich, G. (2010). Use of displacement piles for reinforcement of existing piles. *Proc. of DFI and EFFC 11th Int. Conf. in the DFI series: Geotechnical Challenges in Urban Regeneration* in London/UK, 113–119.

EA Pfähle (2012): *Empfehlungen des Arbeitskreises „Pfähle".* Deutsche Gesellschaft für Geotechnik (DGGT), Arbeitskreis AK 2.1 „Pfähle", 2. Auflage, Berlin: Ernst & Sohn.

Fleming, W.G.K. (1995). The understanding of continuous flight auger piling, its monitoring and control. *Proceedings, Institution of Civil Engineers Geotechnical Engineering, 113*, 157–165.

Franke, J. & Woldt, J. (2009). Die Gründung der Elbphilharmonie Hamburg. Pfahl-Symposium 2009. *Mitteilungen des Instituts für Grundbau und Bodenmechanik, 88,* Technische Universität Braunschweig. Braunschweig, Germany. 147–169.

Grabe, J. & König, F., 2004. Zur aushubbedingten Reduktion des Drucksondierwiderstandes. *Bautechnik 81(7):* 569–577.

Grabe, J., Henke, S., Kinzler, S. & Pucker, T. (2010). Inverse determination of soil density and stress state using dispersion wave measurements and cone penetration tets in non-layerd soil. *Soil Dynamics and Earthquake Engineering, 30*, 481–489.

Hannink, G. & van Tol, A.F. (2005). Reduction of the Cone resistance caused by the installation of CFA piles. *Proceedings of the XVIth international conference on Soil Mechanics and Geotechnical Engineering*, Osaka. 1983–1987.

Hird, C. C., Ni, Q. & Guymer, I. (2008). Physical modelling of displacements around continuous flight augers in clay. *Proceedings of the 2nd British Geotechnical Association International Conference on Foundations 2008*, Dundee, UK. pp. 565–574.

Jamiolkowski, M., Ghionna, V.N., Lancelotta, R. & Pasqualini, E. (1988). New correlations of penetration tests for design practice. *Proceedings of the first International Symposium on Penetration Testing*. Orlando, USA. 263–296.

Prezzi M. (2005). Overview of construction and design technologies of auger cast-in-place and drilled displacement piles. *Proc. of 30th Annual Conf. on Deep Foundations (DFI),* Chicago/USA: 497–512.

Van Impe W.F. und Peiffer H. (1997): Influence of screw pile installation on the stress state in the soil. *In 'Design of Axially Loaded Piles – European Practice'*, ed. by De Cock and Legrand: 3–20. Rotterdam: Balkema.

Vermeer P., Bernecker O. & Weirich T. (2008): Schraubpfähle: Herstellung, Tragfähigkeit und numerische Modellierung. *Bautechnik, 85(2)*:133–139.

Viggiani, C. (1993). Further experiences with auger piles in Naples area. *Deep foundations on bored and auger piles II:* 445–455. Rotterdam: Balkema.

Baltic Piling – Mets & Raudsepp (Eds)
© 2013 Taylor & Francis Group, London, ISBN 978-0-415-64334-4

Comparison of integrity and echo tests in the assessment of defects in pile foundations

R.P. Cunha, A.F. Cordeiro & J.C.R. Cárdenas
University of Brasília, Brasília, Federal District, Brazil

ABSTRACT: Pulse Echo Integrity Testing, also known as sonic test, is a non destructive pile testing method that evaluates the integrity of foundation piles, being normally performed after pile installation and curing. If major defects exist, the test estimates their magnitude and location. This test has two distinct (main) equipments, from major worldwide suppliers. It can be named or known as "Pile Integrity Test (PIT)" or "Pile Echo Test (PET)", depending on the hardware. The University of Brasília (UnB) has both systems, one much older than another. This paper compares the results from them when used side by side in real scale foundations of Brazil and assess the interpretation criteria that was exclusively developed to categorize the level of "integrity" of bored piles, which are commonly executed in the region.

1 INTRODUCTION

The sonic test is used to evaluate the post construction "integrity" conditions of foundation piles, i.e., their continuity, density homogeneity, cross section & length variations, and overall quality in subjective terms. If major "defects" (deviation from nominal construction original standards) exist, the test estimates their magnitude and approximate location. This test is normally performed some days after pile execution. It requires minimal pile preparation, and because of its simplicity, speed of execution and low cost, it is now widely used in Brazil under ASTM (2007) given the lack of a national standard on this subject.

The University of Brasília (i.e. GPFees Group, www.geotecnia.unb.br/gpfees) has two testing systems from two distinct suppliers, both respectively related to the PIT and PET equipments. The PIT is an old version equipment acquired in 1995 by the university and which is about to reach its (deserved) "retirement" age. It uses a simple interpretation software (on time domain) to depict the curves of wave velocity versus pile length. The PET was recently acquired in 2010 to replace the old system, and operates in a similar way but has slightly distinctive interpretation software as well as accessories (accelerometers, hammer, etc.). Nevertheless, the usage scope, modus operandi, and results of both systems are quite the same. Both systems are used for research and teaching.

In order to validate and certify that the new equipment could be employed similarly as the old one, PIT and PET integrity tests were carried out on existing piles of the well known Experimental Site of Foundations and In Situ Testing of the University of Brasília, where the typical porous, collapsible and tropical clay of Brasília prevails (Cunha et al. 1999, 2001 and Marques et al. 2004). Moreover, the cross-comparison was also done in distinct construction sites of Brazil, where different foundation and soil types from those of the research site are found. Some of the gathered data is presented and discussed herein.

This paper therefore shows some few results from this comparative exercise, concluding on the similarity and validity of the obtained results. It provides a quick overlook on previously published information which relates to the interpretation criteria that was exclusively developed by the first author to categorize (according to eventual damage) the tested piles (Cunha 2008). It explains the criteria and classifies the results with data from the same foundations, sequentially tested by both integrity systems.

The paper neither has the objective to lecture and state fundamental principles of interpretation of such integrity tests, nor to teach their usage in practice and interpretation in office by existing techniques (time domain, frequency domain, etc.). It is rather a more practical publication focused in comparing the results of the main existing systems, so that any doubts of eventual differential interpretative outcomes between them are exposed and cleared. It nevertheless quickly addresses some key (theoretical/practical) aspects of interest on the following item.

2 GENERAL ASPECTS OF THE SONIC TEST

2.1 *Scope*

According to ASTM (2007) "low strain impact integrity testing provides acceleration or velocity and force (optional) data on slender structural elements (that is, structural columns, driven concrete piles, cast in place concrete piles, concrete filled steel pipe piles, timber piles, etc.). These data assist evaluation of pile integrity and pile physical dimensions (that is, cross-sectional area, length), continuity, and consistency of the pile material, although evaluation is approximate and not exact". However, the test method will not give information regarding the pile bearing capacity.

Also according to ASTM (2007), there are two methods of doing the sonic test in the field:

- The "Pulse Echo Method (PEM)": The pile head motion is measured as a function of time. The time domain record is then evaluated for pile integrity;
- The "Transient Response Method (TRM)": The pile head motion and force (measured with an instrumented hammer) are measured as a function of time. The data are evaluated usually in the frequency domain.

Therefore, the sonic test allows a fast assessment of a large, and statistically valid, number of executed piles within a particular tested site, under quality assessment criteria which depends on pre-established rules plus the experience of the performer. Construction problems as pile ruptures, "necking", "bulges", cracks and any other sorts of impedance variations of the pile shaft can be readily detected in the site, allowing corrective measures to be readily taken. Besides, even if no problems are detected, the test also serves as a quality control and assurance for the tested piles, allowing the foundation contractor plus any other involved parties with the construction (owner, surveillance team, designers) to positively rely on the good quality of the pile execution.

Nevertheless, one should remember that this testing technique was initially developed for pre-cast driven piles, which, on the other hand, has not prevented its further usage with other types of foundations, executed under distinct conditions from the original assumptions. Hence, the sonic test has been adapted since then to be used with deep bored cast-in-place foundations, as drilled piers, drilled and auger piles (with or without bentonite mud), continuous flight auger piles, and other pile types.

2.2 *History and components*

Non destructive tests performed on deep foundations with the exclusive aim to evaluate their physical "integrity" started to become routine solely in the end of the 70's, in particular within the United States. Part of this large expansion is owed to the fact that portable computers could (and were) employed in this new testing technique. The use of digital techniques in construction sites leads to a fast evaluation of the testing results, allowing the performer to have an initial idea of potential problems while still present in the site, thus enabling a fast decision on post testing procedures, i.e., from the piles' retest under other conditions (as for instance using high strain dynamic test of the pile).

A comprehensive and detailed explanation of this test (historical grounds, wave theory, interpretation and usage) is well exposed in Amir (2009), and it has served as basis for the contents of this sub item, given his extremely didactical approach for this question.

Hence, as pointed out by this author, the sonic method for the integrity testing of piles is aimed at routinely testing complete (100%) piling sites. To perform this test, a sensor (usually accelerometer) is pressed against the top of the pile while the pile is hit with a small hand-held hammer. Output from the sensor is analyzed and displayed by a suitable computerized instrument, the results providing meaningful information regarding both length and integrity of the pile. To have a better understanding of the inner workings of the method one should go through the basics of 1-D wave propagation in prismatic rods.

According to Amir (2009), Steinbach & Vey (1975), who used a makeshift system consisting of an oscilloscope and amplifiers, were probably the first to investigate the sonic test. Although their results look pretty crude by modern standards, they were still able to get a rather convincing reflection from the toe. The first-generation of commercial pile-testing equipment, which soon followed, still consisted of purely analog components, based on an oscilloscope and a Polaroid camera.

The early eighties saw the transition to the second-generation: Digital systems based on purpose-built computers and some proprietary operating system. Third-generation equipment appeared a few years later, once "ruggedized" laptop computers became commercially available, and used some version of DOS. Today, practically all testing systems in use are computerized, belonging to either second or third generation. Fourth-generation testing equipment, which recently became available, makes optimum use of the accelerated progress in both computing power and software capabilities, with a heavy accent on the software aspect. It runs under the worlds' most popular operating system, with all the resulting advantages.

Thus, a modern system for sonic testing of piles consists of the following components:

- A suitable wave generator;
- A transducer which is pressed against the top of the pile and is sensitive to motion;
- An analog-to-digital (A/D) converter;
- A portable computer;
- Dedicated software that handles the input and displays the results. The usual presentation is of pile-head velocity vs. time, or time-domain presentation. As a rule, the time axis is multiplied by c/2 and transformed to length base. This form is also called a "reflectogram";
- An optional component is the instrumented hammer that contains an internal force transducer.

2.3 *North American and British integrity testers*

Non destructive testing procedures, as the sonic test, are recommended by the Brazilian foundation standard ABNT (2010) to assure the quality of the pile construction, regardless of the equipment manufacturer, testing setup at site or specific interpretation approach.

The pile integrity tester, known as PIT, comes from a North American manufacturer. It can use a 450 g or 900 g instrumented hammer coupled to a rugged data acquisition system unit that serves A/D converter, data collector, data reduction (dedicated software), and interpretation module (site computer for preliminary analyses). In this case, similar testing characteristics are followed from those already mentioned before, but the equipment owned by the University of Brasília is ruled by procedures put forward by the PIT (1988) manual version.

The pile echo tester, known as PET, comes from a British manufacturer. It also comes with a lightweight handheld hammer, but distinctively as the American product, needs a PDA (pocket PC) computer to support field operation of signal acquisition, reduction and interpretation. Given that particular feature, equipments are pricewise distinctive, besides of similar accessories, testing and interpretation schemes. Both make use of accelerometers to capture the reflected wave, and both comply with the sonic standard ASTM (2007). PET software has two versions, i.e., for PC (laptop) and for PDA (pocket PC), which operate in full compatibility. Besides, PET hardware has two models, i.e., USB and bluetooth. The USB is more suitable for portable PCs and the bluetooth model is more adequate for pocket PC. University of Brasilia's model is a bluetooth system coupled to a PDA computer which was acquired separately. It is ruled by procedures put forward by the PET (2010) manual version.

For both testers, site acquired signals are chosen or discarded, averaged, magnified, filtered and interpreted in the time domain, in accordance to standard values recommended by aforementioned references and personal experience of the authors. The procedure to manipulate the signals, interpret, classify the possible "anomaly" observed with each pile curve (primary, secondary and tertiary reflected waves), and to finally write the Report (condemning or not the tested piles, or suggesting further alternative investigative approaches), is quite similar for both equipments. Hence, direct cross-comparison of results is possible, and will be addressed herein with the chosen Brazilian sites.

3 INTERPRETATION CRITERIA

Cunha (2008) used a large data bank of integrity tests in Brasília, which were statistically interpreted, to refine the already existing (from previous works) quality control and assurance (QCA) classification system for bored piles in tropical soils. Under this modified criteria, bored piles in this region started to be classified accordingly, leading to additional (and recommended) QCA procedures to be taken by the foundation contractor.

The adopted classification criteria is based on grades "A" to "D", being A related to perfect piles without integrity problems or other sort of defects, and D related to questionable piles which are liable to condemnation. The "C" criterion is subdivided into 3 sub criteria, to better detail it.

The criteria follow below:

A. "No significant defects are found in the pile, and there is a perfect reflection of the acoustic signal in its toe. Hence the foundation can be considered as physically intact";
B. "No significant defects are found in the pile, but apparently there was no acoustic signal reflected in its toe. The pile can be considered as partially tested, with integrity confirmed in its initial meters, and possibly all extension, although this fact can not be conclusively defined with the acquired signal";
C. "Impedance variations are noticed in the pile, but there is a reflection of the acoustic signal in its toe. Most probably, such impedance variations reflect sectional shaft variations and/or density variations of the pile material." Further sub-classifications (C1 to C3) are required with this grade, and can be directly found in the aforementioned reference.

They were used to compare the similarity between equivalent signals from both integrity testers.

4 COMPARISON OF SIGNAL RESULTS

4.1 *Research site of the University of Brasília*

The city of Brasília is surrounding by a special quadrilateral zone of the federation denominated as the "Federal District". Within this district it is common the occurrence of extensive areas (more than 80% of the total area) covered by a weathered laterite of the tertiary-quaternary age. This "latosol" has been extensively subjected to a leaching process and it presents a variable thickness throughout the District, varying from few centimeters to around 40 meters. It is basically a residual soil leached of silica that contains concentrations particularly of iron oxides and hydroxides and aluminum hydroxides, developed in humid, tropical and subtropical regions of good drainage. The studied latosol constitutes into a "collapsible" sandy clay with traces of silt, that overlays a saprolitic/residual soil with a strong anisotropic mechanical behavior and high standard penetration resistance, which is originated from a weathered slate. Most of the bases of the deep foundations of the city are founded in this lower stratum.

The Experimental Site of the University of Brasília is located within its Campus, where a typical section of the geotechnical deposit of the District exists. The initial comparison between both systems started in this site given its convenient location and abundance of (load test & in situ test) results. Figure 1 presents the initial setup of both systems, where the first author (at

Figure 1. Assemblage of both PIT and PET systems in the Experimental Site, and details.

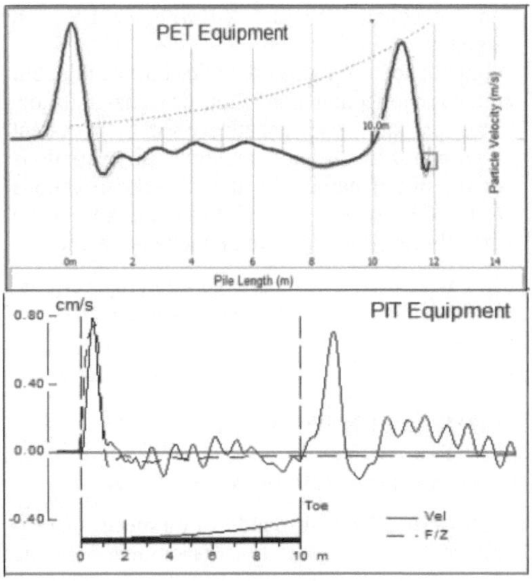

Figure 2. PET × PIT comparison for a 10 m–50 cm dia bored pile in UnB Experimental Site.

right in picture) is assembling both units with one of his doctoral students (herein third author). A schematic detail from both systems is also shown.

The results from the comparison of both systems in a 10 m length bored pile of this site are shown in Figure 2. This figure depicts the particle velocity measured at the pile head interpreted by the pulse echo method, as previously commented. The typical interpreted output is given, where it can be noticed that:

- Both convey similar results and can be classified by the same interpretation criteria according to Cunha (2008), which in this case would be of the type "C". Both also yield the same interpreted pile length;
- PIT signal seems to be more sensitive to minimal changes in particle motion than PET data, but this may relate to the internal characteristics of the accelerometer, and particular details of the systems, as their sampling and integration processes. On an overall comparison, and for engineering purposes, both results are remarkably close when adopting similar manipulation (filters, amplification, etc.) variables. In other words, both testers do provide similar output data and interpretation analyses.

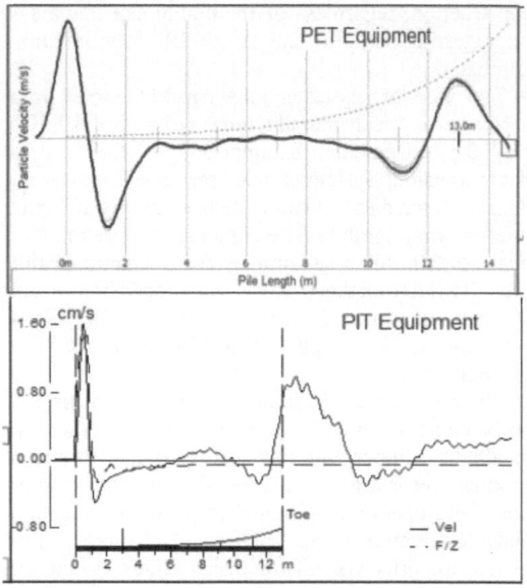

Figure 3. PET × PIT comparison for a 13 m–40 cm CFA pile in Águas Claras city (Federal District).

4.2 *Construction sites of Brazil*

The comparison was also undertaken in real case jobs where distinct soil profiles and foundation types can be found, when compared to the Research Site (where the study initiated). Given space limitations, only the results from two sites will be shown, one of them also located within the Federal District (but slightly distinct from the Research Site in geotechnical terms). This site is located in the city of Águas Claras, around 30 km from Brasília, where the collapsible sandy clay also prevails, but with distinct thickness and underneath bedrock. In this one, a continuous flight auger (CFA) pile was tested.

The other site is located in Manaus, in the Amazon region, a city more than 3000 km from Brasília. It has a deposit of alluvial fine silty sand, with low to high compacity levels depending on the depth. Water table is high and concretions of sandstone are usually found within the profile. In this site, a CFA pile was also tested, as this pile type is getting common in Brazil – given higher productivity and financial aspects that set out in this millenium (where CFA equipments became available at better prices).

Figures 3 and 4 respectively show the results, where it can be noticed that similar observations, as given before, can also be provided. In Figure 4 it is also seen that both systems were able to detect a secondary reflection of the signal, caused by variations of impedance. However, one recognizes a slight divergence in the depths where this change is marked in the

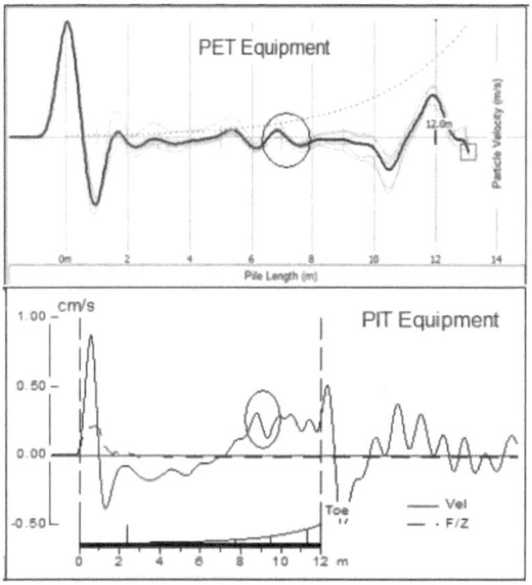

Figure 4. PET × PIT comparison for a 12 m–40 cm CFA pile in Manaus city (Amazonas).

reflectograms, probably caused by differences in the sensitivity of the equipments and difficulties in precisely defining the corresponding (at each signal) ups and downs of the reflected waves.

5 CONCLUSIONS

It is concluded that both integrity-testing systems are equivalent and give reasonably comparable results that are analogous in overall (time domain) manner, and lead to similar engineering judgments solely based upon them, besides of distinct operational modes, accessories and particular facilities.

ACKNOWLEDGEMENTS

Authors thank the Engineering Companies Engesol Engenharia Ltd. and Direcional Engenharia S.A. for allowing this study to take place in their construction sites.

Grant provided by MCT/CNPq 14/2009 Research fund proposal (Process 470531/2009-6), to buy the PET equipment, is also acknowledged and valued.

REFERENCES

ASTM Standard D5882. 2007. Standard Test Method for Low Strain Impact Integrity Testing of Deep Foundations. *ASTM International*, West Conshohocken, PA, www.astm.org.

Cunha, R.P., Jardim, N.A. & Pereira, J.H.F. 1999. In Situ Characterization of a Tropical Porous Clay via Dilatometer Tests. *Proceedings, Geo-Congress 99 on Behavorial Characteristics of Residual Soils, ASCE Geotechnical Special Publication 92, Charlotte*, pp. 113–122.

Cunha, R.P., Pereira, J.H.F., Soares, J.M., Mota, N.M.B. & Poulos, H.G. 2001. Backanalyses of field loading tests on deep foundations in a tropical clay. *Proceedings, XV International Conference on Soil Mechanics and Geotechnical Engineering – XV ICSMGE, Istanbul*, Vol. 2, pp. 869–872.

Marques, F.E.R., Almeida e Sousa, J., Santos, C.B., Assis, A.P. & Cunha, R.P. 2004. In-situ geotechnical characterisation of the Brasília porous clay. *Proceedings, International Site Characterization – ISC'2, Porto*, Vol. 2, pp. 1301–1309.

Cunha, R.P. 2008. Quality evaluation criteria for pile integrity test on deep foundation bored in tropical soil. *Proceedings, 8th. International Conf. on the Application of Stress Wave Theory to Piles – SWAVE'08, Lisboa*, Vol. 1, pp. 465–470.

Amir, J.M. 2009. *Integrity Testing*. Web download, www.piletest.com.

Steinbach, J. & Vey, E. 1975. Caisson valuation by stress-wave propagation method. *ASCE Journal of Geotech. Division*, 101(4), April.

ABNT Standard NBR6122. 2010. Design and Execution of Foundations. *Brazilian Association of Technical Norms, ABNT*, São Paulo, www.abnt.org.br (In Portuguese).

PIT. 1988. *Pile Integrity Tester – Collector Model: User's Manual*. Pile Dynamic Inc., 38 p.

PET. 2010. *Pile Echo Pulse Tester: Users Manual*. Piletest.com Ltd., CD Rom.

Baltic Piling – Mets & Raudsepp (Eds)
© 2013 Taylor & Francis Group, London, ISBN 978-0-415-64334-4

Pile bearing capacity of the new boiler house at cogeneration power plant Sisak

I. Matković, K. Bolanča, M. Bišćan & L. Bolfan
Institut IGH d.d., Zagreb, Croatia

ABSTRACT: Turbines and boiler house for the new cogeneration power plant block in Sisak are founded on reinforced-concrete piles. Piles are embedded in layers of soft clays, silty sands and dense clays. Dynamic pile analysis was performed using CAPWAP analysis. Pile bearing capacity was obtained at the toe and shaft. Pile static testing was performed and total bearing capacity was obtained. Trial boreholes and geophysical profiling were executed on site. CPTU testing, or static penetration testing was performed up to and under the pile toe depth. According to the results of investigation works and CPTU testing, pile bearing capacity was analyzed from the correlations. Comparison of the obtained results was done for boiler house example, and numerical calculation of boiler house foundation structure is shown. Monitoring results of the boiler house foundation slab settlement are also presented.

1 INTRODUCTION

The Thermal Power Plant Sisak (TPP Sisak) is situated in the central part of the Republic of Croatia in Sisak-Moslavina County, in the south industrial zone of Sisak town.

According to the Physical Development Plan of Sisak Municipality, the production of electric energy in Sisak area started in 1970 by putting into operation Block 1 of TPP Sisak, power 210 MW. Already at that time, the construction of the system in three phases of total power 600 MW was planned.

The construction of the new combi block in TPP Sisak has been harmonized with the physical development plans at the state level, i.e. it is in compliance with the reconstruction priority program and the existing capacities of the electric energy system given within the Strategy and the Physical Development Plan of the Republic of Croatia.

2 GEOTECHNICAL INVESTIGATIONS

For the purposes of construction of the new block of TPP Sisak, trial boring was carried out, one borehole at 80 m of depth and three boreholes at 30 m of depth. Afterwards, the location was tested using CPTU probes at three positions; the depth of probes was around 30 m. Subsequently, probing was carried out at four more locations, and the depth of additional probes was 14 m. At nine positions the soil was examined by dilatometer down to the depth of 11–15 m. Geophysical investigations included down hole testing and borehole logging in the borehole of 80 m in depth. Also, geophysical profiling of the terrain was carried out by testing the velocity of shear seismic waves per depth using the ReMi method.

The investigated terrain is characterized by deposits which originated as a shallow lake, wetland and flood-plain river deposits. The terrain has the following structure:

Fill material is of 1–1,5 m in thickness and it covers the entire area of construction. The embankment is made of a mixture of clay, silt, sand, cobbles and rubble from the remaining building material.

Low plasticity, sandy clays and silty sands are the autochthonous flood plain deposits, from soft to medium consistency i.e. loose to medium dense state. The layers extend to the depth of around 11 m. The undrained strength of these clays is $c_u = 50–70$ kPa.

Silty sands constitute the following layer but they have greater compaction, i.e. these sands are moderately dense. The layers are located at the depth between 11–18 m. Sands and silty sands have the number of blows in SPT 10–20.

Wetland clay deposits i.e. a layer of high-plasticity clay, of semi-stiff to stiff consistency within which thin elongated lenses of fine grained silty sand and organic soil – peat appear. The clays have undrained strength $c_u = 200–300$ kPa. Laterally, the clays become silts and silty sands with lenses of silty and sandy gravel. The gravels have the number of SPT blows 20–25. The bottom of this layer consists of low-plasticity silts. The layers are registered at 18–25 m of depth. At 25 m from the terrain surface, shallow lake sediments – silty sands appear again.

Alternating of deposits of clay and silt of stiff consistency are registered down to the depth of 80 m.

Inside this layer, there are thin elongated lenses of silty sands and silty gravels and rarely organic clays and peat.

The TPP Sisak is situated in a seismotectonically active area. The input parameters have been determined for the calculation of the impact of seismic forces at the investigated location. The assessment of amplification in surface layers was made, nonlinear ground response analysis was carried out, as well as the calculation of expected maximum accelerations and earthquake intensity.

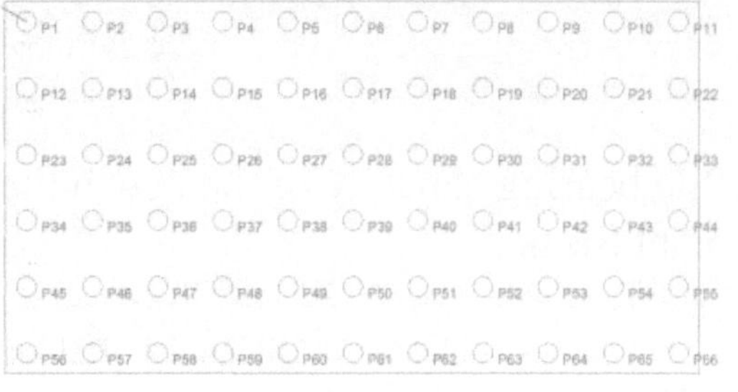

Figure 1. Layout of piles at the boiler house.

3 FOUNDING OF STRUCTURES IN THE NEW POWERPLANT BLOCK

A boiler house, bases for the steam turbine and gas turbine, which are founded on piles, were made inside the new power block. Other structures in the power block are executed on strip foundations and slabs.

For the boiler house, a group of reinforced-concrete bored piles was made using CFA technology i.e. it was executed by means of continuous auger. Total 66 piles of 100 cm in diameter and 15,0 m in length were made as well as the pile cap for the placement of main boiler.

For the steam turbine, a group of reinforced concrete bored piles including borehole protection by steel casing was executed. Total 60 piles of 100 cm diameter and 23,0 m in length and a pile cap for placing of the turbine plant were made. The foundation soil under the foundation slab, and down to the depth of 4,0 was reinforced by jet grouting.

For the gas turbine, a group of reinforced concrete bored piles was also executed, including borehole protection by steel casing. Total 63 piles of 90 cm in diameter and 18,0 m in length were executed, and the pile cap for the placing the gas turbine. The foundation soil under the foundation slab and down to depth of 4,0 m was also reinforced by jet grouting.

The analysis for the boiler house, which was executed on reinforced concrete foundation slab of 1,5 m in thickness and with CFA piles of 100 cm in diameter, is shown in continuation.

Settlement of the foundation structure was analyzed using the software Plaxis Foundation 3D v2 (finite element method for 3D state of deformations). In the software, the pile cap is modeled using plane elements, the piles using line elements, and the ground using solid elements.

The settlement analysis was carried out for the action of permanent and long-term loads.

Geotechnical parameters for the drained condition were used. Stress and strain relation is defined according to the hyperbolic soil model (hardening soil). The analysis was carried out for the following phases during construction:

1. Analysis of the initial geostatic state in the ground including the hydrostatic pressure.
2. Activation of the foundation slab and piles.
3. Load activation.

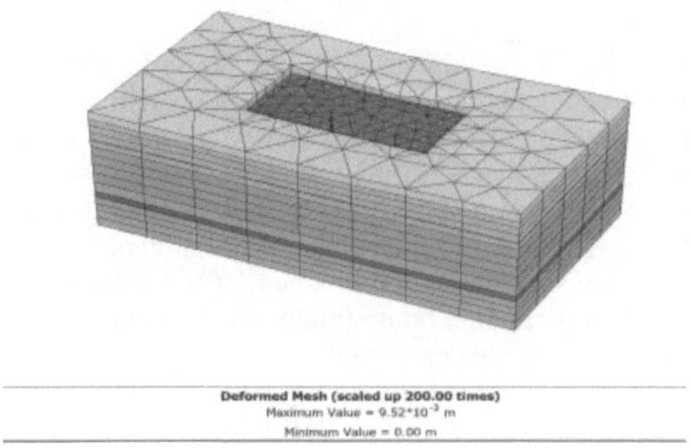

Figure 2. 3D model.

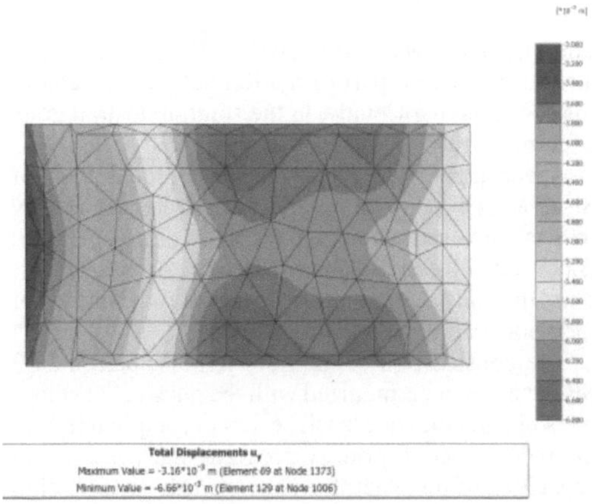

Figure 3. Isolines of the settlement of boiler house slab.

The following distribution of vertical displacements of the slab was obtained for the permanent and long term load.

Vertical displacements of the slab are 3 to 7 mm, i.e. differential settlement amounts to 4 mm. The distribution of longitudinal loads in the piles is also presented, as well as the vertical pile displacements for the outer row of piles (P1–P11).

Settlement of pile heads is also graphically presented (P1–P11).

Because of the arrangement of bearings along the slab edges and the distribution of load on the piles, the distribution of displacement is not uniform. Displacement of pile heads for the outer row of piles (P1–P11)

Table 1. Longitudinal forces and pile displacements.

Pile No	Longitudinal force on the pile head N (kN)	Longitudinal force on the pile toe N (kN)	Displacement of pile head uy (mm)	Displacement of pile toe uy (mm)
1	1040	101	3,6	3,2
2	928	75	4,4	4,1
3	972	87	4,9	4,6
4	1090	99	5,6	5,2
5	1280	117	6,3	5,9
6	1300	121	6,6	6,2
7	1130	108	6,4	6,1
8	1160	105	6,4	6,0
9	1160	105	6,4	6,0
10	1190	108	6,1	5,7
11	1510	145	5,7	5,2

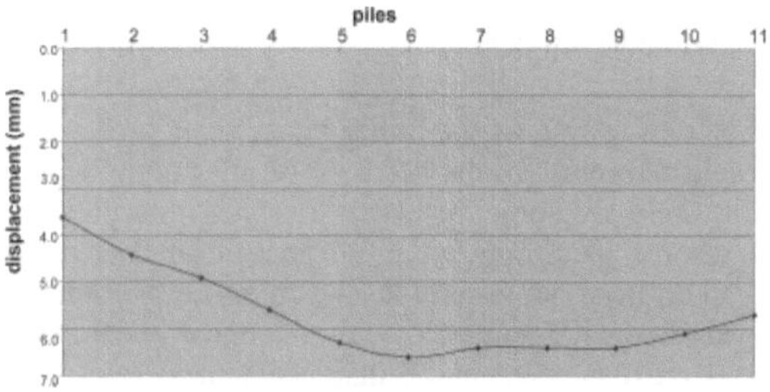

Figure 4. Settlement of pile heads along the slab.

is 3,6–6,6 mm and for the inner row of piles (P23–P33) it is 3,1–6,1 mm. Actual displacement values can be expected to reach up to 1 cm.

Greatest longitudinal force in the piles for dead load is N = 1500 kN.

Foundation soil mostly consists of sand particles, thus the subject overall and differential settlement shall largely occur in the course of the construction works.

The slab displacements during dead load, long-term and seismic load are between 2 and 8 mm.

The greatest longitudinal force in the piles, during dead load and seismic load is 1900 kN.

Bearing capacity analysis for individual, 15 m long pile was done according to the semi-empirical method from O'Neill, M.W. and Reese, L.C. Bearing capacity of pile toe $R_b = 1178$ kN and bearing capacity of pile shaft $R_s = 4032$ kN was obtained, i.e. overall bearing capacity of $R_{uk} = 5210$ kN.

Bearing capacity was also analyzed according to the static penetration test results. Semi-empirical analysis was applied according to Eslami-Fellenius, and in this case the bearing capacity of pile toe is $R_b = 1212$ kN, bearing capacity of pile shaft is $R_s = 2348$ kN, i.e. overall bearing capacity is $R_{uk} = 3560$ kN.

4 TESTING AND MONITORING OF EXECUTED PILES

A trial pile was constructed next to the boiler house, diam. 100 cm and 15 m long, primarily to test the static bearing capacity. The trial load program was executed in accordance with suggested ISSMGE method.

Maximum pile load was defined on the basis of given design pile resistance (up to 3948 kN), and it is 4000 kN. The main breaking criterion is the vertical pile displacement of 10%, pile diameter i.e. ≈10 cm.

The pile has two stages of load:

In the first stage the pile is loaded with up to 50% of the testing force, i.e. up to 2000 kN, which corresponds to nearly maximum forces in the designed piles. The load was applied in four steps, 500 kN each, after which the pile was relaxed of load. Each load increment was maintained for 1 hour or until the displacement velocity falls under 0,1 mm/h. The force under load relaxation was maintained for 10 minutes. In order to determine the permanent settlement of piles, before repeated load, the pile was kept at zero load for 30 minutes.

In the second stage, the pile was loaded up to 4000 kN. The load was applied in increments of 500 kN.

Every load increment was maintained for 1 hour, i.e. until the settlement velocity dropped under 0,1 mm/h. The force under load relaxation was maintained for 10 minutes.

The following was measured during the test: force at pile cap, displacement of pile cap, displacement of reference beam.

The force at soil failure was determined according to the behavior at pile creep, i.e. according to the criterion which proved to be more conservative than achievement of a 10 cm displacement. Measurements on piles with trial load undertaken until now show that the resistance at pile shaft becomes completely active even at very small pile settlement, while full resistance at pile toe requires a significantly bigger pile settlement, app. 10% of the pile diameter. In order to avoid uncertainties when interpreting the curve which shows the load-settlement relation, the limit load estimate is based on identifying the creep force Q_c from the diagram where the ordinate shows a settlement increment of $\Delta_{si,30min}$ during the last 30 minutes of the load

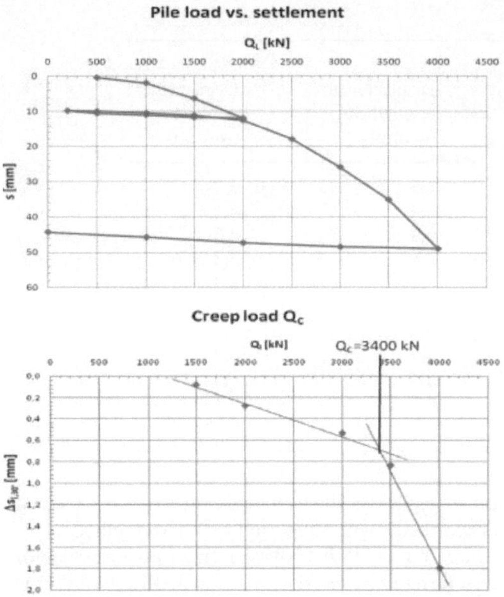

Figure 5. Settlement of pile heads and determining the failure force in relation to pile creep.

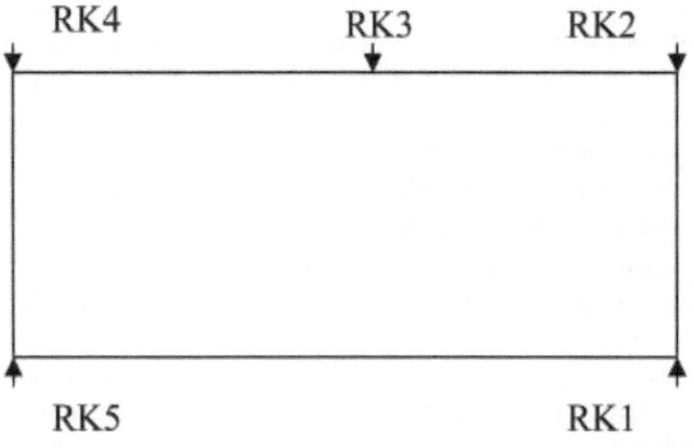

Figure 6. Position of bench-marks on the boiler house foundation slab.

level and in relation to the applied force shown on the abscissa. A fracture is usually visible on the obtained curve and some other formal definition of failure is not required.

The following results were obtained after trial testing of piles.

Load bearing capacity of one pile was also tested by dynamic procedure, according to recommendations given in ASTM D 4945-96. The dynamic load test on piles was done using twelve tons heavy weights placed at a defined height and released by free-fall. Strain meters and velocity meters were placed at the pile head. The testing was done using Pile Dynamics Analyzer-a (PDA), and numerical analysis was done using the CAPWAP program (CAse Pile Wave Analysis Program). CAPWAP analysis, by interactive procedure, determines the limit structural bearing capacity of piles. The pile toe bearing capacity $R_b = 1255$ kN and bearing capacity of pile shaft $R_s = 2258$ were obtained, i.e. overall bearing capacity of $R_{uk} = 3513$ kN.

After the boiler house construction was completed, settlement was monitored during a period of one year. Five bench-marks were placed on the foundation slab. The following picture shows the position of bench-marks on the slab.

The following settlement values were obtained in Table 2.

Table 2. Settlement of bench-marks on the boiler house.

Benchmark	RK1	RK2	RK3	RK4	RK5
Settlement (mm)	9,2	9,7	11,9	0,77	5,0

5 CONCLUSIONS

Structural testing of the boiler house piles presented a realistic picture of the behavior of piles under load. Still, there are some uncertainties when interpreting the test results. Strain at soil failure is defined according to the behavior of piles during creep, i.e. according to criteria which proved to be more controversial than the 10 cm displacement, i.e. displacement corresponding to 10% of the pile diameter. Having in mind that a large number of tests undertaken until now are based on the second criterion which was a basis for the semi-empirical methods of analyses, there are certain discrepancies between the calculated load bearing values and the values obtained by structural and dynamic testing. The following load bearing values of piles were obtained:

	Rs shaft resistance	Rb toe resistance	Rt total resistance
Static test			3400 kN
Dynamic test	2258 kN	1255 kN	3513 kN
Reese-O'Neill	4032 kN	1178 kN	5210 kN
Eslami-Fellenius	2348 kN	1212 kN	3560 kN

Numerical modeling was also presented of the complete axially loaded foundation structure and surrounding soil. Monitoring of the foundation slab settling showed that the settling is in accordance with the calculated values, i.e. that they are about 1,0 cm.

REFERENCES

ASTM D 4945-96 – Standard Test Method for High-Strain Dynamic Testing of Piles.

Eslami, A., Fellenius, B.H. *Pile capacity by direct CPT and CPTU methods applied to 102 case histories*. Canadian Geotechnical Journal, Vol. 34, 1997, pg. 886–904.

Reese L. C., Insenhower W. M., Wang Shin-Tower., *Analysis and Design of Shallow and Deep Foundation.* John Wiley and Sons, N.Y. USA., 2006., pg. 323–345.

Suggested method ISSMGE, Subcommittee on Field and Laboratory Testing: "Axial Pile Loading Test-Part 1: Static Loading", Geotechnical Testing Journal, GTJODJ, Vol. 8, No. 2, June 1985, pg. 79–90.
Institut građevinarstva Hrvatske d.d., *Glavni građevinski i geotehnički projekt temeljenja kotlovnice,*/Detailed Construction and Geotechnical Designs of foundation works for the boiler house / Cogeneration Combi power plant in Sisak, 2009.
Elektroprojekt d.d., Glavni građevinski projekt parne turbine/Detailed Construction Design for the power turbine Cogeneration Combi power plant in Sisak, 2009.
Elektroprojekt d.d., Glavni građevinski projekt plinske turbine/Detailed Construction Design for the gas turbine, Cogeneration Combi Power Plant in Sisak, 2009.
Građevinski fakultet Zagreb, *Izvješće o statćkom ispitivanju pilota*/Report on the Static pile test, Cogeneration Combi Power Plant in Sisak, 2010.
Institut IGH d.d., *Izvještaj o ispitivanju nosivosti pilota dinamickim postupkom*/Report on the pile bearing capacity by dynamic testing, Cogeneration Combi Power Plant in Sisak, 2010.

Baltic Piling – Mets & Raudsepp (Eds)
© 2013 Taylor & Francis Group, London, ISBN 978-0-415-64334-4

Numerical investigations on vibratory sheet piling in embankments

T. Hamann & J. Grabe
Institute of Geotechnical Engineering and Construction Management,
Hamburg University of Technology, Hamburg, Germany

ABSTRACT: When driving piles in the area of embankments the soil is dynamically stressed. Temporary and locally limited instabilities of the embankment can occur resulting in a horizontal loading of the driven pile due to the earth pressure from the embankment. In practice usually pile driving guides are used to prevent large horizontal deformations of the pile during the installation process. The horizontal loading of a pile driving guide depends on many parameters and is difficult to estimate. Dimensioning a pile driving guide is normally based on the experience of the developing engineer.

In this contribution the obliquity of a sheet pile wall due to the vibratory driving in the area of an embankment is back-calculated numerically by the use of the finite element method. The calculated deformations of the sheet pile wall are compared to measurements. Further different parameters influencing the development of the earth pressure and the supporting forces of a pile driving guide are investigated in a parametric study to determine possibilities to prevent damages.

1 INTRODUCTION

For the construction of waterfront structures or railroad embankments sheet piles often have to be installed in the area of embankments. Due to the vibratory driving of sheet piles the surrounding soil is dynamically stressed and a temporary and locally limited reduction of the stability of the embankment can occur. Pile driving guides are usually used to prevent large deformations of the embankment or the sheet piles. The horizontal loading of a pile driving guide depends on many parameters and is difficult to estimate. In case of an undersized pile driving guide damages or deficits of the construction can occur.

In this contribution the obliquity of a sheet pile wall due to the installation process of vibratory driving in an embankment is numerically simulated. The arising deformations and the development of the earth pressure as well as the reaction forces of a pile driving guide are investigated. In a parametric study different influences like the initial bulk density of the soil, the vibratory frequency, the stiffness of the sheet pile and the position of the lower support of the pile driving guide are investigated.

2 VIBRATORY DRIVING IN EMBANKMENTS – BACKGROUND

When installing piles by vibratory driving or impact driving the soil in the near-field of the pile is dynamically loaded due to the propagation of shear waves. The soil, dissipating the energy of the shear waves, gets compacted and the horizontal stress state is reduced. Due to the compaction of the soil in the near-field, soil of the surrounding areas slides into the compacted area of the near-field and settlements are arising at the surface. In case of a horizontal ground level surface extensive numerical investigation of the penetration process of piles in drained sand have been carried out by Mahutka (2007) and Henke (2008) to estimate the compaction of the soil or the altitude of settlements on the surface. A numerical model to calculate settlements due to the vibratory driving of sheet piles in case of drained and undrained conditions has been proposed by Meijers (2007).

In case of an embankment, the soil compaction and the reduction of the horizontal stress state can reduce the stability of the embankment. Locally limited parts of the embankment can slide downwards. An imbalance of the horizontal earth pressure on the active and passive side of the pile can develop resulting in horizontal deformations of the installed sheet pile wall. In practice pile driving guides are usually used to prevent large horizontal deformations of the sheet pile wall due to the installation process. Constructive recommendations for the use of pile driving guides and guidelines for dimensioning are given for example in TESPA (2001) and Buja (2001). Approaches to estimate the arising reaction forces especially for the installation of piles in embankments are not given.

3 CONSIDERED CASE OF DAMAGE

For the enlargement of a power plant a new waterfront consisting of a sheet pile wall of Z-Sections of

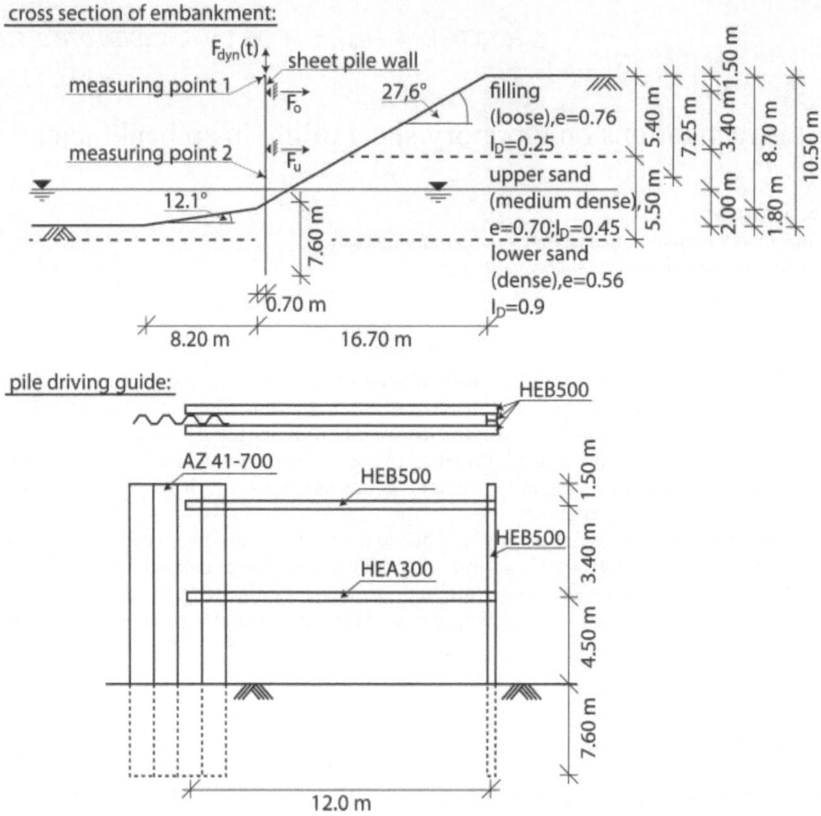

Figure 1. Top: cross section of the embankment at the considered case of damage; bottom: engaged pile driving guide.

the type "AZ 41-700" has been built up in the area of an embankment. The geometry and dimensions are shown in Figure 1. The sheet piles were installed as double piles with a "Müller-Vibrator MS-48 HFV" with a vibratory frequency of $f = 36$ Hz by the use of an upper and lower pile driving guide as shown in Figure 1. During the vibratory driving a horizontal deformation and obliquity of the sheet piles occurred into the direction of the waterside. At the first measuring point (head of the sheet piles) a horizontal deformation up to $u_h = 13$ cm was detected and at the second measuring point a horizontal deformation up to $u_h = 9$ cm.

4 NUMERICAL MODELING

The numerical simulations are carried out with the finite element software Abaqus/Explicit, Version 6.9 (Dassault Systemes 2009). For the subsoil a sand under drained conditions is assumed. A possible development of excess pore water pressure can not be taken into account due to restrictions of the finite element software. First approaches of the numerical modeling of the subsoil as a three-phase material under dynamic loading are proposed in Holler (2006) and Schümann (2010).

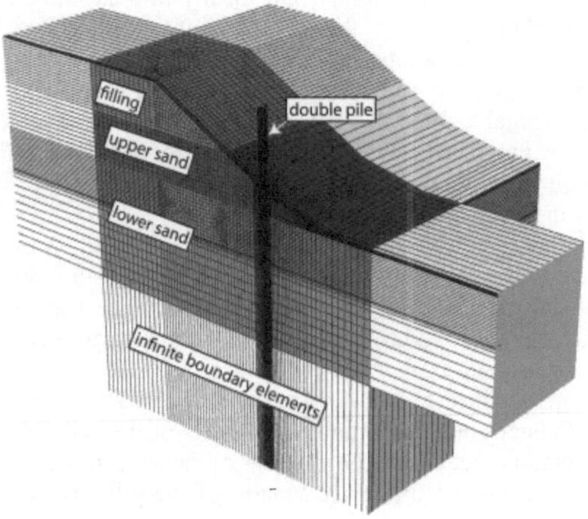

Figure 2. Numerical model for the back-calculation of the considered case of damage with discretisation.

4.1 *Numerical model*

For the simulation of the vibratory driving a plane of symmetry is introduced in the three-dimensional numerical model (Fig. 2). The back-calculation of the considered case of damage is done with the cross section depicted in Figure 1. The

Table 1. Hypoplastic parameters of "Karlsruher Sand".

Material	φ_c [°]	h_s [MPa]	n [–]	e_{d0} [–]	e_{c0} [–]	e_{i0} [–]	α [–]	β [–]	m_T [–]	m_R [–]	R [–]	β_R [–]	χ [–]
Karlsruher Sand	30	5,800	0.28	0.53	0.84	1.00	0.13	1.05	2.0	5.0	0.0001	0.5	6.0

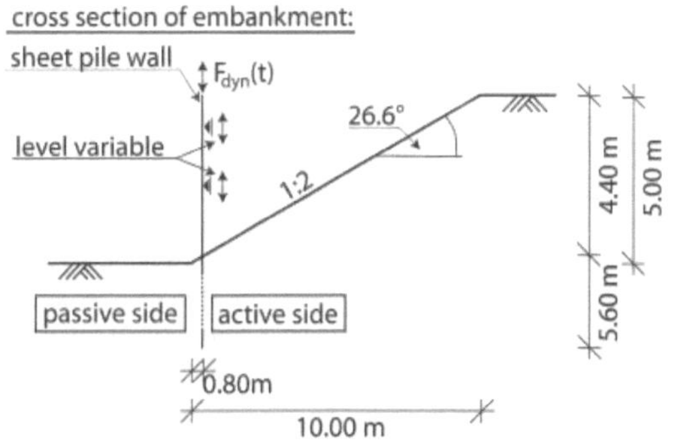

Figure 3. Cross section of a simplified embankment for a parametric study.

parametric study is carried out on a simplified cross section shown in Figure 3. At the boundaries of the numerical model infinite boundary elements are located to prevent a reflection of waves induced by the vibratory driving. The double pile is modeled as a deformable body with linear elastic material behavior. A Young's modulus of $E = 210{,}000\,\text{MN/m}^2$ and a Poisson's ratio of $\nu = 0.3$ is assumed for the steel. The double pile, which has to be installed, is modeled using the wished-in-place-technique. Using the so-called zipper-technique (Cudmani 2001) is not possible due to the elastic material behavior of the double pile. The underlying assumption regarding the penetration process is, that the surrounding soil is mainly influenced by shaft friction. Toe resistance plays a minor role. The point of load application is at the head of the double pile. Due to the elastic material behavior a realistic wave propagation within the double pile can be simulated as well as lateral oscillations of the double pile.

4.2 *Constitutive model of the sand*

For the constitutive modeling of the sand a hypoplastic model is used to simulate the dynamic compaction of the subsoil realistically. The hypoplastic model in the version of Gudehus (1996) and von Wolffersdorff (1996) in the formulation of vonWolffersdorff (1996) with the extension of intergranular strains by Niemunis & Herle (1996) is used. The hypoplasticity is suitable to model the non-linear and anelastic behavior of granular materials like sand. Typical characteristics like dilatancy, contractancy, different stiffness for loading and unloading as well as the dependency of the stiffness from the void ratio and the mean pressure are modeled.

For the carried out calculations the parameters of "Karlsruher Sand" are assumed because no hypoplastic parameters of the sand in situ were available (Tab. 1). The bulk density of the sand is given by the definition of an initial void ratio as depicted in Figure 1.

4.3 *Contact formulation*

The contact between double pile and soil is modeled with the "general contact algorithm" implemented in Abaqus (Dassault Systemes 2009). An angle of wall friction of $\delta = \frac{2}{3}\varphi$ with a friction angle of $\varphi = 30°$ is assumed.

4.4 *Discretisation*

The numerical model is discretised with app. 50,000 three-dimensional 8-node linear brick elements with reduced integration and hourglass control.

4.5 *Loading*

In a first step a surface pressure of $p = 1\,\text{kN/m}^2$ for a better numerical stability and gravity of $g = 10\,\text{m/s}^2$ is applied to the model, except the double pile to prevent a penetration during the simulated vibratory driving in the next step. In the second step the vibratory driving is simulated. The simulation of the vibratory driving is done displacement-controlled to prevent a penetration of the double pile into the soil and a failure of the mesh, as it occurs in a force-controlled simulation. For a period of $t = 15$ s a harmonic oscillating vertical displacement is applied to the head of the double pile as an external loading. The magnitude of the oscillation is determined in a short foregoing force-controlled simulation. The influence of a neighboring double pile is neglected.

5 INVESTIGATION OF THE CONSIDERED CASE OF DAMAGE

The back-calculation of the investigated case of damage is carried out by consideration of two limit cases:

1. Free-riding vibratory driving without a pile driving guide
2. Assumption of a rigid pile driving guide

The behavior of the real pile driving guide is to be expected between these two limit cases. The case of a free-riding vibratory driving represents the upper limit regarding the deformations and the case of a rigid pile driving guide the lower limit.

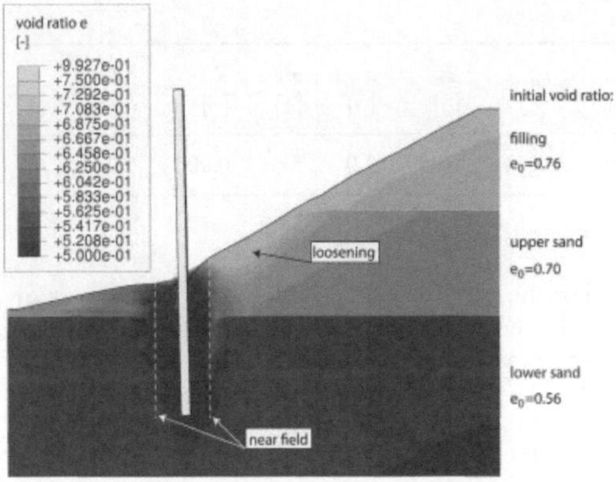

Figure 4. Void ratio after a vibratory driving of 10 s, free-riding vibratory driving.

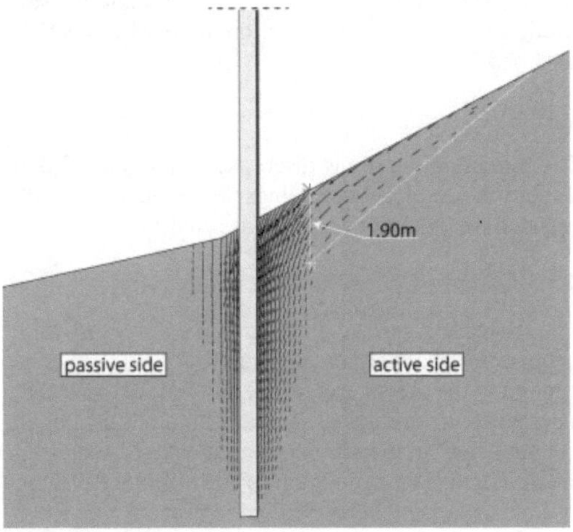

Figure 5. Deformation mechanism of the embankment after a vibratory driving of 10 s, free-riding vibratory driving.

5.1 *Dynamic compaction of the soil*

Due to the vibratory driving a cyclic shearing of the soil occurs in the near-field of the double pile with the result of soil compaction as depicted in Figure 4. In accordance with Mahutka (2007) the near-field has a radius of approximately 2 to 2.5 times the profile height. On the surface a marginal dilatant behavior can be observed due to a wedge of soil which slides downwards the slope into the area of soil compaction.

5.2 *Deformation mechanism*

The deformation mechanism can be divided into two zones. In a surface zone at the active side (down to a depth of 1.90 m) a wedge of soil is sliding downwards the slope into the near-field due to the occurring soil compaction (Fig. 5). In a deep zone (depth of 1.90 m and deeper) the soil is moving in approximately vertical direction due to the occurring soil compaction with the result of surface settlements.

5.3 *Horizontal deformation of the double pile*

The simulated development of the deformations of the double pile with respect to time is shown in Figure 6 for both limit cases. In both cases the deformations are increasing asymptotically to a limit which is not reached after a simulation time of 15 s. A longer period of simulation time is not possible due to numerical instabilities. In case of a free-riding vibratory driving an obliquity of the double pile into the direction of the waterside occurs. The horizontal deformations reach a maximum of $u_{h,lim1,P1} = 37$ cm at measuring point 1 and $u_{h,lim1,P2} = 21$ cm at measuring point 2 (see Fig. 1). In case of a rigid pile driving guide only small horizontal deformations are possible in the upper area of the double pile due to the two rigid supports of the pile driving guide. In this case a little different deformation mechanism consisting of a marginal oblique into the direction of the embankment occurs. The main deformations arise at the lower part of the double pile with a horizontal deformation of $u_{h,lim2,toe} = 4.4$ cm at the toe of the pile.

The deformation mechanism of the embankment due to the vibratory driving can be simulated qualitative correctly. The measured deformations are within the range of the two considered limit cases and thus can be recalculated quantitative correctly:

measuring point 1:

$$u_{h,lim\ 1} = 37 \geq u_{h,meas} = 13 \geq u_{h,lim\ 2} = 0.1 \quad [\text{cm}]$$

$$u_{h,lim\ 1} = 21 \geq u_{h,meas} = 9 \geq u_{h,lim\ 2} = 0.8 \quad [\text{cm}]$$

5.4 *Earth pressure and reaction forces*

During the simulation of the vibratory driving the development of the earth pressure and the reaction forces are investigated (Figs 7, 8). The resulting earth pressure forces at the active and the passive side of the double pile decrease abruptly to approximately 50% after the beginning of the vibratory driving due to the cyclic shearing. During the vibratory driving the earth pressure forces are remaining nearly constant. In case of a rigid pile driving guide the earth pressure force at the active side decreases less than at the passive side due to the wedge of soil sliding from the embankment into the compacted near-field (see Fig. 7). The difference between the earth pressure force of the active and the passive side is the resulting reaction force of the pile driving guide.

The investigation of the distribution of the earth pressure with respect to the embedment depth of the double pile shows, that a reduction of the earth pressure at rest occurs especially for the deeper areas of the embedment (Fig. 8). At the active side down to a depth of approximately $z = 2.5$ m an increase of the earth pressure at rest occurs. This is caused by the developing wedge of soil which is sliding downwards and is held by the double pile.

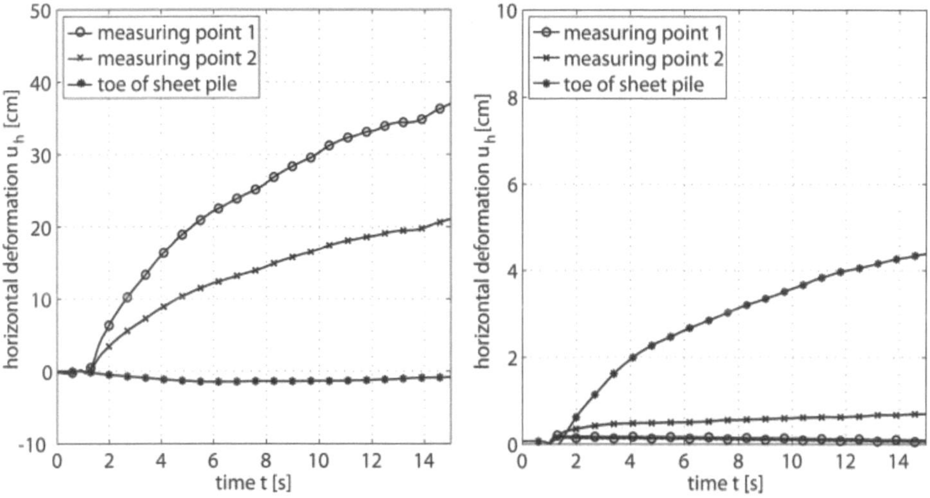

Figure 6. Development of the horizontal deformations of the double pile with respect to time, left: free-riding vibratory driving; right: rigid pile driving guide.

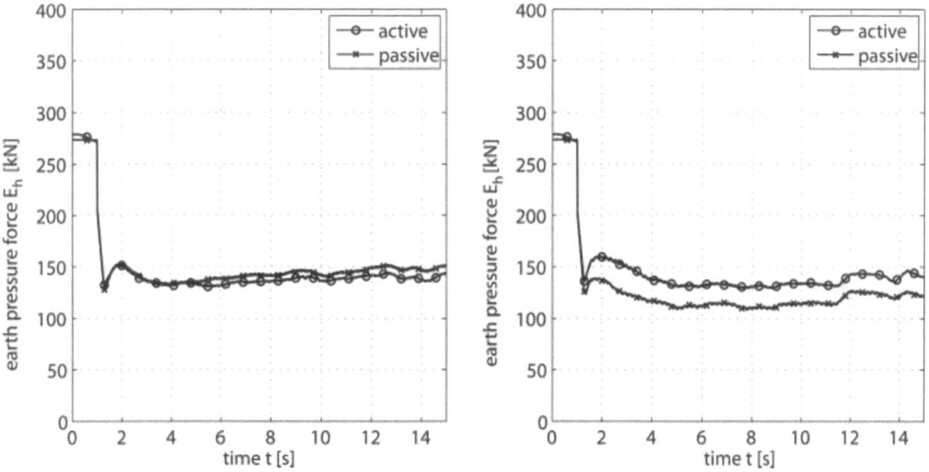

Figure 7. Development of the horizontal earth pressure force at the active and passive side of the double pile with respect to time, left: free-riding vibratory driving; right: rigid pile driving guide.

The development of the reaction forces with respect to time shows a similar behavior to the earth pressure forces (Fig. 8). The reaction forces increase very fast due to the abrupt decrease of the earth pressure force at the beginning of the vibratory driving. During the vibratory driving only a small increase of the reaction forces occurs, although the resulting earth pressure forces are remaining nearly constant. This is caused by the development of settlements on the surface of the embankment. The depth the resulting earth pressure force is acting on the double pile is moving deeper resulting in a larger momentum, which has to be absorbed by the pile driving guide. Reaction forces of $F_{upp} = -34$ kN at the upper support and $F_{low} = 56$ kN at the lower support have developed after a simulation time of 15 s.

6 PARAMETRIC STUDY

A parametric study is carried out on a simplified cross section of an embankment to investigate the influence of single parameters on the reaction forces of a pile driving guide and the development of the resulting earth pressure force (Fig. 3). Parameters like the position of the lower support of the pile driving guide, the initial bulk density of the soil, the vibratory frequency and the stiffness of the double pile are investigated.

A variation of the position of the lower support shows no effect on the development of the earth pressure force (Fig. 9). As expected, the position of the lower support shows a significant influence on the reaction forces. The smaller the distance between both

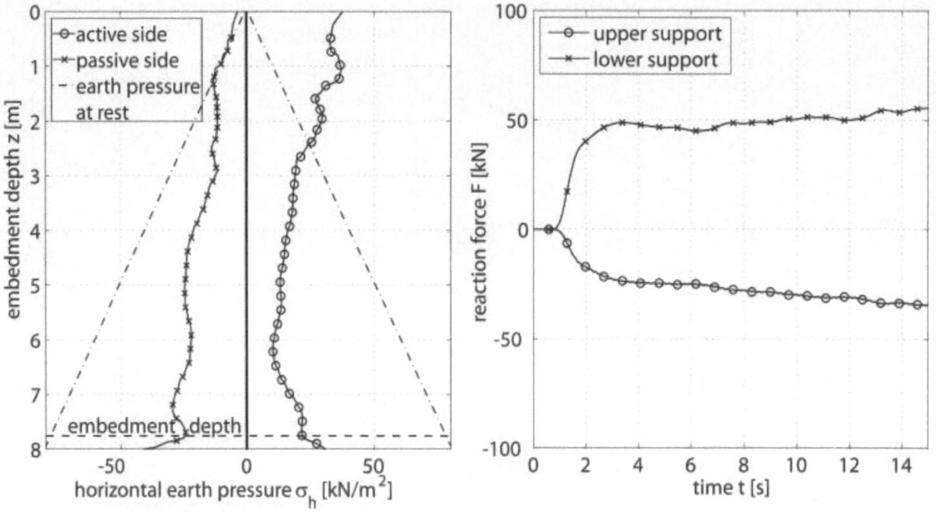

Figure 8. Left: distribution of the horizontal earth pressure with respect to the embedment depth after the vibration driving; right: development of the reaction forces of the rigid pile driving guide with respect to time.

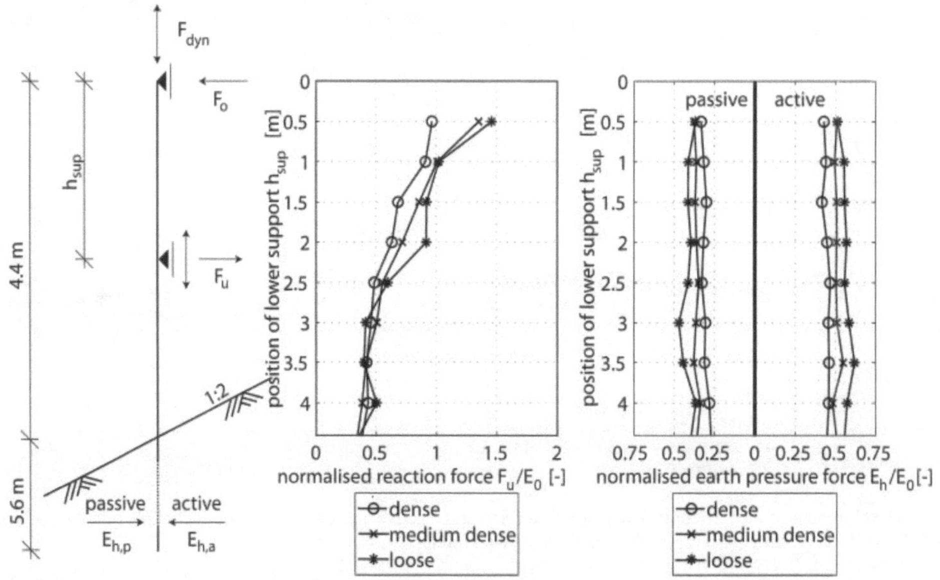

Figure 9. Left: cross section of the investigated embankment; middle: reaction force normalized with respect to the earth pressure force at rest; right: earth pressure force after the vibratory driving of the double pile normalized with respect to the earth pressure force at rest; 30 s of vibratory driving, dynamic force of $F_{dyn} = 1200$ kN, vibratory frequency of $f = 25$ Hz, bulk density of the soil and position of the lower support variable.

supports is, the bigger are the reaction forces to absorb the resulting moment acting on the double pile due to the earth pressure force.

A variation of the bulk density shows a significant influence on the reaction forces and the development of the earth pressure force (Fig. 9). The denser the soil is initially layered, the smaller are the reaction forces and the higher is the reduction of the earth pressure force at rest. The reason of this behavior is the deformation mechanism during the vibratory driving. The soil of the near-field is densified due to the cyclic shearing and the horizontal earth pressure force decreases. The soil outside the near-field is sliding into the compacted near-field and the earth pressure force is increasing little again. On the active side a wedge of soil on the surface is sliding against the double pile additionally resulting in a larger increase of the earth pressure force than on the passive side. The looser the soil is initially layered, the bigger is the potential of compaction and the more soil slides into the compacted near-field resulting in a higher increase of the earth pressure force.

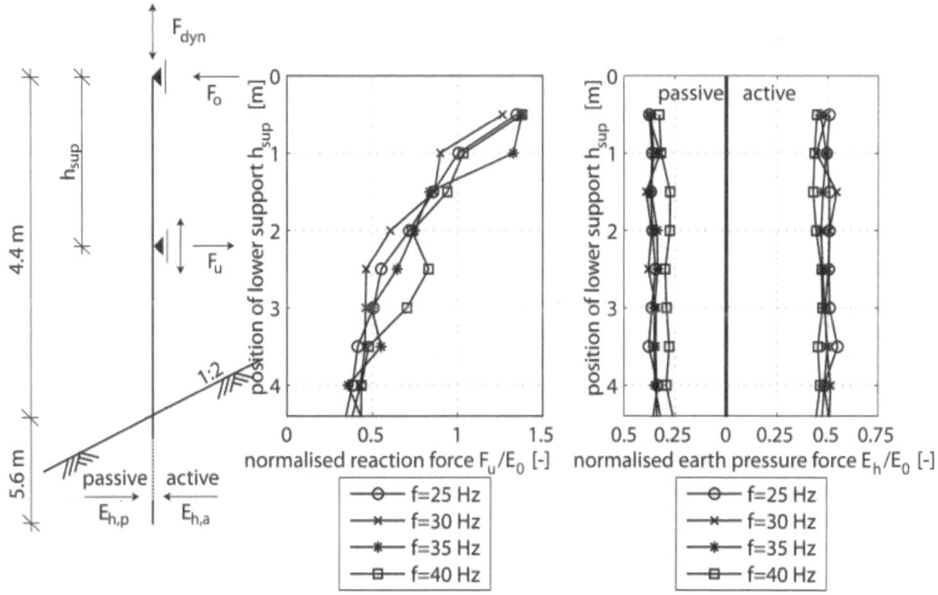

Figure 10. Left: cross section of the investigated embankment; middle: reaction force normalized with respect to the earth pressure force at rest; right: earth pressure force after the vibratory driving of the double pile normalized with respect to the earth pressure force at rest; 30 s of vibratory driving, dynamic force of $F_{dyn} = 1200$ kN, medium dense layered sand, vibratory frequency and position of the lower support variable.

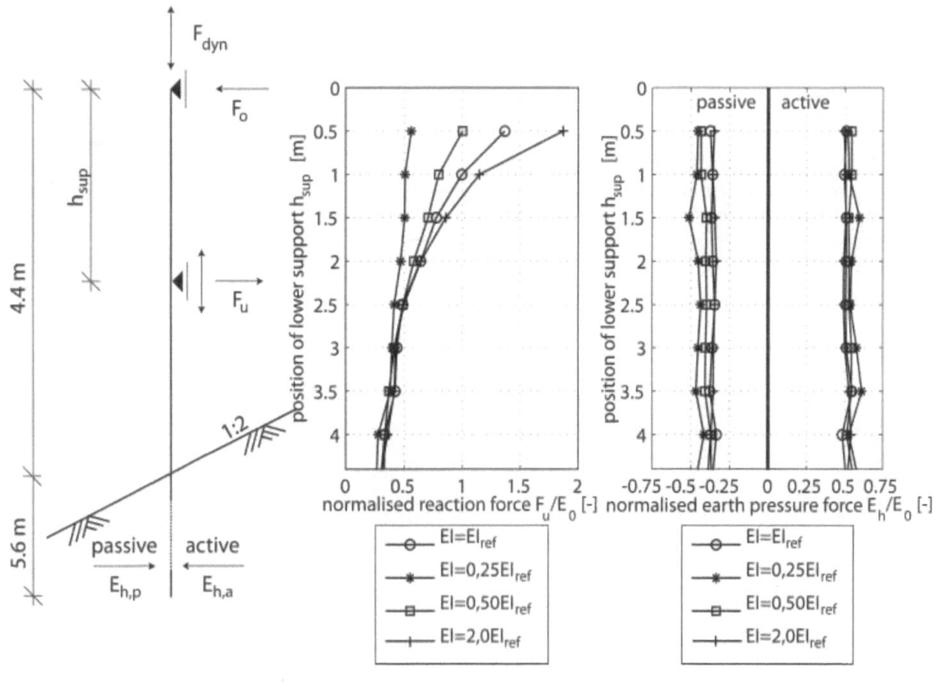

Figure 11. Left: cross section of the investigated embankment; middle: reaction force normalized with respect to the earth pressure force at rest; right: earth pressure force after the vibratory driving of the double pile normalized with respect to the earth pressure force at rest; 30 s of vibratory driving, dynamic force of $F_{dyn} = 1200$ kN, vibratory frequency of $f = 25$ Hz, medium dense layered sand, stiffness of the double pile and position of the lower support variable.

The vibratory frequency shows a marginal influence on the reduction of the earth pressure force and the reaction forces (Fig. 10). When the vibratory frequency is reduced, the shear waves, induced by the vibratory driving, and thus the area of soil compaction have a larger range. The larger the area of soil compaction is, the more soil can slide into the compacted near-field and thus lead to a higher increase of the earth pressure force again. When reducing the vibratory frequency, the initially reduced earth pressure force (caused by soil compaction) is later further increasing again than in the case with high frequency vibration.

The stiffness of the double pile indicates a distinct influence on the earth pressure force and the reaction forces: the lower the stiffness of the double pile is, the smaller is the reduction of the earth pressure force at rest and the reaction forces. This is caused by the horizontal deformation of the pile toe into the direction of the waterside. Reducing the stiffness of the pile, it needs less force to deform the pile. If soil slides into the compacted near-field of the active side and thus leads to an increase of the reduced earth pressure force, the toe of the pile deforms into the direction of the waterside reducing the earth pressure force again. The horizontal deformation of the pile toe enlarges the area of soil sliding into the compacted near-field of the active side with the result of a renewed increase of the earth pressure force. Simultaneously the horizontal deformations of the pile toe lead to a higher increase of the reduced earth pressure force at the passive side.

7 CONCLUSIONS

When driving piles in embankments locally limited instabilities of the embankment can occur due to the dynamic loading caused by vibratory driving or impact driving for example. Parts of the embankment can slide downwards against the pile to be installed. An imbalance of the horizontal earth pressure forces on the active and the passive side of the pile can develop resulting in horizontal deformations of the installed pile. Pile driving guides are usually used to prevent large deformations due to the installation process. By use of the finite element method a case of damage, arised during the construction of a new waterfront consisting of a sheet pile wall, is back-calculated. The installation process of a double pile in the embankment is simulated by investigating two limit cases: a free-riding vibratory driving and the use of a rigid pile driving guide. The calculated deformations of the double pile are compared to in situ measurements. In a parametric study different influences on the development of the earth pressure forces acting on the double pile and the reaction forces of a pile driving guide are investigated.

Back-calculating the case of damage the occurred deformation mechanism of the embankment can be reproduced qualitative correctly. By the use of numerical simulations the mechanical processes in the soil, like the development of the horizontal earth pressure due to the vibratory driving, can be explained. Further the measured deformations can be back-calculated quantitative correctly. The measured deformations are in the range of those obtained by the two investigated limit cases.

The parametric study shows that the magnitude of the reaction forces and the development of the earth pressure is strongly influenced by the initial bulk density of the soil, significantly influenced by the stiffness of the pile and marginal influenced by the vibratory frequency. A high vibratory frequency, a dense layered soil or a low stiffness of the pile cause the smallest reaction forces.

ACKNOWLEDGEMENT

The presented investigations were carried out in the framework of the research training group "Ports for Container Ships of Future Generations" funded by the Deutsche Forschungsgemeinschaft (DFG). The DFG is greatly acknowledged for funding this work.

REFERENCES

Buja, H.-O. (2001). *Handbuch des Spezialtiefbaus.* Düsseldorf, Germany: Werner Verlag.

Cudmani, R. (2001). *Statische, alternierende und dynamische Penetration in nicht-bindigen Böden.* Karlsruhe, Germany: Dissertation, Karlsruhe Institute of Technology, Institute of Soil Mechanics and Rock Mechanics.

Dassault Systemes (2009). *Analysis User-Manual. Abaqus 6.9.*

Gudehus, G. (1996). A comprehensive constitutive equation for granular materials. *Soils and Foundations 36(1)*, 1–12.

Henke, S. (2008). *Herstellungseinflüsse aus Pfahlrammung im Kaimauerbau.* Hamburg, Germany: Dissertation, Hamburg University of Technology, Institute of Geotechnical Engineering and Construction Management.

Holler, S. (2006). *Dynamisches Mehrphasenmodell mit hypoplastischer Materialformulierung der Feststoffphase.* Aachen, Germany: Dissertation, RWTH Aachen University, Lehrstuhl für Baudynamik und Baustatik.

Mahutka, K.-P. (2007). *Zur Verdichtung von rolligen Böden infolge dynamischer Pfahleinbringung und durch Oberflächenrüttler.* Hamburg, Germany: Dissertation, Hamburg University of Technology, Institute of Geotechnical Engineering and Construction Management.

Meijers, P. (2007). *Settlement during vibratory sheet piling.* Delft, Netherlands: Dissertation, Delft University of Technology.

Niemunis, A. & I. Herle (1996). Hypoplastic model for cohesionless soils with elastic strain range. *Mechanics of cohesive-frictional materials 2(4)*, 279–299.

Schümann, B. (2010). Modeling of soils as multi-phase materials with Abaqus. In *Proc. of SIMULIA Customer Conference 2010 at Providence/USA*, Providence, USA, pp. 384–399.

TESPA (2001). *Technical European Sheet Piling Association: Installation of Sheet Piles.* Luxembourg: Werner Verlag.

von Wolffersdorff, P.-A. (1996). A hypoplastic relation for granular material with a predefined limit state surface. *Mechanics of cohesive-frictional materials 1*, 251–271.

Baltic Piling – Mets & Raudsepp (Eds)
© 2013 Taylor & Francis Group, London, ISBN 978-0-415-64334-4

Study of bearing capacity of vibratory pile applying acceleration record

K. Kelevišius, J. Amšiejus & A. Norkus
Vilnius Gediminas Technical University, Vilnius, Lithuania

Z. Sikora
Gdansk University of Technology, Gdansk, Poland

ABSTRACT: The method for evaluation of ultimate bearing capacity for a vibratory pile is introduced. It employs test acceleration record and discrete model for evaluation of vibratory pile response The static laboratory test was performed for the installed pile. The Smith rheological model was chosen for determination of bearing capacity of the vibratory pile. A contact FE-interface is used for modeling interaction of the simulated pile bottom and the soil. This contact interface transfers only compression. The spring stiffness in the chosen rheological model is obtained by processing data of the static pile tests. The rheological model of pile installation approximates: I) vibratory pile values of damper, II) ultimate displacements (responses) of shaft and bottom. The measured and simulated accelerations are compared in time intervals for obtaining an agreement.

1 INTRODUCTION

The installation of piles using vibrodriveability techniques recently becomes an important issue in the geotechnical world. The customers expect that the installed vibratory piles sustain bearing capacity for the design loads, while the contactors aim for the smallest difference of pile settlements, not exceeding the limit values. The bearing capacity of installed pile can be verified by applying the below listed methods: static pile test (ASTM D1143/D1143M – 07e1), dynamic pile test (ASTM D4945 – 08, 2003) and the Standard Test Method for Axial Compressive Force Pulse Testing of Deep Foundations (ASTM D7383 – 10). The calculated bearing capacity on the vibratory pile model is determined by applying mathematical procedures for simulating vibrodriving process. The results of the calculated and measured vibratory pile accelerations are expected be in a certain agreement. The following rheological models can be used for the modeling: Randolph and Simons (Randolph, M.F. and Simons, H.A., 1986), Novak (Novak, 1978), Randolph and Worth (Randolph and Worth 1978), Holeyman (Holeyman, 1985), Nguyen (Nguyen, 1988), El-Naggar and Novak (see El-Naggar, M. and Novak, M., 1994), Deeks and Randolph (Deeks, A.J. and Randolph, M.F., 1995), Michaelides (Michaelides OD, 1997 and Michaelides OD., 1998). At present it is possible to determine the bearing capacity of vibratory pile by using, e.g. static pile test (ASTM D1143/D1143M – 07e1), dynamic pile test (ASTM D4945 – 08, 2003), axial compressive force pulse test (ASTM D7383 – 10) after when the vibratory pile is installed. Response measures during installation actually have not been employed for determining bearing capacity for vibratory piles. The aim of this paper is a study of a correlation between the records of a vibratory pile during installation process and the bearing capacity after installation.

2 STATE OF ART

The static pile test is an expensive and time consuming procedure when compared with the dynamic method or the axial compressive force pulse test. To perform the static test, one should install a reaction system or apply the ballast mass (total force of reaction system) being larger compared with the bearing capacity of pile. Note that the reaction system or the ballast is rather expensive technical tools to apply in practice. It requires using the relevant beam and hydraulic jacks in addition. The dynamic and axial compressive force pulse pile test methods are quite accurate and reliable. They do not require the expensive equipment, needed for static pile test as it was noted above. The inaccuracy in interpreting results of dynamic and axial compressive force pulse pile test methods can be significantly reduced by highly skilled and qualified engineer.

Due to the reasons briefly listed above one it is obvious that in the last decade many approaches for investigations of vibratory piles have been proposed.

3 MOTIVATION OF THE SELECTED RHEOLOGICAL MODEL

Having ascertained the quality of the rheological models available in the literature, one can indicate the Smith (Smith, 1960) model for modeling the ultimate base resistance of vibratory pile to be the suitable. However, on trial, if another rheological model would be applied, then any determination of shear modulus and Poisson ratio of soil by processing data of pile test records is difficult to obtain, since the soil strength parameters is very sensitive versus the identified soil properties by CPT.

4 DESCRIPTION OF THE PERFORMED TEST

The installation test of vibratory pile was performed in laboratory. The parameters of the vibratory hammer are as follows: total eccentric weight is 3.88 kg; eccentricity of weight is 0.029 m; common mass of vibratory hammer is 67.7 kg. The inserting pile without static surcharge has been performed with the 20 Hz frequency of rotating mass; that of for the piles loaded with 120 kg surcharge – with the frequency of rotating mass of 30 Hz. The closed end pile of the diameter 0.107 m and 1.8 m length have been employed for tests.

The internal friction angle of sand is 43°. The CPT test data: linear increment of cone resistance up to 1.2 m depth was met, reaching 7.0 MPa and further remaining constant per depth.

5 PREPARATION OF THE TEST

The soil was liquefied using the hydrodynamic forces. Then the soil was compacted by the deep vibrator, below the water level. The additional CPT at the place of pile test has been performed. The testing pile was equipped by the vibratory hammer of the testing mount, allowing the vertical displacements only.

The accelerations have been measured at the pile top with frequency of 2250 records per second.

The pile without surcharge was penetrated by vibration up to 80 cm depth. Subsequently, the test for pile was continued with 120 kg of surcharge the pile and finished at 1.40 m of depth.

The static test has been performed after pile installation. The pile has been loaded step-by-step with 7 kN load increment. The displacements of pile top have been measured at 3 points outside the pile, i.e. at point's located 120 degrees to each other.

The output of the measurements is the acceleration plot versus time. A typical record for the pile subjected to 20 Hz vibratory loading during the settlement process is given in Fig. 1.

Typical record for pile subjected to 20 Hz vibratory loadings, in case when settlement process was stopped, is given in Fig. 2.

Typical records for the pile subjected to 30 Hz vibratory loadings during the settlement process are given in Fig. 3.

Typical records for the pile subjected to 30 Hz vibratory loading and 120 kg static load surcharge when settlement process stopped, is given in Fig. 4.

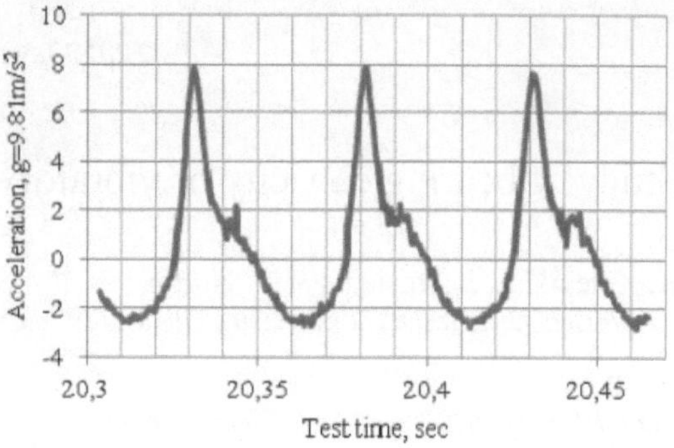

Figure 1. Typical values of accelerations for the pile subjected to 20 Hz vibratory loading during the settlement process.

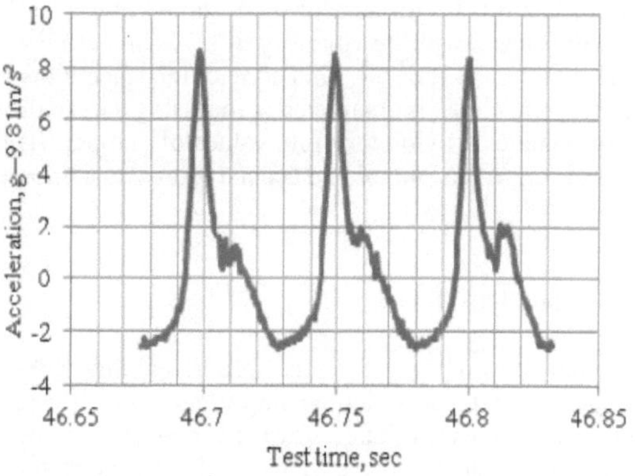

Figure 2. Typical values of accelerations for the pile subjected to 20 Hz vibratory loadings after settlement process has been stopped.

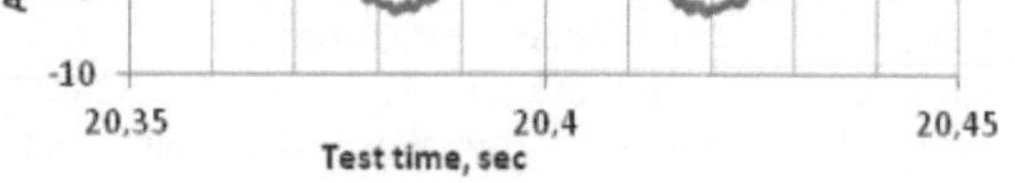

Figure 3. Typical values of accelerations, for pile with settlements of 30 Hz loadings and 120 kg static load surcharge during the settlement process.

6 STATIC PILE TEST DATA

The static test data is given in Fig. 5. The determined pile bearing capacity is equal to 48.9 kN corresponding the 10% relative pile diameter settlement.

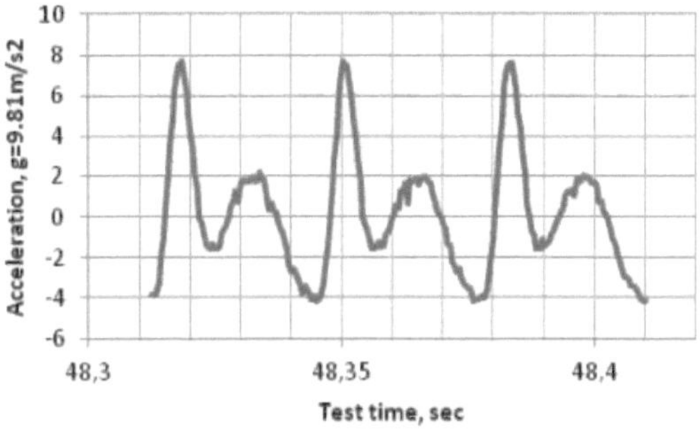

Figure 4. Typical records for the pile subjected to 30 Hz vibratory loadings and 120 kg static load surcharge after settlement process has been stopped.

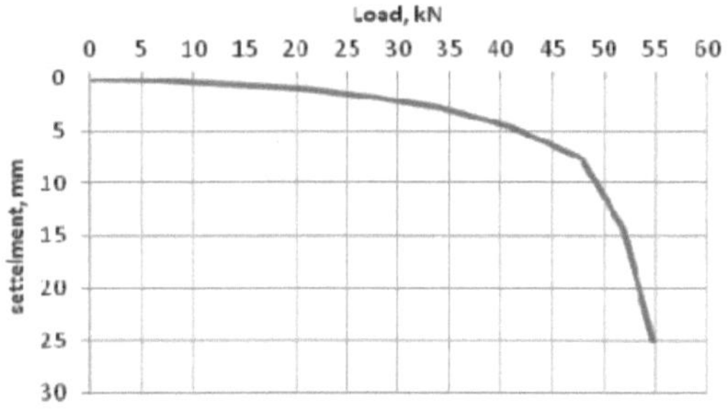

Figure 5. Pile load versus settlement.

7 PILE RESPONSE MODELING

The mathematical modeling of the installed vibratory pile has been performed using the Smith (Smith, 1960) rheological model. The original computer program was developed applying the MATLAB™ software.

The parameters for calculations procedures were chosen aiming to receive the best agreement between the calculated accelerations of the vibratory response and the measured data. The data processing interface is given in Fig. 6, the simulation results is given in Fig. 7.

The Newmark integration method for the time integration was employed at developed computer program. The computation procedure started at zero time point and was continued by a constant increment of the design load magnitude to reach the calculated eccentric vibratory and surcharge force magnitudes. The discrete model for vibratory pile following the Smith rheological model is as given in Fig. 8.

According to Cudmani (Cudmani, 2002) the interface element was introduced into the rheological model. The theoretical ultimate pile bearing capacity was chosen on the basis of static test results. The soil ultimate strength in calculations was determined from the static test results assuming that the ultimate bearing capacity corresponds to 10% of relative pile diameter settlement. The determined pile ultimate bearing capacity was 48.9 kN. The soil stiffness modulus (computed) was 4570 kN/m. In the model, by assumption, the stiffness of all spring-finite elements were analogous to the calculated pile-soil stiffness modulus.

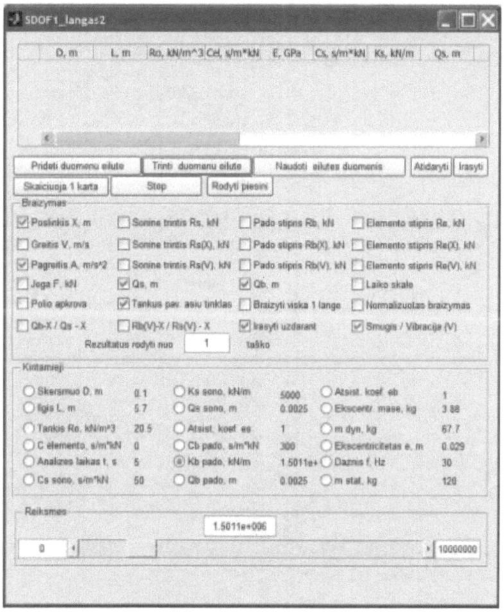

Figure 6. Data processing interface.

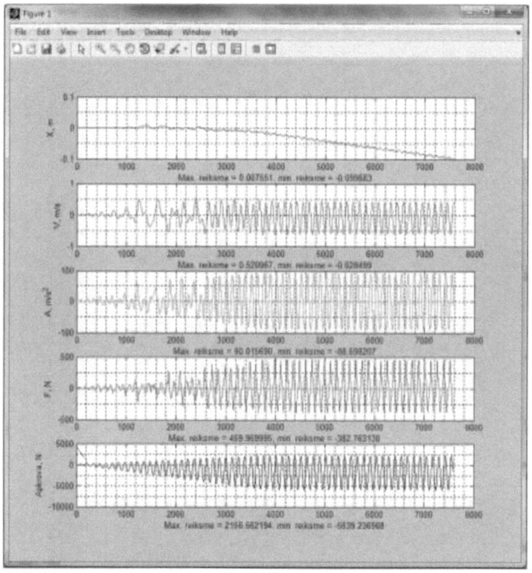

Figure 7. Result plot window.

The pile discrete model data are given in Fig. 9. The comparison of measured accelerations during the pile installation with the ones, obtained via mathematical modeling is given in Fig. 10.

The damping of shaft is 0.1 s/m*kN, and that of at bottom (1.4 m depth) – 15 s/m*kN. Therefore the stiffness of spring finite element was chosen as 4500 kN/m and 7 kN/m for the bottom and the shaft, respectively.

The friction at finite elements was activated at 2.5 mm quake at the pile shaft and at 10 mm quake at the pile base respectively.

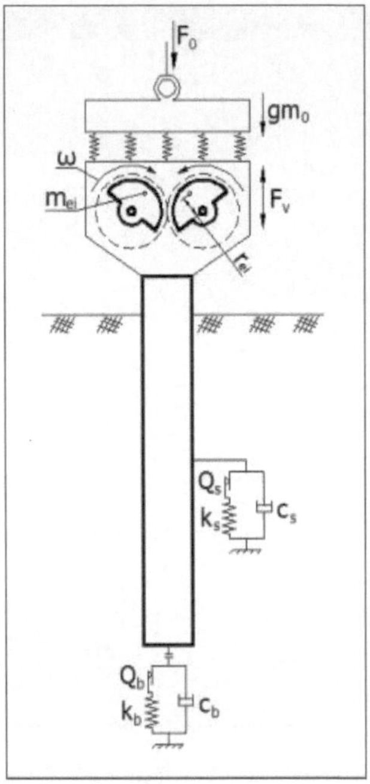

Figure 8. Discrete model for vibratory pile.

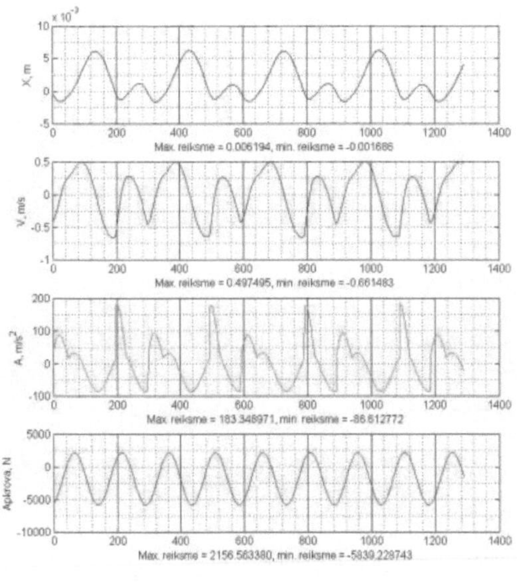

Figure 9. Data processing interface.

8 CONCLUSIONS

1) The computational results match with sufficient accuracy the test-measurements.
2) Aiming to reduce a discrepancy of acceleration curve (see Fig. 10) one should add Randolph and Simons (1986) developed rheological models, corresponding the vicinity of soil of the analyzed pile.
3) One should perform more tests of pile installation in order to evaluate bearing capacity via mathematical modeling. The field experiments for benchmarking analysis are also needed to obtain better accuracy.

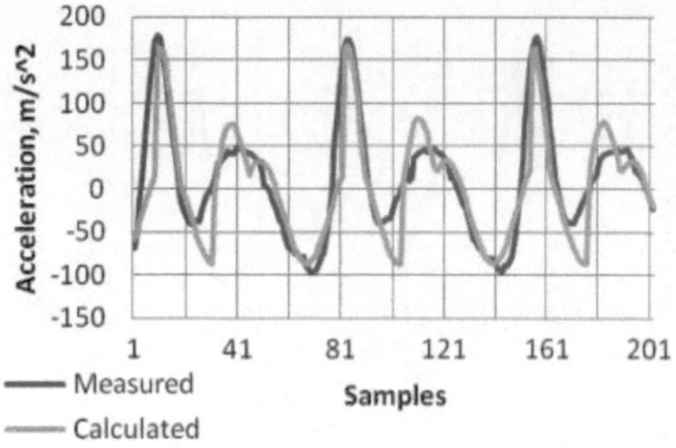

Figure 10. Comparison of the measured accelerations during installation with the ones obtained via mathematical modeling.

ACKNOWLEDGEMENTS

An equipment and infrastructure of Civil Engineering Center of Vilnius Gediminas Technical University was used for the investigation. Geotechnical laboratory facilities from the Department for Geotechnics, Geology and Maritime Eng. in Gdansk University of Technology were also used in these test-investigations.

REFERENCES

ASTM D1143/D1143M – 07e1. (n.d.). Standard Test Methods for Deep Foundations Under Static Axial Compressive Load.

ASTM D4945 – 08. (2003). Standard Test Method for High-Strain Dynamic Testing of Piles. Standard Test Method for High-Strain Dynamic Testing of Piles.

ASTM D7383 – 10. (n.d.). Standard Test Methods for Axial Compressive Force Pulse (Rapid) Testing of Deep Foundations. (10).

Deeks, A.J. and Randolph, M.F. (1995). A Simple Model for Inelastic Footing Response to. International Journal for Numerical and Analytical Methods in, 109(5), 307–329.

El-Naggar, M. a. (1994). Non-Linear Model for Dynamic Axial Pile Response. J. Geotech. Eng. Div., Am. Soc. Civ. Eng.; (United States), 120(2), 308–329.

Holeyman, A. E. (1985). Dynamic Non-Linear Skin Friction of Piles. International Symposium on Penetrability and Driveability of Piles, 1, p. 173–176. San Francisco.

Holeyman, A. E. (1988). Modeling of Dynamic Behaviour at the Pile Base. 3rd International Conference on the Application of Stress-Wave Theory to Piles, (p. 174–185). Vancouver, Canada.

Ishibashi, I. and Zhang, X. (1993). Unified Dynamic Shear Moduli and Damping Ratios of. Soils and Foundations, 33(1), 182–191.

Kondner, R. (1963). Hyperbolic stress-strain response: Cohesive soil. Soil Mechanics and Foundations Division, 189(1), p. 115–143.

Lysmer, J. a. (1966). Dynamic Response Of Footing To Vertical Loading. ASCE, 92(1), 65–91.

Michaelides OD, G. G. (1997). Approximate Non-Linear Dynamic Axial Response of Piles. Géotechnique, 48(1), 33–53.

Michaelides OD., a. B. (1998). Non-Linear Soil Properties. Soils and Foundations, 129–142.

Nguyen, T. T. (1988). A New Soil Model for Pile Driving and Driveability Analysis. Third International Conference on Application of, (p. 353–367). Vancouver, Canada.

Novak, M. a.-E. (1978). Dynamic Soil Reactions for Plane Strain. Engineering Mechanics Division, 104(4), 953–959.

Randolph M.F. and Simons H.A. (1986). An improved soil model for one dimensional pile driving analysis. 3rd International Conference of Numerical Methods in Offshore Piling, (p. 3–17). Nantes.

Randolph, M. (2003). Science and Empiricism in Pile Foundation Design. 43rd Rankine Lecture, Geotechnique, 54(1).

Randolph, M.F. and Simons, H.A. (1986). An improved soil model for one-dimensional pile. 3rd International Conference of Numerical, (p. 3–17). Nantes France.

Randolph, M.F. and Wroth, C.P. (1978). Analysis of deformation of vertically loaded piles. J. Geotech. Eng. Div., Am. Soc. Civ. Eng.; (United States), 104(GT12), 1–17.

Smith, E. (1960). Pile-Driving Analysis by the Wave Equation. Soil Mechanics and Foundations Division, pp. 35–61.

Baltic Piling – Mets & Raudsepp (Eds)
© 2013 Taylor & Francis Group, London, ISBN 978-0-415-64334-4

Post installation dynamic pile testing of 29 vibratory driven piles in the Neretva river delta

O. Mirković
Institut IGH d.d., Zagreb, Croatia

ABSTRACT: This paper presents a case study of post installation dynamic pile testing of 29 steel open end vibratory driven piles in the Neretva river delta, Republic of Croatia. The total of 46 dynamic load tests was made, as a number of piles have been tested two or three times, with the minimum of 11 day, and the maximum of 61 day interval between subsequent tests. A custom made 12 ton weight drop hammer was used to perform the tests. Besides from obtaining bearing capacity values, the important purpose of the project was to obtain the time dependence of capacity increase.

The tests were monitored by Pile Driving Analyzer (PDA), used to collect strain and acceleration measurements and to convert them to force and velocity curves. These curves were then analyzed with the signal matching software CAPWAP (Case Pile Wave Analysis Program).

1 INTRODUCTION

As a part of a motorway project in Croatia, a junction with two viaducts and one bridge was constructed in the Neretva river delta. The total of 29 piers is built on open-ended steel pipe piles (six for each pier), 40 meters long, with the diameter of 120 cm, which were vibratory driven into the ground. The soil in the alluvial river delta is mostly high and low plasticity silty clay.

The main scope of the pile testing project was to establish the time dependence of pile capacity change. It was expected that the long term capacity would be greater than the initial one, because of the soil strength change and the so called "set-up" phenomenon in clay soil.

Since the piles were vibratory driven, it was impossible to perform testing during driving, and all the tests were performed as restrike tests. Some piles have been tested two or three times, with the minimum of 11 day, and the maximum of 61 day interval between subsequent tests.

2 TEST PROCEDURE

Custom made 12-ton ram, with a specially designed steel construction for holding and leading the weight, was used for restrike testing. The weight mass was about 1.8–2.6% of the expected final pile bearing capacity. The free fall height of the ram is 0.5–3.0 m.

Four strain transducers and two accelerometers were attached to the test pile for measuring deformation and acceleration. The sensors were placed axially and symmetrically on 4 sides of the pile, approximately 1.2 m below the pile head. The data collected by the sensors were sent to the PDA device (Pile Driving Analyzer), for recording and processing.

Before the beginning of the project, the precise testing time-table was defined, in order to obtain the most relevant capacity time dependence curve.

All the tests were performed as restrike tests, with the minimum period of 6 days, and the maximum period of 81 days after installation. The minimum and the maximum time interval between two subsequent tests on the same pile were 11 and 61 days.

The total number of tested piles is 29 (one pile on each pier), 18 piles were tested once (with varying time intervals after installation), 5 piles were tested twice, and 6 piles were tested three times.

Permanent sets per each blow were measured independently with a leveling instrument, and were typically in the range of 1–5 mm.

3 TEST RESULTS

Upon carrying out the dynamic load testing, the obtained data are digitalized for computer analysis on a PC by way of the CAPWAP Program (Case Pile Wave Analysis Program), with wave equation as its basis. The pile and soil model is presented with one of the measured curves, whereas iteration (the change of pile and soil model) is used to find the best fit between the measured and computed curves. Once a satisfactory overlap of curves is achieved, the soil model properties and the ultimate bearing capacity of the tested pile are established.

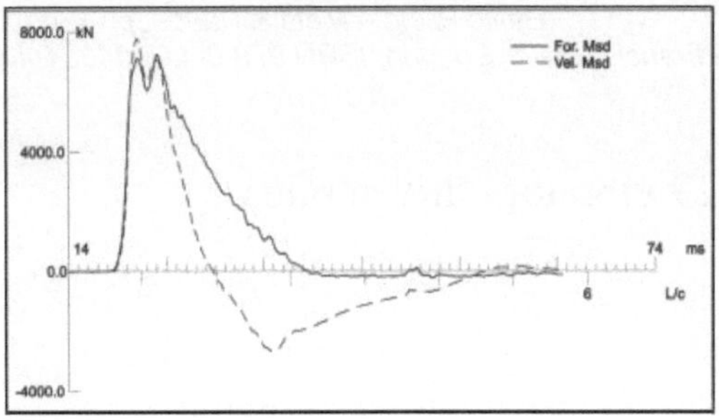

Figure 1a. Force and velocity curves for the pile tested 20 days after installation.

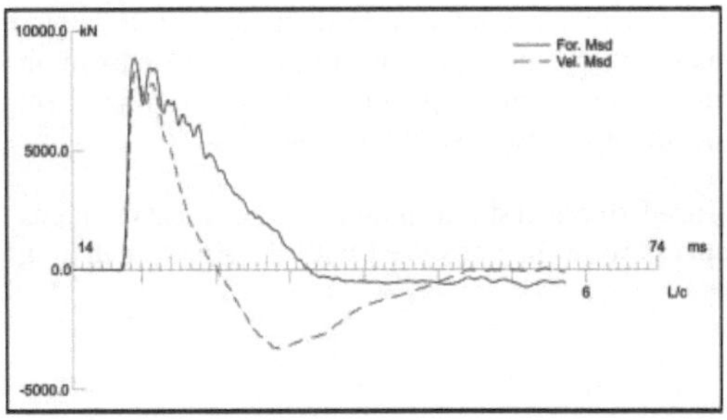

Figure 1b. Force and velocity curves for the pile tested 81 days after installation.

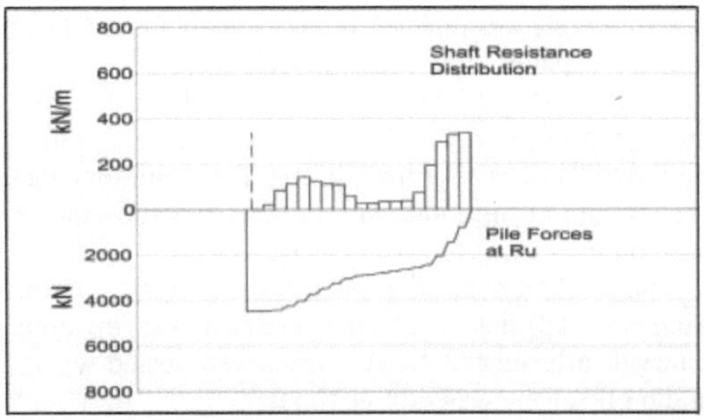

Figure 2a. Shaft resistance distribution for the pile tested 21 days after installation.

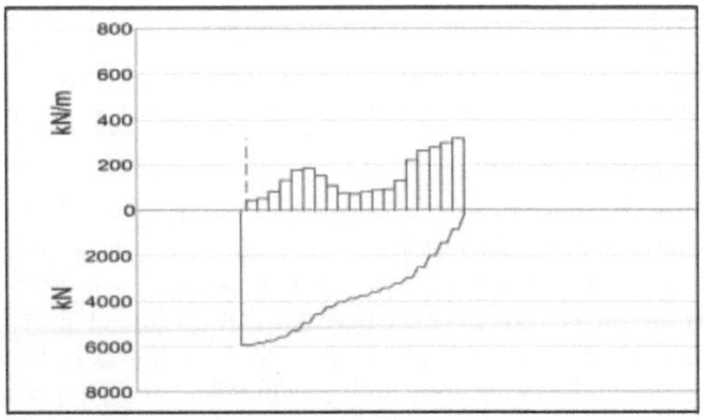

Figure 2b. Shaft resistance distribution for the pile tested 81 days after installation.

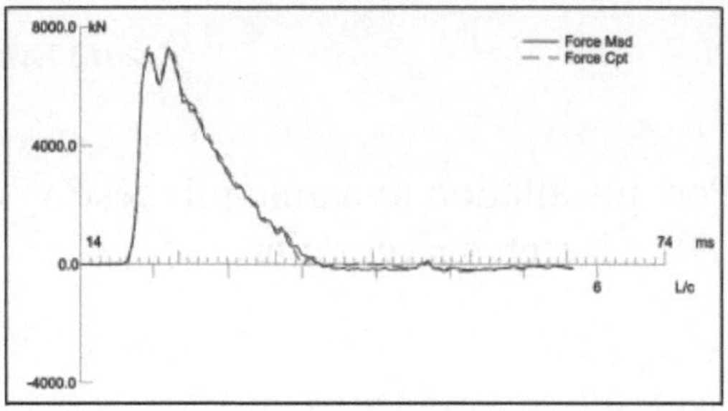

Figure 3a. Force curve match for the pile tested 21 days after installation.

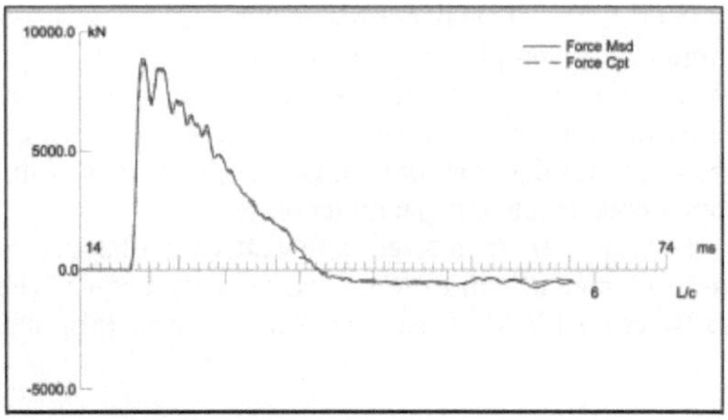

Figure 3b. Force curve match for the pile tested 81 days after installation.

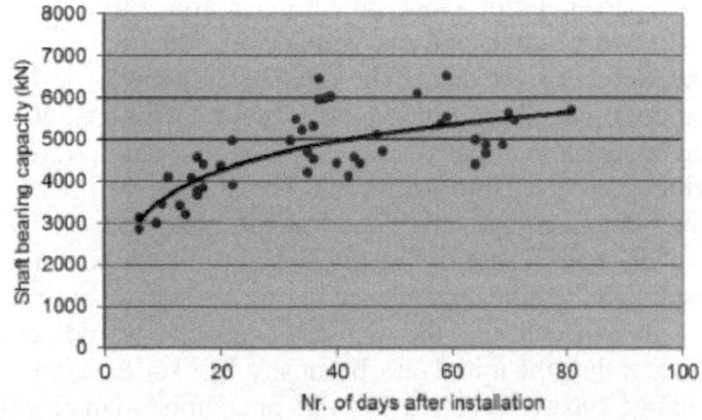

Figure 4. Time dependence of shaft bearing capacity.

Since the piles were open-ended pipes driven by vibrating, it was expected that the base bearing capacity would be relatively small compared to the shaft capacity. The design predicted rate of the base capacity was 10–20% of the total capacity. Test results for base capacities show even smaller rates, 3–8% of the total bearing capacity.

Typical examples of force and velocity curves, shaft resistance distribution and force curve match for two successive tests on the same pile are shown in Figures 1, 2 and 3. The first test was performed 20 days after pile installation (1a, 2a and 3a), and the second one 81 days after installation (1b, 2b and 3b).

After the processing of all the data, it was clear that the base, shaft and total bearing capacity increase with time, as expected. Since the soil charactions, and pile dimensions were the same, the results could be well compared to each other.

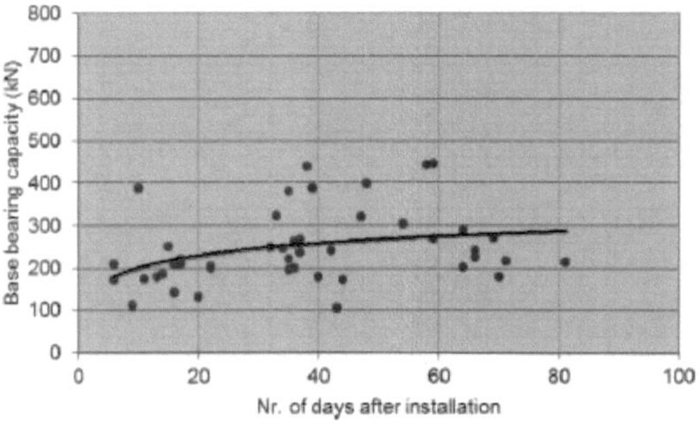

Figure 5. Time dependence of base bearing capacity.

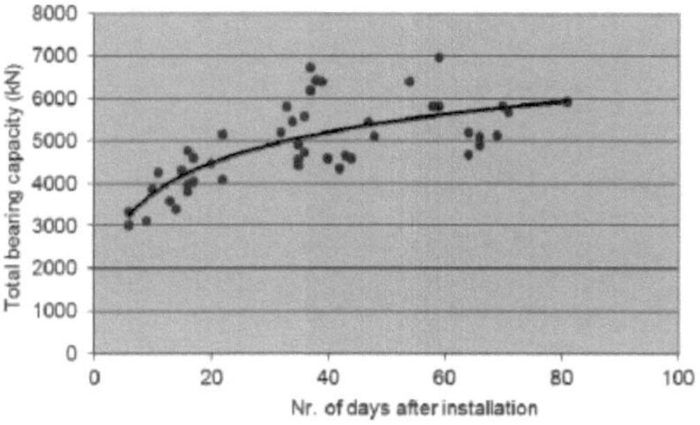

Figure 6. Time dependence of total bearing capacity.

Time dependence of the capacity increase is shown in Figures 4 (shaft bearing capacity), 5 (base bearing capacity) and 6 (total bearing capacity).The dots in the graphs present test results, and the curve is the function interpolation of the increase.

4 CONCLUSION

46 dynamic load tests were performed on 29 vibratory driven piles in the Neretva river delta in Croatia. Since the soil is mostly silty clay, it has been predicted by the design that the pile bearing capacity would increase in time due to set-up phenomenon.

Dynamic load tests were performed in various time intervals after pile installation and repeated two or three times for some piles, according to the predefined time-table.

The results show the capacity increase in time for both shaft and base bearing, and confirm the prediction that base bearing would be small (almost insignificant) compared to the shaft bearing.

REFERENCES

Bolanča, K., Ultimate resistance of vibratory driven piles in the Neretva river delta, *master theisis, 2009.*

Bosscher, P.J., Menclova, E., Russel, J.S. & Wahl, R.E., 1998. Estimating Bearing Capacity of Piles Installed with Vibratory Drivers, *Technical Report CPAR-GL-98-2, July 1998, U.S. Army Corps of Engineers, Washington, DC 20314-1000, USA.*

Kavur, B., Vrkljan, I. & Bolanča, K., 2006., Terenska CPTU i FVT ispitivanja u delti Neretve. *Priopcenja 4. Savjetovanja HGD-a, Opatija 2006., str. 205–214.*

Morgano, C.M., Whit e B.A. & Allin, R.C, 2008., Dynamic testing in sensitive & difficult soil conditions, *The 8th International Conference on the Application of Stress Wave Theory to Piles, Lisbon, 2008.*

Baltic Piling – Mets & Raudsepp (Eds)
© 2013 Taylor & Francis Group, London, ISBN 978-0-415-64334-4

Behavior models of plate-pile foundations under cyclic loading

I.T. Mirsayapov & M.I. Shakirov
Kazan State University of Architecture and Engineering, Kazan, Russia

ABSTRACT: The paper presents results of experimental studies of bearing capacity and deformation of base composite plate-pile foundation. According to test results obtained graphic dependences of the stresses and forces in the soil between the piles and piles.

At high load levels on the ground or adverse ground conditions, one way to increase the carrying capacity is the use of plate-pile foundations.

The study of plate-pile foundations in a large number of works. However, these studies are largely limited to the establishment of the basic theoretical patterns and features the work of plate-pile foundations under the short-term static loading.

However, buildings and structures and their foundations, along with static exposed to various types of cyclic loads, which in many cases are the main determinants security and integrity of the building elements.

The joint deformation of the "pile – slab grillage – ground between pile" with regard to redistribution of effort between the individual elements in the process of cyclic loading is practically unknown.

In connection with the above, it is necessary to study the behavior of plate-pile foundations under cyclic loading.

Experimental studies were conducted in a laboratory bulk tray with dimensions 1,0 × 1,0 × 1,0 m (Figure 1). As a foundation of grillage model was used reinforced concrete slab with dimensions of 400 × 400 × 40 mm, reinforced with a wire armature Ø3 Bp-I. Total were produced three experiments.

Figure 1. Skin test stand.

The piles were modeled by hollow plastic tubes with diameters of 7 mm, with a length of 400 mm and wall thickness of 1 mm. Deformation of the piles were determined by using strain gages glued along the length.

Installation of piles was carried out by the layer-dumping and compaction soil between pile.

Sandy loam soil was a semi-solid foundation (modulus of deformation E = 4,1 MPa, internal friction angle $\phi = 15°$, specific cohesion C = 3,3 kPa, the density $\rho = 14\,kN/m^2$, humidity W = 11%). The pressure in the soil was determined using an array of pressure sensors.

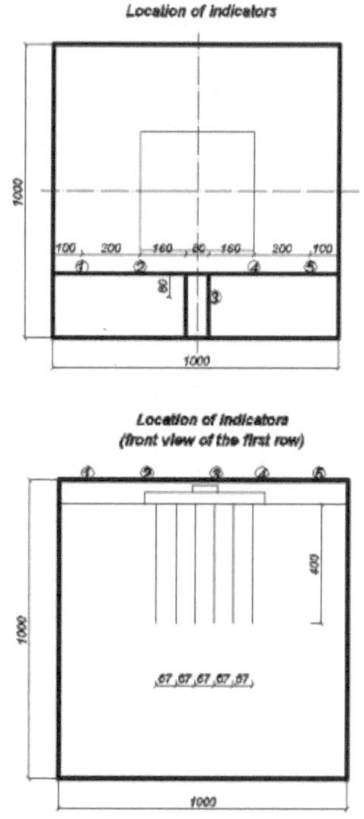

Figure 2. Arrangements of indicators.

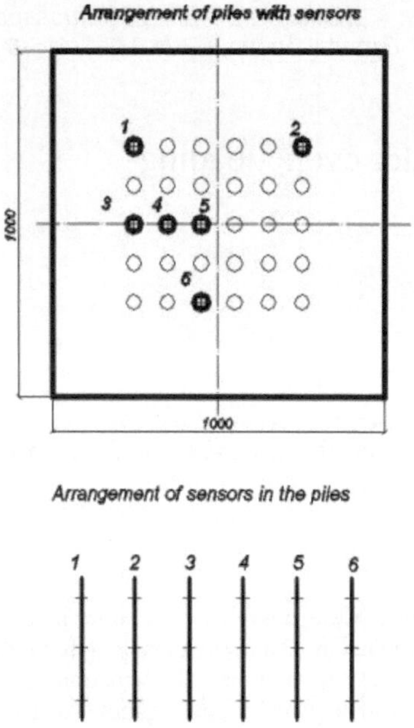

Figure 3. Schemes of arrangement of strain gauges in the piles.

The experimental setup of measuring devices to determine the ground surface deformation is shown in figure 2.

Schemes of plate-pile foundation and arrangement of strain in the piles and pressure sensors in the soil are shown in figures 3–4.

In carrying out experimental investigations were recorded moving slab foundation, vertical and horizontal displacement piles, deformation in the subgrade, in piles and grillage slab reinforcement.

Our studies have established the basic laws of the stress-strain state of the base of plate-pile foundation during cyclic loading.

Change efforts in the piles are located in specific zones of plate-pile foundation are shown in figures 5–7. As can be seen from the figures cyclic loading does not lead to an increase in effort in piles. Nature of the change effort in piles shows the redistribution of effort from the piles into the ground between piles.

Figures 8, 9 and 10 are graphs of the stresses in different zones of soil between piles. As can be seen from the figures, there is an increase of stresses in the ground in all areas of soil with an increase in the number of loading cycles. It should be noted that the greatest increase occurs under stress of plate grillage.

Cyclic loading caused increased sediment in the base plate within a grillage, and beyond, and the intensity of their development was dependent on the coordinates of the point. Figures 11–16 are graphs of increasing sediment basis, depending on the number of loading cycles. As can be seen from the graphs, the intensive development of the sediment, regardless of the origin points considered, the base occurred within

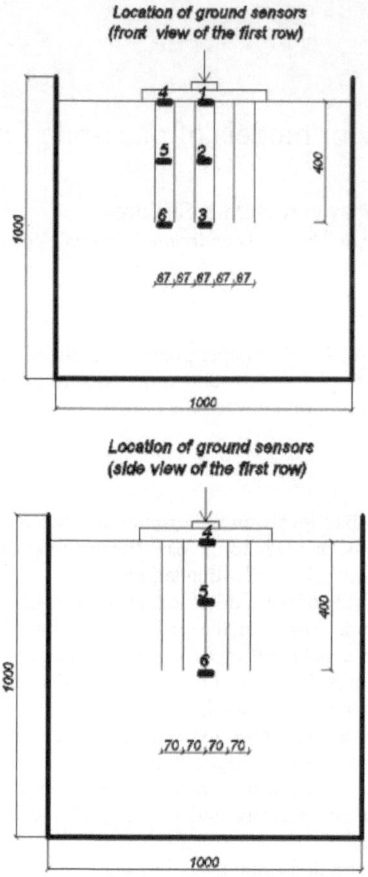

Figure 4. Arrangements of ground sensors.

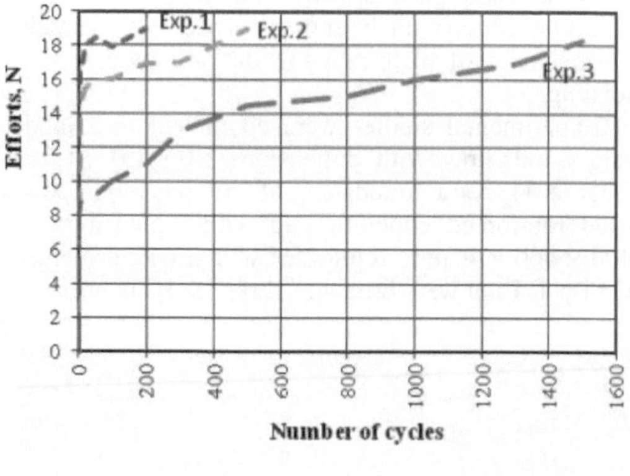

Figure 5. Change efforts in the pile 5 under cyclic loading.

the first 200 cycles of repeated loading. For example, at experiment number three, for points 1, 2, 3, 4, 5 full value of the base cake for the first 200 cycles with respect to the initial loading of the first increased by 417%, 395%, 377%, 381%, 411%, and subsequently for the entire period until the end of loading 121%, 117%, 131%, 119%, 124%.

This shows that after 200 cycles of repeated loading rate of increase sediment decreased significantly,

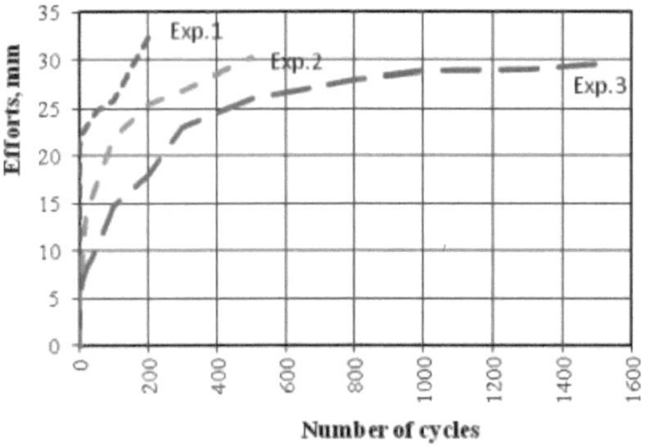

Figure 6. Change efforts in the pile 6 under cyclic loading.

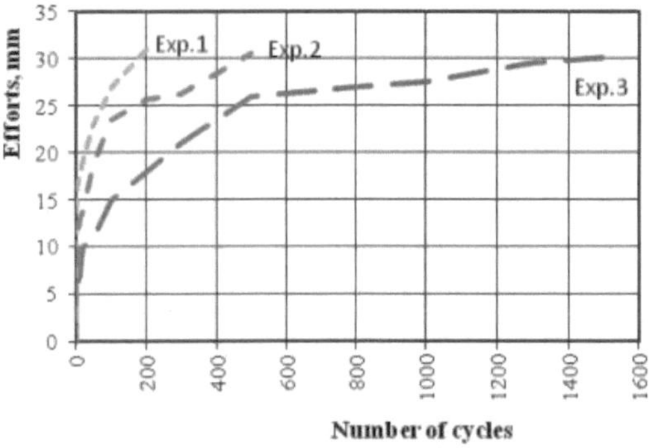

Figure 7. Change efforts in the pile 1 under cyclic loading.

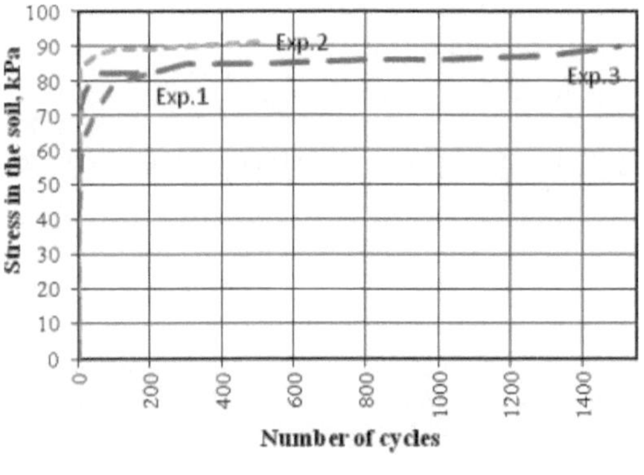

Figure 8. Changing stresses in the soil under cyclic loading at the bottom of the plane of pile.

but their complete stabilization is not observed. This pattern of sediment due to the fact that in general the changes in sediment depends on the development of ground deformation between piles under the grillage, the shear strain between the ground and piles of soil and deformations in the area under the lower end of the pile.

In the initial stages of deformation of the seal being implemented in almost all areas. The increment of strain compression after 200 cycles of loading almost stopped, causing a decrease in the intensity increase of the sediment.

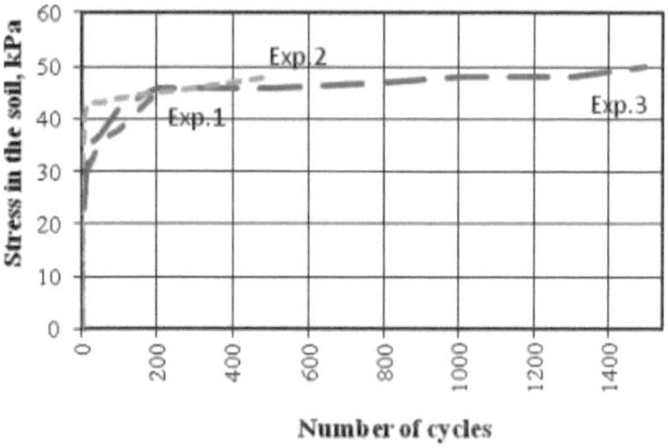

Figure 9. Changing stresses in the soil under cyclic loading at 20 cm from the bottom of the grillage.

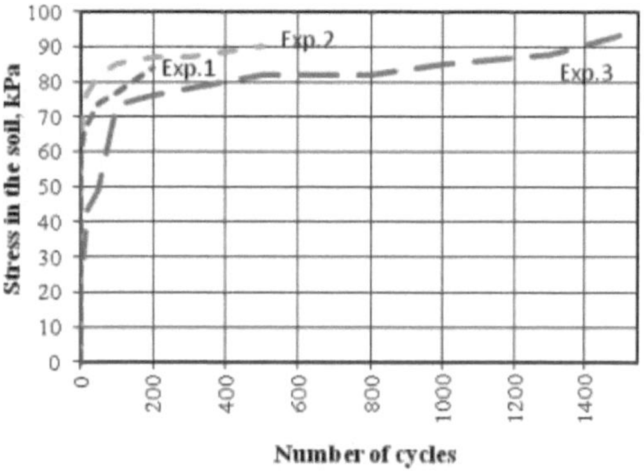

Figure 10. Changing stresses in the soil under cyclic loading at a depth of 2 cm from the bottom of the grillage.

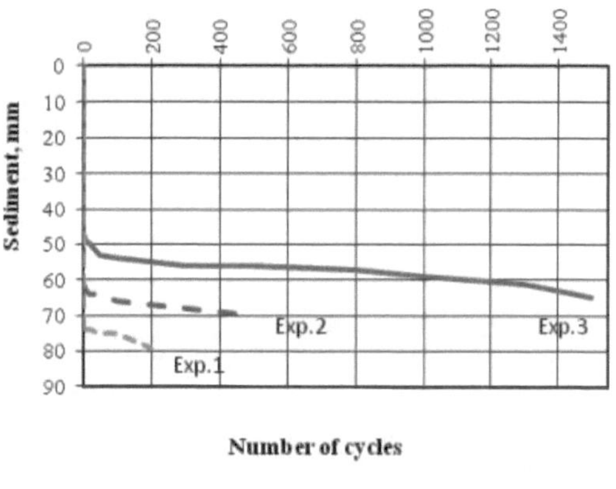

Figure 11. Sediment of base of plate-pile foundation with a maximum load.

Sediment base, measured in static loading step after re-loading different amounts of change similar to strains of soil between piles. As an example, in figures 11–16 shows the change in sediment within the base plate and grillage plate faces with stepwise static loading.

As can be seen from the figures 13–14, is a qualitative transformation of the diagram P – S, t. Is transition

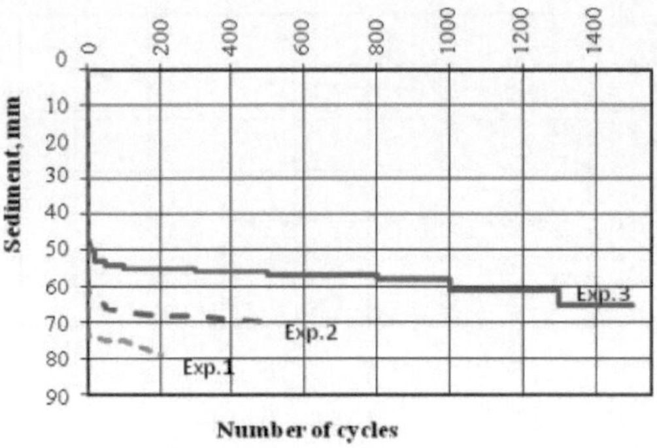

Figure 12. Sediment of base of plate-pile foundation with minimum values of load.

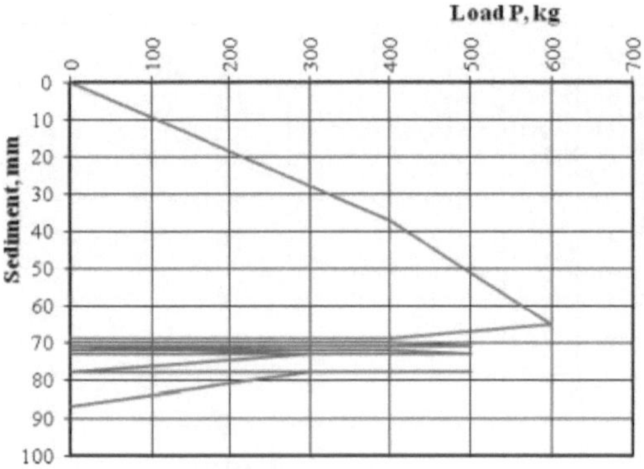

Figure 13. Sediment of base slab grillage foundation after 200 cycles of step loading (experiment 1).

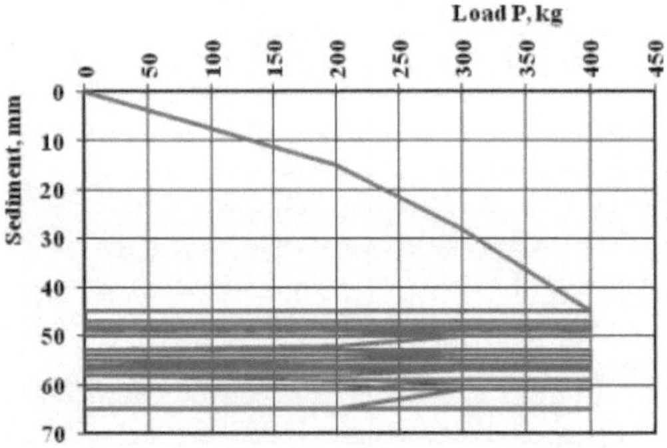

Figure 14. Sediment of base slab grillage foundation after 1500 cycles of step loading (experiment 3).

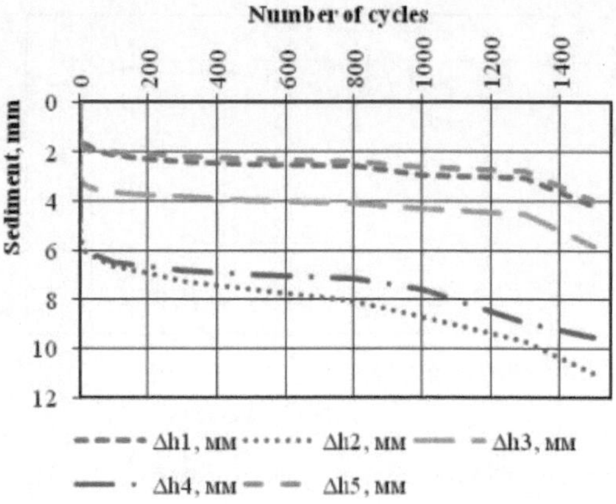

Figure 15. Changing the sediment at various points in the base at maximum load in a cycle from number of cycles (experiment 3).

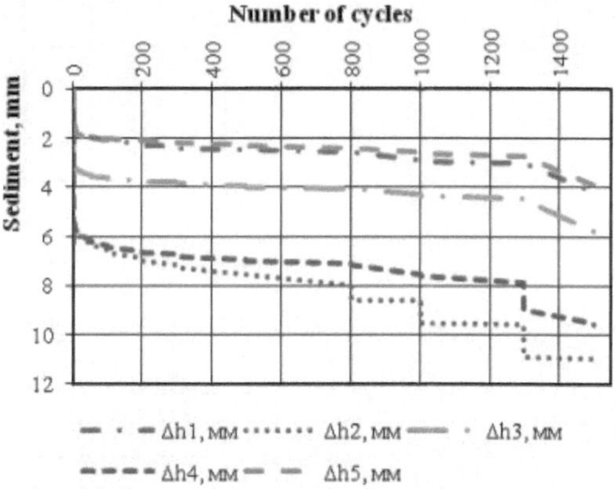

Figure 16. Sediment at the minimum loads of cycle (experiment 3).

from curved diagram in a straight, indicating that the transition of soil deformation in the linear stage.

Analysis of changes in sediments after step loading on bases with a different number of cycles shows that the increment of the sediment is mainly due to the increase of the residual part (figures 11–14).

The quantity of sediment during one cycle varies slightly. In testing these changes are fixed ("elastic") deposit with the increase in the number of loading cycles. During the first 20 cycles, these sediments are reduced. Reduction of "elastic" sediment can be explained by compaction due to the decrease of the pore volume of soil, which decrease in intensity faster than the shear modulus of shear strain between the piles and surrounding soil. Since the intensive soil compaction occurs during the initial period of repeated loading, and the change in coupling between the soil and piles is a lengthy process, then after 20 cycles of loading "elastic" sediments are beginning to increase. If the limit is not reached the state of the base, at the time of 1200 cycles of loading occurs relatively stable, ie dependence of F-S is nearly linear (figures 13–14).

The causes of change in the "elastic" and the residual sediment is likely to vary.

"Elastic" part of the sediment associated with the destruction of the structural links between the solid soil particles and a decrease in the modulus of cohesion between piles and surrounding soil, and the residual part – with the inelastic deformation of soil under the due soil vibro creep of plate grillage and the development of fatigue cracks in a plastically deformed local areas.

REFERENCES

E.A. Voznesensky. 1997. The behavior of soils under dynamic loads. Moscow: Moscow University Press.

Y.K. Zaretsky. 1989. Lectures on Modern Soil Mechanics. Rostov-on-Don.

A.L. Kryzhanovsky. 1983. Mechanical behavior of soils under conditions of spatial stress // Foundations, foundations, soil mechanics. In Number 1. P. 23–27.

I.T. Mirsayapov & I.V. Koroleva. 2009. The study of strength and deformability of clay soils with long-term triaxial compression // Proceedings of the Kazan State Architectural University, Number 2 (12). P. 167–172.

I.T. Mirsayapov & I.V. Koroleva. 2010. Features of the deformation of clayey soils under cyclic triaxial compression // International Journal of Geotechnics. In Number 6. P. 64–67.

Baltic Piling – Mets & Raudsepp (Eds)
© 2013 Taylor & Francis Group, London, ISBN 978-0-415-64334-4

Geosynthetic reinforced load transfer platforms for infrastructural projects – Actual design codes and verification by in-situ measurements

L. Vollmert & O. Doygun
BBG Bauberatung Geokunststoffe GmbH & Co. KG, Espelkamp, Germany

ABSTRACT: With the help of a parameter-study and a comparison of the results from an in-situ measurement, the arching effect within a fill embankment over pile foundations and the supporting of this arch through a geosynthetic reinforced load transfer platform were proved. The load distribution and the settlement behavior of the construction are proved to be enhanced with a geosynthetic load transfer platform. A parametric analysis has been performed comparing different calculation methods, especially for the German EBGEO and associated models. Modifications and additional methods such as considering the pile-soil-geosynthetics-interaction with the help of a finite element analysis, corrections for analytic bedding approach and load regime are discussed. It can be shown that slight modifications have significant influence and allow a further improvement of the EBGEO calculation method, giving hints for the best practice in design.

1 INTRODUCTION

The design of the geosynthetic load transfer platforms above pile foundations is regulated according to the new published German technical guideline EBGEO 2010 (Recommendation for the design and the calculation of soil bodies reinforced with geosynthetics). The mentioned construction above piles is made of a combined body from soil and geosynthetics. The design of this construction is based on the main principle that the geosynthetic reinforced soil construction bridges the subgrade with poor bearing capacity among the vertical piles and transfers the vertical loads directly to the vertical piles. The design is significantly influenced by the subgrade bedding and the chosen load figure as design input parameters. In order to make a safe and accurate assumption for the bedding modulus of the subgrade, it is important to take the pronounced interactions among the elements of pile-soil-geosynthetics into consideration. According to EBGEO 2010, the bedding modulus as an input parameter is to be calculated with the help of an average subgrade stiffness modulus depending on the underlying soil layer thicknesses. The scope of the work is to analyze the mentioned pile-soil-geosynthetics interactions on the basis of numerical calculations (Vollmert & Doygun 2011) and to investigate the influence of the chosen two load figures based on in-situ investigation results from a realized structural project (Vollmert et al. 2006), with the help of which a better understanding of the subgrade bedding influence on the design is achieved.

2 PARAMETER STUDY

2.1 *Finite-element model (FEM)*

The detailed boundary conditions of the performed finite-element analysis were given in (Vollmert & Doygun 2011). The input parameters of pile-, subgrade- and the geogrid-stiffness modulus were varied. The bedding modulus of the subgrade has been calculated with the help of vertical stress and settlement values which were determined from the axial symmetrical Plaxis model analysis as follows:

$$k_s = \frac{Vertical\ stress\ (\sigma_0)}{Settlement\ (s)} \tag{1}$$

To be on the safer side, the maximum settlements between the vertical piles and the average vertical stress on the geogrid layer have been taken into account by the determination of bedding modulus as illustrated in Figure 1.

2.2 *Results of the parameter study*

The results of the parameter study have shown that the subgrade bedding effect increases with an increase in pile stiffness under constant geogrid stiffness for all investigated soil types as illustrated in Figure 2. This mentioned increase is more pronounced for the pile stiffness values between 50 MN/m² and 26000 MN/m². For the pile stiffness values greater than 26000 MN/m² the growth in advantageous

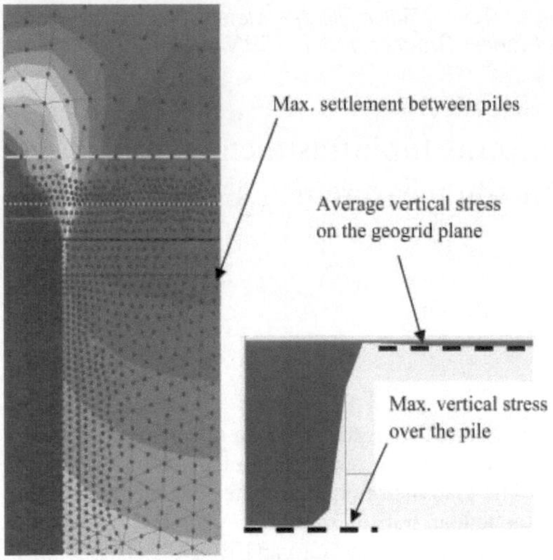

Figure 1. Settlement and vertical stress paths from FEM.

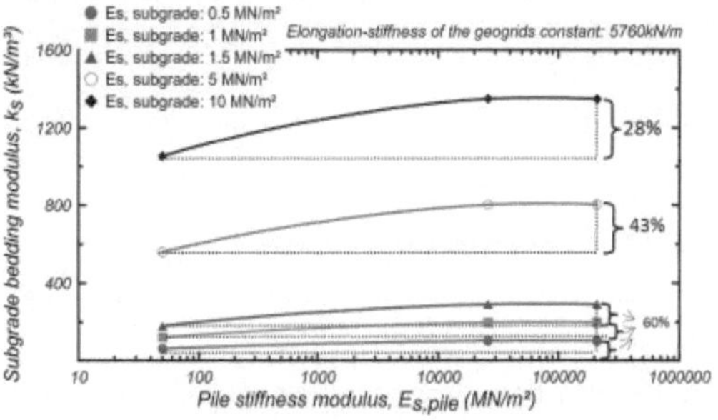

Figure 2. Effect of pile stiffness on the bedding modulus depending on the subgrade stiffness.

bedding effect is nearly constant. In other words, under mentioned constant boundary conditions, the pile stiffness plays a more essential role up to a threshold value, whereas this role is afterwards negligible. This threshold value is to be searched among the corresponding pile stiffness values greater than those which are attained by usual soil improvement techniques without any use of concrete.

The effect of the pile stiffness for smaller values than the threshold value of $E_{s,pile} = 26000\ MN/m^2$ indicates different tendencies depending on the subgrade bearing capacity. For the soil types with a poor bearing capacity ($E_{s,subsoil}$: 0.5–1.5 MN/m^2), the mentioned growth in the bedding effect is about 60%, whereas this is smaller for the soil types with a better bearing capacity (The increase in the bedding effect for $E_{s,subsoil}$: 5 MN/m^2 is approximately 43% and for $E_{s,subsoil}$: 10 MN/m^2 is about 28%). In other words, this tendency indicates the fact that the importance of the soil-geosynthetics-pile interaction with an increase in pile stiffness under other constant boundary conditions is greater for the soil types with poorer bearing capacity (e.g., peat, $E_{s,subsoil}$: 0.5–1.5 MN/m^2) in comparison to the soil types with a relatively better bearing

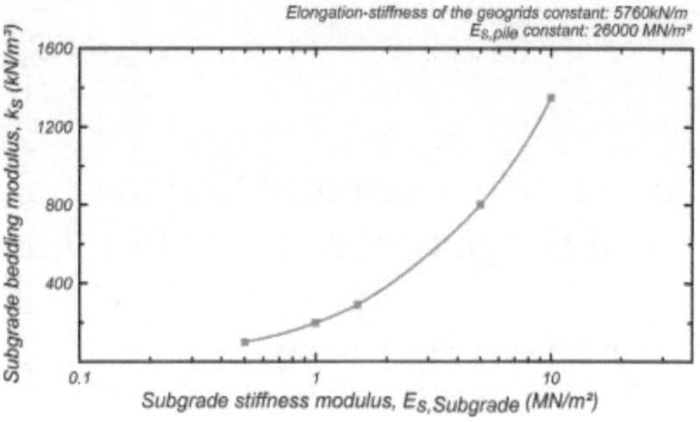

Figure 3. Effect of subgrade stiffness on the bedding modulus.

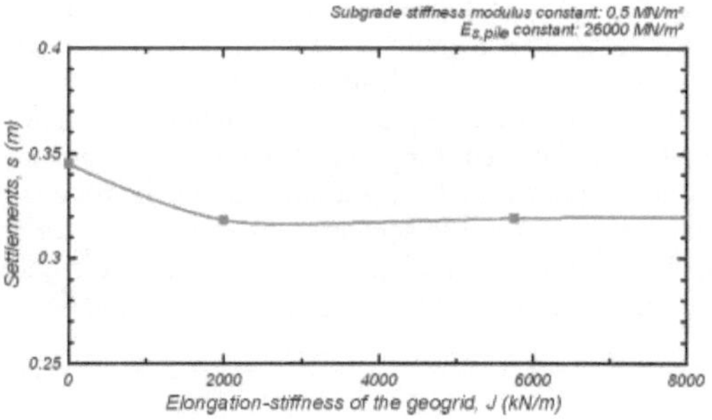

Figure 4. Effect of geogrid's stiffness on the settlements.

capacity (e.g., clay (CL), $E_{s,subsoil}$: 5–10 MN/m^2). The correlation based on the effect of subgrade stiffness modulus is illustrated in Figure 3 again under other mentioned constant boundary conditions.

The research results of the soil-geosynthetics interaction have also exhibited (see Figure 4) that the maximum settlements below the geosynthetic load transform platform go down under constant subgrade stiffness- and pile stiffness modulus with an increase in geosynthetics' stiffness. This decrease in settlements was apparently more pronounced for the geosynthetics stiffness range of between 0–2000 kN/m. It means, with constant subgrade stiffness and pile stiffness, the advantageous soil-geogrid interaction which is mobilized through an increase in geogrid stiffness can lead to a reduction in settlements down to a certain amount and a compensation of differential settlements, whereas the settlements cannot be avoided completely.

As a consequence of the performed parameter study, it is important to note that the subgrade bedding can be determined with the help of FEM more properly due to its advantages such as being able to consider the influence of the arching effect of the load transfer platform or the pile skin friction on the bedding. A design based on EBGEO can be combined with FEM in terms of a more realistic subgrade bedding approximation.

3 ACTUAL DESIGN CODES

3.1 *Overview of the calculation methods*

The first calculation method was developed by (Kempfert et al. 1997). In this method, the bedding

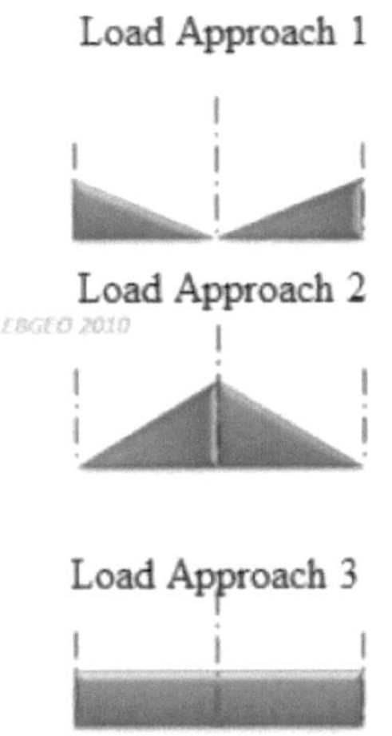

Figure 5. Load approaches.

of the subgrade was considered through its bearing capacity. In this method, a uniformly distributed load was taken into consideration as shown in Figure 5 (Load approach 3).

Following the mentioned Kempfert calculation method, Zaeske (2001) developed a design method based on the publication of Emde (1995) where the subsoil bedding could be considered via catenary differential equations for the geosynthetics between the piles. With this method, it is possible to consider not only the deformation below the geosynthetics but also the pile settlements. This method allows all three load approaches as illustrated in Figure 5. Zaeske (2001) has mainly analyzed the load approach 3 and developed an analytical solution method for this one. He also modified the design into a simplified graphical form, which is the basis for the EBGEO (2010) calculation method, where the load figure is assumed as "Load Approach 2" and the piles will be considered as rigid elements in comparison to the soft bedding behavior of the subsoil. In other words, after fulfilling the mentioned bedding requirement, the deformation of piles will be ignored and the design will be focused on the deformation behavior of the geogrids between the piles, which enables a design method on the safe site. This simplified graphical calculation method can be used according to EBGEO after fulfilling a bedding requirement that the bedding modulus of the piles has to be 75-times greater than the subsoil bedding modulus.

After Zaeske (2001), Geduhn & Vollmert (2005) developed and simplified the Zaeske Method (2001) into a new form, called "Zaeske, simplified", where the piles are considered as rigid elements analogous to EBGEO calculation method. In addition, the "Zaeske, simplified" calculation method enables to consider all three load approaches (see Figure 5), which were already described by Zaeske (2001) and are commented on by van Eekelen (2011). A brief comparison of the mentioned calculation methods is summarized and illustrated in Table 1.

Table 1. The comparison of the calculation methods.

Method	Pile stiffness	Load function	Calculation
Kempfert (1997)	not considered (rigid)		from free hanging catenary
Zaeske (2001)	considered		from DGL for bedded system
Zaeske, simplified (2005)	not considered (rigid)	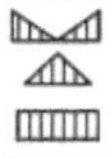	from DGL for bedded system
EBGEO (2010)	not considered (rigid)		from DGL for bedded system

Heitz (2006) extended the calculation method of Zaeske (2001) in terms of cyclic loading. Based on the results of the performed model tests, the soil arching effect of the geogrids tends to be reduced due to cyclic loading. This reduction can be considered by a calculation method which was developed in the mentioned publication of Heitz (2006).

The effect of the chosen calculation method (Tab. 1) has a big influence on the design results. The research of van Eekelen (2011) indicates that the shape of a deformed geosynthetic reinforcement between the piles does not follow a power law function lower than the third order in any of the performed tests. The inverse triangular load (load approach 1) also gives a third order power law function, which demonstrates the most similarity (among the other load approaches) to the real deformation form of the geosynthetic reinforcement according to the test results. For the same geometry considered in the publication of van Eekelen (2011), the same total load with the inverse triangular form (load approach 1) results in a tensile force in the geosynthetics that is 75% of the tensile forces calculated by a triangular load distribution (load approach 2), which shows the big importance of the load distribution choice on the design. To put in a nutshell, the load approach 1 leads to a minimum stress on the geosynthetics and a maximum stress on the piles and the load approach 2 shows exactly the opposite tendency, whereas the load approach 3 ends up somewhere in the middle. These mentioned tendencies were also confirmed by the parameter study of (Vollmert et al. 2012).

3.2 *The modified bedding model*

Besides the load approaches discussed above, the influence of the bedding model on the calculation results is very significant. (Vollmert et al. 2012) introduced a modified bedding model where a bedding adjustment was undertaken. This adjustment bases on the following realistic assumption: According to the Zaeske (2001) calculation method, the resulting total load on the geosynthetics area (A_L) is to be considered in a redistributed form on a reinforcement strip with a width of (b) and a clear length of (L_W) between the piles. However, there is still a need of an analogous adjustment for the bedding modulus as well for the reason that the resulting settlements between the piles have to remain the same within the same calculation model. Therefore, the subsoil bedding modulus is to be corrected as follows:

$$Settlements = \frac{\frac{\sigma_o * A_L}{L_W * b}}{k_{s,corrected}} = \frac{\sigma_o}{k_s} \tag{2}$$

$$k_{s,corrected} = k_s * A_L/(L_W * b) \tag{3}$$

with σ_o: resulting normal stress on the geosynthetics.

After this minor adjustment, the bedding modulus is considered to be corrected by taking the settlement equality on the reinforcement strip between the piles into account. In the following chapter, the appropriateness of introduced bedding determination through FEM (see 2.1) combined with the bedding correction (see (3)) will be investigated by considering two loading approaches (LA1 and LA2, see Figure 5) with the help of a case study.

4 CASE STUDY – SCHLEUSE NEUER HAFEN/BREMERHAVEN/GERMANY

(Vollmert & Doygun 2011) calculated the bedding modulus of a realized project as an example in the same fashion to the parameter study (see 2.1) with the help of the FEM program Plaxis. The prognoses resulting from this calculation were compared with the in situ-measurements for the project, called "Schleuse Neuer Hafen, Bremerhaven" in Germany. The considered boundary conditions and soil parameters are given in (Vollmert et al. 2006) and (Vollmert & Doygun 2011) more in detail. Further to the mentioned research, the effect of the described bedding correction and the influence of the load approach 1 (LA1) have been investigated with new back-calculations and the results are illustrated in the following figures. In order to assess the results, the degree of utilization, the elongation of the geogrids, the bedding modulus on the reinforcement plane and the maximum settlement below the load transfer platform have been chosen as comparison criteria. By the determination of the degree of utilization, partial safety factors were taken into consideration. To make a reasonable comparison, the utilization degree for the in-situ measurements was also calculated by taking reduction factors for installation damages, creep and a partial safety factor of γ_M based on DIN 1054:2010 and EBGEO into account to determine the tensile strength of the geogrid. All the rest of comparison parameters (strains, settlements, bedding modulus) and the corresponding in-situ measurements were calculated by the characteristic values.

The results illustrated in Figure 6 indicate that the recalculated design with EBGEO 2010 (LA2) and

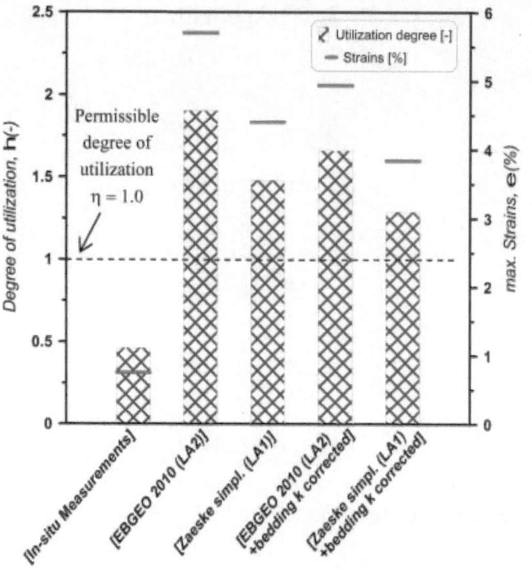

Figure 6. Effect of load approach & bedding correction.

"Zaeske simplified" (LA1) for the given boundary conditions provides significantly deviating prognoses in terms of the utilization degree and the strains in comparison to the in-situ measurements. Nevertheless, the load approach 1 improves the results compared to load approach 2 a bit and the correction of the bedding leads to a slight enhancement of the results, which shows a good agreement with the tendencies determined by (van Eekelen et al. 2011) and (Vollmert et al. 2012). Yet, the results clearly show that the calculation of the subsoil bedding according to the mean stiffness modulus of the subgrade layers depending on their thicknesses as described in EBGEO 2010 (without paying attention to the pile-soil-geosynthetics-interaction) leads to no satisfactory results. Due to this determined fact that the mentioned way of calculation of the bedding modulus results in an underestimating of the real subsoil bedding, a further investigation has been carried out based on the settlements and bedding modulus values as shown in Figure 7.

As can be seen in Figure 7, the described way of bedding modulus determination according to EBGEO 2010 results in a divergence factor of 14 in comparison to the in-situ measurements. A better approximation was obtained by taking the pile-soil-interaction into consideration with the help of finite element method. The maximum in-situ settlements were about 2.3 cm, which shows a good agreement with the results of the back-calculation based on the bedding modulus determined from axial-symmetric FE analysis with a settlement of 3.5 cm.

It is observed in Figure 8 that both the EBGEO calculation method (LA2) and Zaeske simplified method (LA1) combined with the bedding modulus from FEM allow a better approximation in terms of strain and utilization degree compared to the results shown in

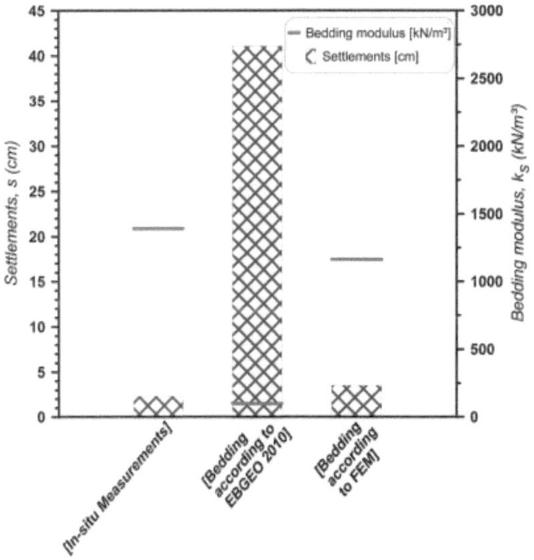

Figure 7. Effect of the way of bedding modulus determination.

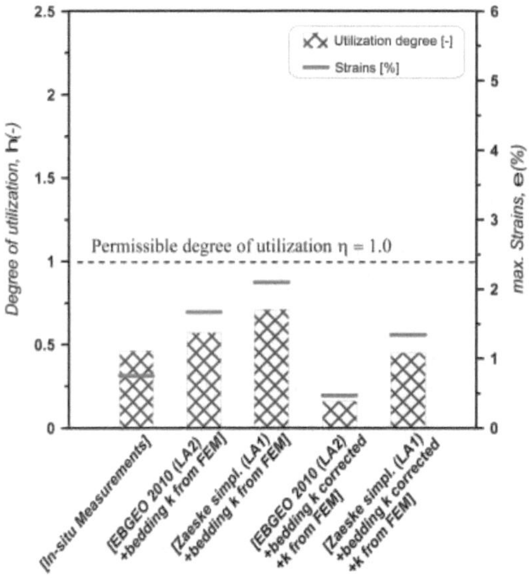

Figure 8. Effect of load approach, FEM & bedding correction.

Figure 6. On the other hand, the EBGEO calculation method (LA2) combined with the bedding modulus from FEM and having a bedding correction according to (3.2) leads to an underestimation of the strain and the utilization degree, whereas strains from the Zaeske simplified method (LA1) indicate a good agreement with the measured strains without an underestimation of the utilization degree.

Another noteworthy point to be mentioned about Figure 8 is the contrast in tendencies between the two applied load approaches. Contrary to the previously observed tendency (see Figure 6), the load approach 1 leads to a larger degree of utilization and greater strains

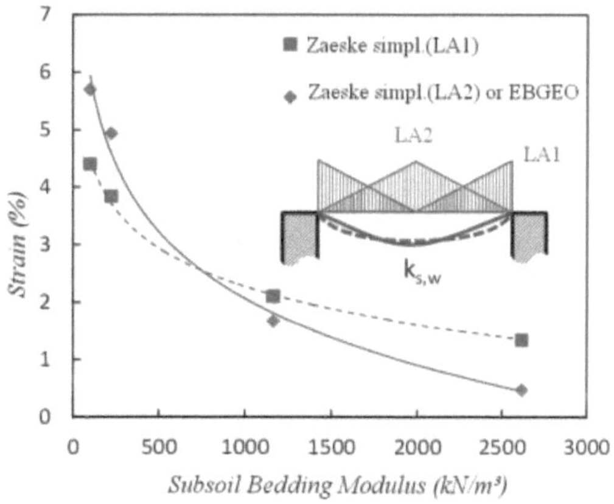

Figure 9. Effect of subsoil bedding on two load approaches.

compared to load approach 2. It is evident from these two figures (Figure 6 & Figure 8) that the reason for this contra tendency is attributed to the large difference between the applied subsoil bedding modulus values. In order to clarify this phenomenon, the effect of the subsoil bedding on the mentioned two loading figures has been investigated with the help of Figure 9.

It is apparent in Figure 9 that load approach 2 leads to larger deformations below a threshold value for the subsoil bedding modulus. Beyond this threshold value (e.g. subsoil types with a better bearing capacity), load approach 1 leads to larger strains. It is assumed that the reason for this is dependent on the supposed location of the load maximum. For the soil types with poor bearing capacity (small bedding modulus), a load maximum in the middle (LA2) results in larger settlements in comparison to load approach 1 (LA1), where the maximum loading takes place in the vicinity of the pile (under a good supporting effect of the pile compared to the subsoil in the middle). In contrast to this, for the soil types with a good bearing capacity (large bedding modulus), a load maximum in the middle (LA2) takes more advantage out of the good subsoil-bedding with much smaller settlements in total compared to load approach 1, where the maximum loading in the vicinity of the pile still chooses the subsoil as decisive bearing element and leads to a punching-like effect at the pile edge and results in larger strains on the geosynthetics. In summary, depending on the subsoil type, LA1 can case-specifically cause a greater utilization degree on geosynthetics than LA2 can do.

5 CONCLUSIONS

A realistic estimation of bedding modulus and the choice of the decisive load approach play the key role to make safe and still economic designs for geosynthetic reinforced load transfer platforms. Among the available design methods, the calculations following the current EBGEO code without any correction but considering the pile-soil-geosynthetics-interaction with the help of a finite element analysis leads to satisfactory results.

Yet, it has certain restrictions. One restriction of EBGEO method is that it solely considers the most unfavorable load approach (LA2). In contrary to this, the simplified Zaeske method (2005), on which EBGEO is based, enables all three load approaches known in the literature. The simplified Zaeske method (2005) supported with FEM to estimate the subsoil bedding, combined with bedding correction and the loading approach 1 (LA1) leads to very plausible results with respect to in-situ measurements. Therefore, the authors recommend a slight improvement on the current EBGEO code to advance our understanding of this specific structure, which leads to economic and still safe designs.

Consequently, for daily practice, a careful design by EBGEO, as published taking a bedding modulus as estimated on the safe side but still realistic (e.g., determination by Finite-Element-Analysis) will give satisfactory results. For the best practice, the design should take

- load approach 1 (LA1, inverse triangle)
- soil support calculated by finite-element-analysis
- corrected analytic bedding calculation of the EBGEO-model into consideration.

REFERENCES

Deutsche Gesellschaft für Geotechnik e. V., Arbeitskreis 5.2. 2010. *Empfehlungen und die Berechnung für den Entwurf und die Berechnung von Erdkörpern mit Bewehrungseinlagen aus Geokunststoffen (EBGEO)*. Berlin: Wilhelm Ernst & Sohn.

Emde, O. 1995. *Der unterspannte schubsteife Balken-ein mechanisches Modell für geotextilbewehrte Tragschichten. Bericht-Nr.15.* Wuppertal: Bergische Universität Gesamthochschule Wuppertal, Grundbau, Bodenmechanik und Unterirdisches Bauen.

Geduhn, M. & Vollmert, L. 2005. Verformungsabhängige Spannungszustände bei horizontalen Geokunststoffbewehrungen über Pfahlelementen in der Dammbasis. *Bautechnik Nr. 82(9)*, P.657–662.

Heitz, C. 2006. *Bodengewölbe unter ruhender und nichtruhender Belastung bei Berücksichtigung von Bewehrungslagen aus Geogittern*. Kassel: Schriftenreihe Geotechnik Universität Kassel.

Kempfert, H.G., Stadel, M. & Zaeske, D. 1997. Berechnung von geokunststoffbewehrten Tragschichten über Pfahlelementen. *Bautechnik Nr. 74(12)*, P.818–825.

Van Eekelen, S.J.M., Bezuijen, A., Lodder, H.J., van Tol, A.F. 2011. *Model experiment on piled embankments. Part 1 & Part 2* Netherlands: Elsevier.

Vollmert, L., Doygun, O. 2011. *Geosynthetic Reinforced Load Transfer Platforms for Infrastructural Projects and Wind Energy Plants, Lectures on 18. Darmstädter Geotechnik. Kolloquim on 17.03.2011, Heft Nr. 88.* Darmstadt: Mitteilungen des Institutes und der Versuchsanstalt für Geotechnik der Technischen Universität Darmstadt.

Vollmert, L., Kahl, M., Giegerich, G. & Meyer, N. 2006. *Schleuse Neuer Hafen, Bremerhaven – In-situ Verifizierung eines erweiterten Berechnungsverfahrens für geogitterbewehrte Gründungspolster über vertikalen Traggliedern. Beitrag zur Baugrundtagung*. Bremen: DGGT.

Vollmert, L., Körlin, R., Herold, A. 2012. *Comparison of design methods for geosynthetics spanning over columns.* Valencia: EUROGEO.

Zaeske, D. 2001. *Zur Wirkungsweise von unbewehrten und bewehrten mineralischen Tragschichten über pfahlartigen Gründungselementen*. Kassel: Schriftenreihe Geotechnik Universität Kassel.

Baltic Piling – Mets & Raudsepp (Eds)
© 2013 Taylor & Francis Group, London, ISBN 978-0-415-64334-4

Instructions for preparing the paper for 10th international conference "modern building materials, structures and techniques" displacement of horizontally loaded piles

Kęstutis Tumosa
Company "Vilniaus Rentinys", LT Vilnius, Lithuania

Vincentas Stragys
Vilnius Gediminas technical university, LT Vilnius, Lithuania

ABSTRACT: In the article results of bearing capacity of horizontally loaded piles are presented and analyzed. Three groups of piles consisting of 2 or 3 piles have been tested in the construction site of Mariampoles multifunctional sports palace. Opening distance of the building was equal to 83 m. Soil conditions – week, soft soils. Type of piles – vibrated cast-in-place. Method of installation – vibrating the steel pipe into the ground using the vibrator. Length of piles – around 9 meters, diameter – 380 mm. Design vertical load on foundation having 18 piles – Nk = 772,3 kN, design horizontal load on the same foundation – Qk = 1650 kN and Mk = 9805.0 kN * m. In the article horizontal load – horizontal displacement relationships are presented and conclusions on bearing capacity of pile foundation for action of horizontal loads given.

Keywords: piles, horizontal loading of piles, horizontal displacement of piles, pile–soil interaction.

1 INTRODUCTION

A rational treatment of ground behavior and its interaction with structures is indispensable condition for limit state design of buildings and civil engineering works [1].

Analysis of stress-strain distribution in soil samples during laboratory tests allows to apply results of findings to field conditions for soil-structure interaction [2].

Vibropiles for construction of foundations were used since the thirties of the last century. With the help of vibrations it is possible to reduce the resistance of soils during the deepening. Since 1993, vibropiles are used in Lithuania also. Results of research of deformations of till around the vibropiles were presented and analyzed in [3].

There are many other publications presenting results of displacement developments of horizontally loaded piles [4, 5, 6, 7]. Different types of piles were tested under horizontal, however there were not observed publications about the performance of horizontally loaded vibropiles.

Lymon C. Rose and William F. Van Impe even published a book [4], thrust of which is detailed presentation of methods of analysis for single piles and groups of piles under lateral loading. The book specifically addresses the analysis of piles of varying stiffness installed into soils with a variety of characteristics, accounting for the axial load at the top of the pile and for the rotational restraint of the pile head, possibly nonlinear, offered by the connection to the superstructure. The text provides example designs as well as the design of pile foundations that support an offshore platform. The book also includes references to a rich body of technical material, including citations of hundreds of relevant publications.

Robert L. Mokwa and Michael Duncan [5] investigated ability of pile caps to resist lateral loads. This resistance is often neglected in design. Published cases involving a variety of pile and cap sizes, soil conditions, and loading conditions. Comparing the response of pile groups with caps fully embedded and with soil removed from around the caps showed that pile caps provide significant resistance to lateral load. This resistance depends primarily on the stiffness.

Article of W.Y. Shen and C.I. Teh [6] reports a variational solution and its spreadsheet calculation procedure for the analysis of laterally loaded piles in a soil with stiffness increasing with depth. The aim of the paper was to provide solutions that can be used simply with recourse only to spreadsheet calculation to solve the displacement and bending moment of laterally loaded piles, so that they can be easily applied in practice as an alternative approach to analyze the response of laterally loaded piles.

Paper of Anderson [7] examines seven case histories of load tests on piles or drilled shafts under lateral load. Since the current design software to estimate lateral load resistance of deep foundations requires *p-y* curves. The *p-y* method is a method of analyzing the ability of deep foundations to resist loads applied in the lateral direction.

The first approach used was correlative whereby soil parameters determined from in situ tests [standard penetration test (SPT) and cone penetration test (CPT)] were used as input values for standard *p-y* curves. In the second approach *p-y* curves were calculated directly from the stress deformation data measured in dilatometer (DMT) and cone pressuremeter tests. The correlative evaluation revealed that, on the average, predictions based upon the SPT were conservative for all loading levels, and using parameters from the CPT best predicted field behavior. Typically, predictions were conservative, except at the maximum load. Since traditionally SPT and CPT correlation-based *p-y* curves are for "sands" or "clays," this study suggests that silts, silty sands, and clayey sands should use cohesive *p-y* curves. For the directly calculated curves, DMT derived *p-y* curves predict well at low lateral loads, but at higher load levels the predictions become unconservative. *p-y* curves derived from pressuremeter tests predicted well for both "sands" and "clays" where pore pressures are not anticipated. Analysis of lateral loading of piles are given also in a many other classical books of foundation engineering [8, 9, 10], however local geotechnical conditions and new methods of pile installation, in particular case vibropiles require to collect and analyze results of in-situ tests.

2 GEOTECHNICAL CONDITIONS OF CONSTRUCTION SITE

Geotechnical conditions of construction site of Mariampoles multifunctional sports palace are presented in the log given in Figure 1. Ground profile of the site consists of different type of soils. Top layer is formed

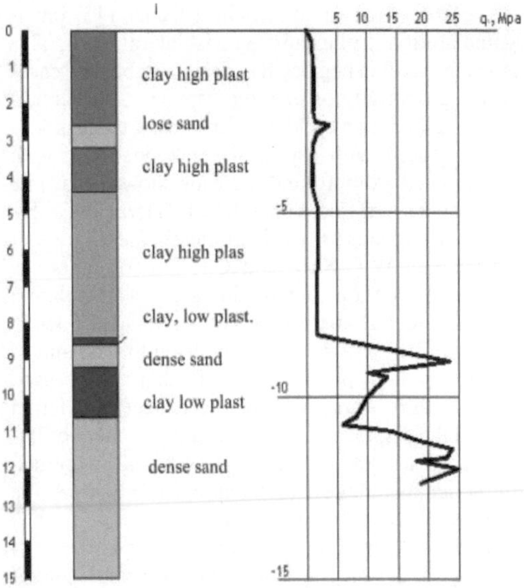

Figure 1. Geological log of site.

from organic deposits of 0.4–1.0 m thickness. Deeper were found layers of clays from low to high plasticity with lenses of loose sand. Below clay layers, deeper than 9.40 m, deposits of dense sands were found.

3 SELECTION OF FOUNDATION TYPE

After evaluation of architecture structure of building and geotechnicalconditions of site it was decided to install foundations having 18 piles each. It was found that design vertical load on foundation having 18 piles is equal to Nk = 772.3 kN, design horizontal load on the same foundation – Qk = 1650 kN and bending moment Mk = 9805.0 kN ∗ m. Installation of vibrated cast in place type of piles was selected. Length of were s equal to 9.00 m. approximately and diameter 380 mm.

It was decided to test piles loding them horizontally. Three group of piles were selected for horizontal loading tests. Two tests were made for pile groups having 2 piles each conected them by cap and one test for group of 3 piles also conected by cap.

4 TECHNOLOGY OF PILE INSTALLATION

Technology of vibrated cast in place type of piles instalation is presented in Figures 2 and 3. Sequence of pile instalation is given below:

1. Positioning the steel pipe on the cover;
2. Vibrating the steel pipe into the ground using the vibrator;
3. Vibrating down to the projected depth and Installing the reinforcement cage into the steel pipe;
4. Filling in the concrete.

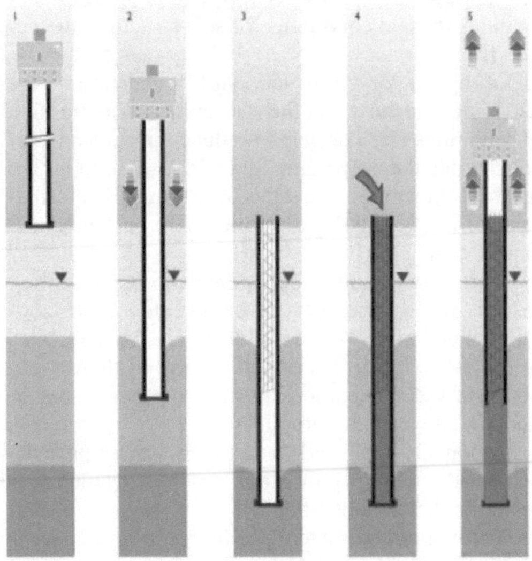

Figure 2. Technology of vibrated cast in place pile installation.

Figure 3. Pile installation in progress.

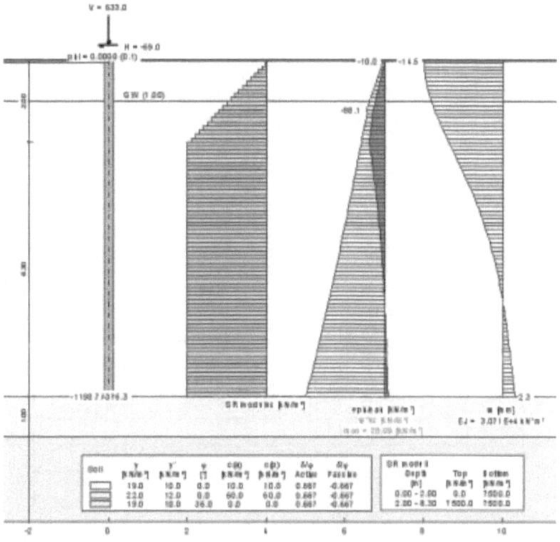

Figure 4. Theoretical displacement graphs.

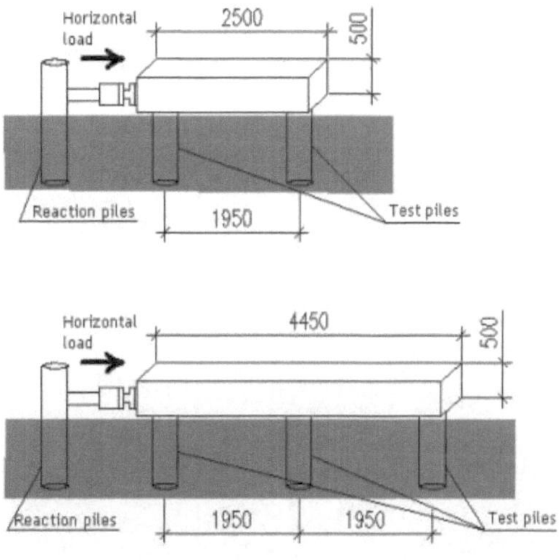

Figure 5. Loading systems for tests of groups of 2 and 3 piles connected by cap.

The concrete is compacted during the extraction. Advantages of vibrated cast-in-place piles are:

- high performance since the pipe is vibrated into ground as one piece,
- due to the sealed designing neither ground water nor soil can get into the casing of the pile to be installed,
- the displacement of soil and the vibration causes an improvement of the ground directly adjoining the pile,
- reinforcement cage and concrete are installed into dry steel pipe in a short time,
- exact determination of the pile length and diameter according to the geological and static
- conditions.

5 THEORETICAL CALCULATIONS DISPLACEMENT OF HORIZONTALLY LOADED PILES

Bearing capacity of displacement piles were calculated using German standard DIN 4085. For this reason software GGU LATPILE was applied. Subgrade reaction modulus was calculation on Vesic formula:

$$k_h = 0.65 \times \sqrt[12]{\frac{E_s \times d^4}{E_p \times I_p}} \times \frac{E_s}{1 - \nu^2}$$

where, k_h – subgrade reaction modulus;
E_p – modulus of elasticity of pile;
I_p – moment of inertia of pile;
E_s – modulus of elasticity of soil;
d – pile diameter
υ – Poisson number.

Using this formula the following coefficients of subgrade reaction were found:

- high plasticity clay, high plasticity 7500 kN/m^3
- loose sand 7500 kN/m^3
- low plasticity clay 40000 kN/m^3
- dense sand 40000 kN/m^3

6 METHODOLOGY OF PILE HORIZONTAL LOADING TESTS

Systems for originating horizontal loading, measuring horizontal force and displacements were installed between reaction pile and test piles cap. Systems included: loading jack, manometers (or load cels) for measuring forces and displacement gauges.

Horizontal loads were icreased stepwise. Each horizontal load was kept constant for 5 minutes before applying next step of horizontal load.

For each step of horizontal load readings of horizontal dispplacements were recorded after −1, −5, −10, −20, ... −40 e.c.t. minutes.

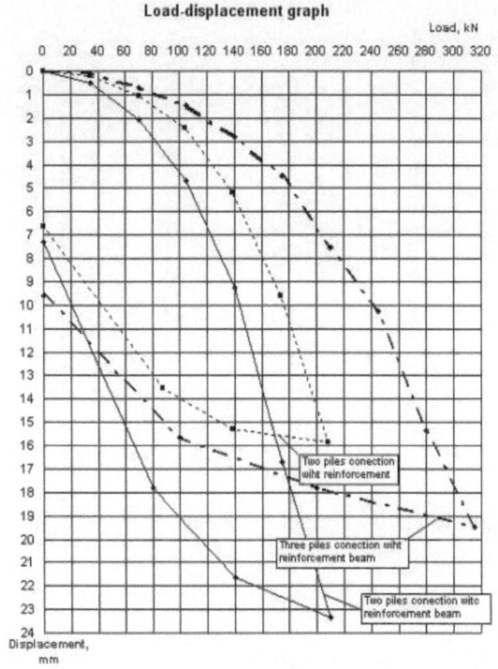

Figure 6. Pile horizontal load-displacement graphs.

Figure 7. Construction of building in progress.

Each load was kept onstant until the relative stabilizaion was reached equal to no more than 0.1 mm per 20 minutes.

Unloading process was done also stepwise recording pile cap horizontal displacements.

Group of two piles (connected by reinforcement beam) test under design static lateral load $H = 69.0$ kN, showed that horizontal displacement value is 5 and 9.5 mm accordingly group of three piles test showed that horizontal displacement value is 7.5 mm. The displacement value 14.5 mm was estimated using theoretical method. Comparing the values obtained by calculation and testing, it can be said that the theoretical estimated displacement values is greater.

7 CONCLUSIONS

1. Analysis of literature sources shows that many theoretical and experimental investigations were carried out in the field of pile lateral loading. However, local geotechnical conditions and new methods of pile installation, in particular application of vibropiles, requires to collect and analyze results of in-situ tests.
2. It was constructed loading systems and 3 horizontal groups consisting of 2 and 3 piles were tested by horizontal applying horizontal loads.
3. Horizontal load-displacement graphs are presented in Figure 4.

REFERENCES

[1]. L. Furmonavičius, [A. Kudzys]. Safety of piles in groups // Modern building materials, structures and techniques: selected papers of the 8th International conference.- ISBN 9986-05-757-4.- Vilnius, Lithuania, May 19–21, 2004, 730–735.

[2]. N. Verveckaite, J. Amsiejus, V. Stragys. Stress-strain analysis in the soil sample during laboratory testing // Journal of Civil Engineering and Management International research and Achievements. Vilnius: Technika, 2007, Vol. 13, p. 63–70.

[3]. Žaržojus G., Kairys T., Gadeikis S., Dundulis K. Results of research of deformations of vibropiles and till interactions.//Modern building materials, structures and techniques: selected papers of the 9th International conference.-ISBN

[4]. Robert L. Mokwa, Michael Duncan. Experimental Evaluation of Lateral-Load resistance of Pile Caps. Journal of Geotechnical and Geoenvironmental Engineering, Vol. 127, No. 2, February 2001, pp. 185–192.

[5]. Single piles and pile groups under lateral loading. Lymon C. Rese, William F. Van Impe. ISBN-10. Taylor and Francis. Jan.1 2000. 480 pages.

[6]. W.Y. Shen and C.I. Teh. Analysis of laterally loaded piles in soil with stifnees increasing with depth. J. Geotech. and Geoenvir. Engrg. Volume 130, Issue 8, pp. 878–882 (August 2004)

[7]. J.B. Anderson, F.C. Towsed, and B. Grajales. Case history evaluation of laterally loaded piles. J. Geotech. and Geoenvir. Engrg. Volume 129, Issue 3, pp. 187–196 (March 2003)

[8]. Pile foundations in engineering practice. Edited by Prakash Shamer, Hari D. Sharma. ISBN 0–471 61653-2 A. Wiley-Interscience publication. 1990. 734pp.

[9]. Laterally loaded deep foundations. Edited by Langer J.A. ASTM Special Technical Publication No 835. ISBN 0-8031-0207-0. 1984. 264 pp.

[10]. Behaviour of deep foundations.Edited by Raymond Lundgren. ASTM Special Technical Publication No 670. ISBN 0-8031-02917. 1979.

Baltic Piling – Mets & Raudsepp (Eds)
© 2013 Taylor & Francis Group, London, ISBN 978-0-415-64334-4

Bearing capacity of piles in strengthened foundations

V.N. Paramonov & P.A. Kravchenko
Petersburg State Transport University (Saint-Petersburg, Russia)

ABSTRACT: The article describes the results of small scale experiments where bearing capacity of short piles in pile-strengthened foundation was studied. Bearing capacity of piles in raft foundations is compared with bearing capacity of piles in pile-strengthened foundations. It is shown that strengthening of the foundation by short piles leads to essential increase of bearing capacity of a foundation and significant reduction of settlements. Bearing capacity of a pile increases about 2 times in comparison with bearing capacity of a single pile and about 1.5 times in comparison with bearing capacity of the a pile in a pile group.

1 INTRODUCTION

There are a lot of historical buildings in St. Petersburg constructed on rubble stone footings. Surveys often show that contact pressure under the foundations exceeds permissible values regulated by contemporary codes of Russia.

A similar problem often takes placed in cases of reconstruction of historical buildings when the load on the foundations is increased. In these cases strengthening of the soils and the foundations is required. One of ways of strengthening of the foundations is strengthening (underpinning) by piles.

There are no directions in modern codes of Russia [1] by which a designer could calculate the part of the load applied to the piles. Besides, bearing capacity of piles in an underpinned foundation is not regulated either. Therefore, as a rule, the designer determines the pile bearing capacity by calculation with the use of formulas of Russian codes or by results of piles static load tests outside of the strengthened foundation.

The part of the load transferred to the soil by the piles is often accepted equal to the full load from a building. The contact pressure under the existing foundation is not taken into account.

In this article we describe our results of laboratory investigation and the effects occurring at strengthened foundations.

2 DESCRIPTION OF LABORATORY CONDITIONS

Some series of small-scale tests were performed in laboratory of St. Petersburg State Transport University for the purpose of investigation of distribution of the load between the strengthened foundation and strengthening piles in the case of increase of the load at the foundation. The soil was modeled by silty sand because it is typical in upper part of the geological condition of the central part of St. Petersburg.

Diameter of the experimental tank was 710 mm, thickness of the sand layer – 1300 mm. Before the beginning of each experiment the soil was totally excavated from the tank and then placed into the tank with layer-by-layer compaction. Then humidity and density of the sand were determined. Experiments were carried out with humidity of the soil in a range from 7,55% to 8,16% and density from 1,38 g/cm^3 to 1,42 g/cm^3 (loose sand). All experiments were repeated three times.

As a criterion of necessity of pile strengthening of the foundation was the load on a model of the foundation at which contact pressure reaches the value of the design strength – the limit of the linear relationship between the pressure under a foundation and its settlement. This load was determined in preliminary series of experiments (plate tests) with a square plate with the sizes 15 × 15 cm corresponding to the small-scale model of the strengthened foundation. The results of the plate tests are shown on fig. 1. It is seen that

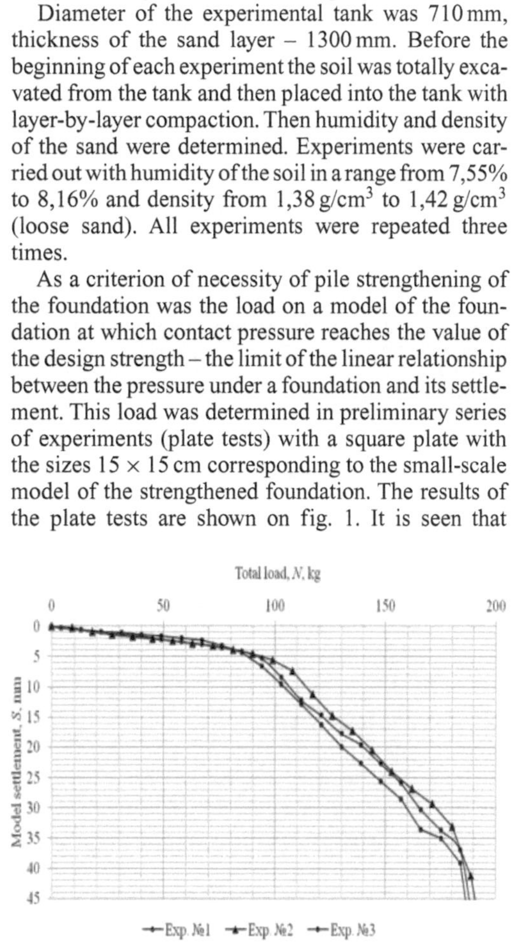

Figure 1. Settlement of the plate versus the vertical load.

the contact pressure corresponded to the value of the design strength reached at the load on the plate equal to about 80 kg.

The models of the strengthening piles were metal pipes with the external diameter of 16 mm and the length of 300 mm. First of all the static load tests of models of single piles with step-by-step loading were carried out. The step of the load was accepted equal to 0,3 kg. The average value of bearing capacity of a single pile in the conditions of small-scale experiment was determined as $F_d = 8{,}4$ kg (see fig. 3).

The same tank with the similar soil conditions was used for modeling of strengthening of the foundation by the piles. The plate was loaded step-by-step up to the load equal to the design strength.

After that four strengthening piles were installed through the plate. Distance between piles was equal to three diameters of the piles, the distribution scheme was 2×2. Technological settlement of the foundation in the process of installation of the piles in all experiments was observed. Then the step-by-step loading of the strengthened plate continued. The load increased until achieving total settlement of the model, equal to the $3/4$ value of the ultimate settlement of non-strengthened plate. It was not possible to reach the real bearing capacity of model of the strengthened foundation. The dependence of the settlements from the load for the plate is compared with the same dependence for model of the strengthened foundation, as it is presented on fig. 2.

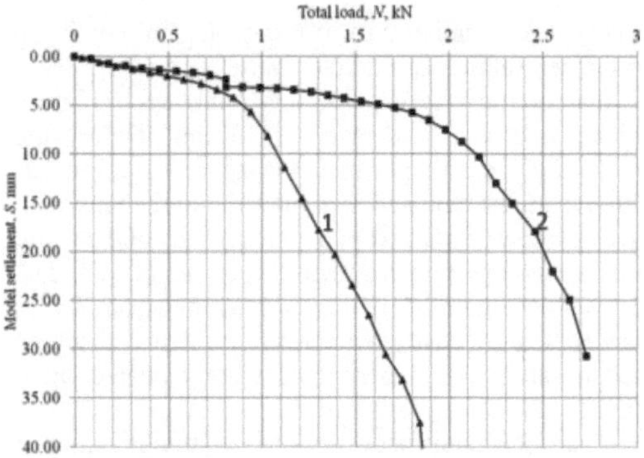

Figure 2. Settlement of the non-strengthened (1) and strengthened (2) plate versus the vertical load.

The diagram shows that the strengthening of the foundation by short piles essentially decreases the settlements under the load exceeding the value of design strength of the soil. During the last loading steps the increase of the settlement increment is similar to increase of the plate settlement increment.

Comparing the diagrams of experiments with the strengthened and non-strengthen foundations we notice that the bearing capacity of the foundation after strengthening increased approximately by 90 kg (1,5 times). Total bearing capacity of 4 strengthening piles is equal to $8.4 \times 4 = 33.6$ kg. Thus the bearing capacity of the strengthened foundation increases by value which exceed bearing capacity of strengthening piles.

Our assumption of this super total effect is in increasing of the bearing capacity of piles in the strengthened foundation in comparison with the bearing capacity of a single pile. The similar effect was described in [2] for buildings on pile rafts. A series of additional experiments on test of piles in the tank was carried out for confirmation of this hypothesis.

3 PILE TESTS

For research of effect of increase of bearing capacity of a pile as a part of the strengthened foundation some series of pile tests in various conditions were

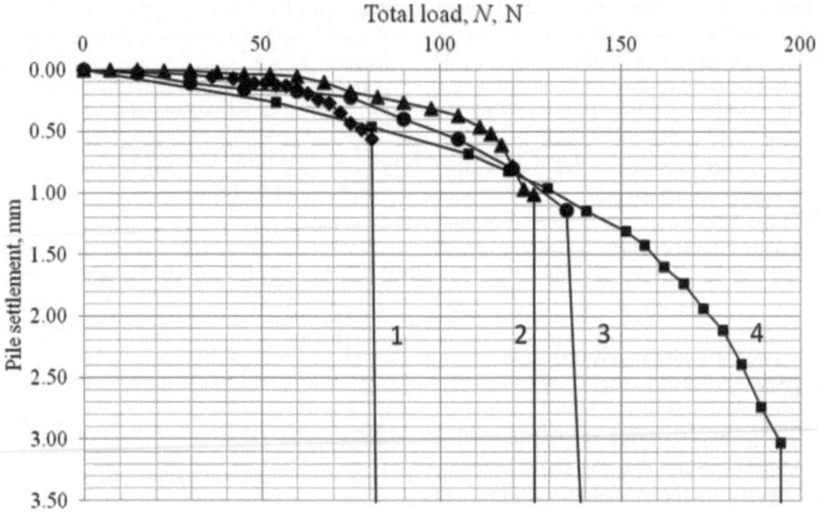

Figure 3. The results of static pile tests. 1 – Single pile, 2 – a pile of a pile group, 3 – a pile in a pile foundation, 4 – a pole in a strengthened foundation.

carried out (all experiments were carried out with triple repeatability):

1. for a single pile;
2. for a pile as a part of the pile group (without contact of the raft and the soil);
3. for a pile as a part of strengthened foundation without saving the contact pressure under the foundation;
4. for a pile as a part of strengthened foundation with saving the contact pressure under the foundation.

The results of the static pile tests are given on Fig. 3.

The results of the tests allow to conclude that the bearing capacity of a pile as a part of the pile group reaches about 12,6 kg after recalculation on one pile that shows increase of bearing capacity of a pile in pile group by 4,15 kg (or for 51%) in comparison with a single pile.

For the third case after the settlement of the strengthened foundation reached about 3/4 of the limit settlement the model was completely unloaded. Then static tests of piles as a part of the foundation were carried out. For this purpose two diagonally located piles were disengaged from work and were loaded to failure. Piles were loaded step-by-step with the loading step 3 kg. In this case the bearing capacity of a pile as a part of strengthened foundation reached 13,5 kg, i.e. it exceeded the bearing capacity of a pile in the pile group by 1,25 kg (10%) and by 61% in comparison with the bearing capacity of a single pile.

Finally, a series of tests was carried out with the strengthened foundation. After the settlement of the strengthened foundation reached the 3/4 value of the limit settlement of non-strengthened foundation the foundation was partly unloaded and the piles were disconnected from the foundation. One pile was loaded step-by-step until failure settlement. Failure of a pile was fixed at the loads in a range from 19,98 kg to 20,52 kg on a pile.

4 CONCLUSION

The series of small-scale experiments, described in this article, showed that strengthening of the foundation by short piles leads to essential increase in bearing capacity of the foundation. Besides, the contact pressure causes the increase of the bearing capacity of a pile by 2,3 times in comparison with the bearing capacity of a single pile and by 1,44 times in comparison with the bearing capacity of a pile in a pile group. That testifies to efficiency of strengthening of the foundations by short piles. The bearing capacity of a pile, calculated with use of formulas and tables of Russian codes, appears to be several times less than the real bearing capacity of a pile of the strengthened foundation. This contradiction, established by the described tests testifies to the possibility of improvement of engineering calculations methods, recommended by Russian codes.

REFERENCES

Paramonov V.N., Dunaevskaya T.A. Modification of bearing capacity of driven piles in time at the opened sites and previously loaded. *Reconstruction of cities and geotechnical engineering.* No. 8, Saint-Petersburg, 2004. c. 102–106. (in Russian).

SP 24.13330.2011 Pile foundations. Moscow, 2011.

Author index